International Conference on Advanced Fibers and Polymer Materials 2017

Next Generation Fibers: Changing Your Life

Shanghai, China
8 - 10 October 2017

ISBN: 978-1-7138-0623-3

Printed from e-media with permission by:

Curran Associates, Inc.
57 Morehouse Lane
Red Hook, NY 12571

Some format issues inherent in the e-media version may also appear in this print version.

Printed with permission by Curran Associates, Inc. (2022)

For permission requests, please contact Donghua University
at the address below.

Donghua University – College of Materials Science & Engineering
2999 North Renmin Road
Songjiang, Shanghai
China, 201600

Phone: 0086-21-67792362
Fax: 0086-21-67792855

http://cmse.dhu.edu.c

Additional copies of this publication are available from:

Curran Associates, Inc.
57 Morehouse Lane
Red Hook, NY 12571 USA
Phone: 845-758-0400
Fax: 845-758-2633
Email: curran@proceedings.com
Web: www.proceedings.com

ICAFPM

2017 International Conference on Advanced Fibers and Polymer Materials

October 8–10, 2017, Shanghai, China

Processdings of 2017 International Conference on Advanced Fibers and Polymer Materials

Organizer

State Key Laboratory for Modification of Chemical Fibers and Polymer Materials
Innovation Base of Advanced Fabrication Technology of Fiber Materials
Center for Advanced Low-Dimension Materials
College of Materials Science Engineering of Donghua University

Co-organizer

Natural Science Foundation of China
Fiber Society of China, CMRS (In preparation)
Polymer Materials and Engineering Society of China, CMRS
China Women's Association for Science & Technology
Shanghai Consulting & Academic Activities Center for Academicians of Chinese Academy of Engineering
State Key Laboratory of Polyolefins and Catalysis

Media sponsor

www.polymer.cn

Contents

3. Abstracts of Oral Presentation 106

1.Abstracts of Plenary Lecture

Advancing Nanocellulose Technologies for Water Purification

Benjamin S. Hsiao

Department of Chemistry, Stony Brook University,Stony Brook, NY 11794, U.S.A.

E-mail: benjamin.hsiao@stonybrook.edu

The fabrication of nanocellulose fibrous scaffolds can be accomplished by chemical and mechanical processes of cellulose from natural resources, where some sources can provide "effortless extractability" with very low energy consumption. Nanofibrous scaffolds have very unique properties including interconnected pores, a very large surface-to-volume ratio and a high capacity for surface modifications, making them ideal candidates for fabrication of high throughput separation membranes (e.g. microfiltration and ultrafiltration) and/or highly efficient sorptive media. Directed water channels in the barrier layer of reverse osmosis and nanofiltration membranes can also be formed through the formation of interface between the nanofibers and barrier layer polymer matrix, while the gap thickness can be regulated by physical interactions or chemical bonding. In the present context, advances in fundamental studies on cellulose microfibrils, the building block from the cell wall of any plant, and their interactions with metal ions by means of synchrotron x-ray and neutron techniques have provided us with new insight into the design and fabrication of highly permeable nanofibrous membranes, having a hierarchically tiered structure, that can enable higher flux, higher retention and lower energy consumption water purification. Nanocelluloses may serve as a revolutionary platform technology for the design and fabrication of low cost, sustainable and highly efficient filtration and sorptive media.

BIOGRAPHY

Benjamin S. Hsiao, Ph.D., is a Distinguished Professor in Chemistry at Stony Brook University. Hsiao has a distinguished reputation in polymer science. His current research interests are mainly focused on the development of sustainable nanostructured biomaterials for water purification applications. He is the recipient of SUNY Distinguished Professor, Chang-Jiang Scholar from Education Ministry of China, Co-operative Research Award from Division of Polymeric Materials Science and Engineering of American Chemical Society, NSF Special Creativity Award and DuPont Young Faculty Award. He holds 21 U.S. patents and 15 foreign patents that have been licensed to 3 companies. He has published over 463 peer-reviewed scientific papers, 45 reviews and chapters in books and encyclopedias, and serves as editorial board member for 6 scientific journals. Hsiao is a Fellow of AAAS, ACS, APS, MRS and NAI, and is an AAAS-Lemelson Invention Ambassador.

Nano Technology Convergence for the Future IT, ET, BT

Jong Min Kim

Department of Engineering, University of Cambridge, 9JJ Thomson Ave, Cambridge, UK

E-mail: jmk71@cam.ac.uk

This speech presents the current and future nanotechnology, especially focusing on the convergence of nano with electronics, photonics, and energy/bio areas. Nano-electronics will cover the graphene and carbon nanotubes, and their applications in flexible, transparent and printable electronics, and transistors for future applications. Nano-photonics will include quantum-dot (QD) displays/lighting and other applications such as including QD solar cell, image sensors and photo detector.

New concept of energy harvesting and flexible battery, e-nose, e-tongue, fragrance generation system, active hologram for future mobile devices, textile electronics including smart lighting energy devices, and e-skin and sensor network will be presented.

These convergence technologies also focuses on new medical imaging areas, non invasive medical detection, and healing patch and else for industrial applications with nano materials, processing and fabrication on this system.

BIOGRAPHY

Professor Jong Min Kim was formerly Senior Vice President and Vice President in Samsung Electronics Corporate R&D Centre, Korea for 13 years. He was the first Samsung Fellow in Samsung Group. Now, he is Professor of Electrical Engineering of Department of Engineering at University of Cambridge. He was Chair of Electrical Engineering at University of Oxford from 2012-2015. Professor Kim had previously held a variety of senior technology positions at the Samsung Group including Display, Materials, Energy and Electronics research/developments. Professor Kim had managed several tens of millions of dollars per annum for major projects in Samsung for 17 years. His research is described in more than 300 journal papers (including 9 Nature/Science, and Nature family journals), more than 300 publications on the Technical Digest and proceedings with around 100 keynote/invited speech at major international conference, and 253 patents (153 international patents). He received a number of awards: Award by Samsung Group Chair, Prime Minister Awards (2001) and many others. He had numerous world first inventions: carbon nanotube (reported variously in Science, Nature, etc. One paper is with more than 1,000 citation); transparent and flexible graphene electrodes (Nature 2009, with more than 3,000 citations) and quantum dot based LEDs and displays (Nature Photonics, Cover Article, 2009 and 2011, Nature Comm'13), LED on glass (Nature Photonics, Cover Article, 2011), CNT network Transistors (Science 2008 and Nature Communications 2011), and many others. He has been leading EC ERC Advanced Grant since 2014 and many other projects. Prof Kim has been leading numerous projects EU ERC-Advanced Program, Horizon-2020, UK EPSRC, and else. He also leads committee members on Int Conferences.

Structural Health Monitoring and Performance of Polymer-based Composites at Space Environment

Alan Kin-tak Lau

Pro-Vice-Chancellor (Research Performance and Development),Faculty of Science, Engineering and Technology, Swinburne University of Technology, Hawthorn Melbourne VIC 3122 Australia

E-mail: Aklau@swin.edu.au

Polymer-based advanced composites always suffer from degradation at extreme temperatures in the range between 220 and 77 K and low atmospheric pressure. Within this temperature range, composite structures behave very brittle and many micro-cracks are easily formed due to differential thermal coefficients of expansion (CTEs) between polymer matrix and high strength reinforcements. Besides, at the Low Earth Orbit (LEO) environment the structures may also be subject to damages due to micro-meteoroid attack, in which many tiny particles left over from the formation of the solar system and they are travelling at very high speed (hyper-velocity) to cause serious impact and abrasion onto the structures. Out-gassing and high oxidation rate are also problems for polymers using at this environment. Therefore, studies on using nano-particles to enhance the anti-cracking resistant properties and prolong the pryrolzing process for ablators are necessary. Besides, due to the increasing use of polymer-based nanocomposites at extreme environment condition, their inspectability becomes a hot topic, at least in coming 5 years to explore more real-time or remote health monitoring techniques to ensure the safety of structures. Embedded sensors, self-healing technology and smart structure designs are most prominent research fields for nanocomposite structures. In this invited lecture, an overview on the nanocomposites, their mechanical, thermal and structural properties at different working environments is given. The potentiality of using different structural monitoring techniques to serve at the extreme environment will also be discussed.

BIOGRAPHY

Professor Lau received his Bachelor and Master degrees of Engineering in Aerospace Engineering from the Royal Melbourne Institute of Technology (RMIT University, Australia) in 1996 and 1997, respectively. Within that period, he also worked for General Aviation Maintenance Pty Ltd, Australia, as an Engineer Trainee, and for the Corporative Research Centre for Advanced Composite Structures (CRC-ACS) Australia, as a Research Assistant designing a repair scheme for composite performs. Afterward, he received his Doctor of Philosophy (PhD) from The Hong Kong Polytechnic University in 2001. Thereafter, he was appointed Assistant Professor in 2002 and promoted to Associate Professor and Professor in 2005 and 2010, respectively. In 2015, he was appointed as Alex Wong/Gigi Wong Professor in Product Design Engineering and Associate Dean (Industrial Relation) in the Faculty of Engineering, PolyU. Currently, he is Pro-Vice-Chancellor (Research Performance and Development) of Swinburne University of Technology, Australia to look after its future research and ranking strategies. Professor Lau has received numerous research and teaching awards since 2002. He has been working in advanced composite, nano-composite and bio-composite materials research for over 20 years, and his published articles have been frequently cited by other researchers. He has received over 12,500 citations with h-index of 52 to date (Google Scholar Citations). Last year, he has been elected as Fellow of the European Academy of Science and Arts. To date only three academics in Australia have received this honour. He has also been elected as International Vice President of The Institution of Mechanical Engineers (IMechE) and appointed as Independent Non-executive Director of King's Flair International (Holdings) Limited (Stock Code: 6822). Currently, he is Fellow of the Institution of Mechanical Engineers; Institute of Materials, Minerals and Mining; Royal Aeronautical Society and Hong Kong Institution of Engineers.

How do we Create and Process Materials for Flexible, Transparent Electronic Circuitry?

Tobin Marks

Department of Chemistry and the Materials Research Center, Northwestern University, Evanston, IL 60208-3113 USA

E-mail: t-marks@northwestern.edu

This lecture focuses on the challenging design, realization, understanding, and implementation of new materials families for unconventional electronics. Fabrication methodologies to achieve these goals include high-throughput, large-area, high-resolution printing techniques. Materials design topics will include: 1. Rationally designed high-mobility p- and n-type organic semiconductors for printed organic CMOS, 2. Self-assembled and printable high-k nanodielectrics enabling ultra-large capacitance, low leakage, high breakdown fields, minimal trapped interfacial charge, and device radiation hardness, 3. Polycrystalline and amorphous oxide semiconductors for printable transparent and mechanically flexible electronics, 4. Combining these materials sets to fabricate a variety of high-performance thin-film transistor-based circuitries, 5. The relevance of these advances to unconventional photovoltaics.

BIOGRAPHY

Tobin Marks is Ipatieff Professor of Catalytic Chemistry, Professor of Materials Science and Engineering, Professor of Applied Physics, and Professor of Chemical and Biological Engineering at Northwestern University. His recognitions include the U.S. National Medal of Science, the Spanish Principe de Asturias Prize, the MRS Von Hippel Award, the Dreyfus Prize in the Chemical Sciences, the National Academy of Sciences Award in Chemical Sciences, and the ACS Joseph Priestley Medal. He is a member of the U.S., German, and Indian National Academies of Sciences, the U.S. National Academy of Engineering, the American Academy of Arts and Sciences, and the U.S. National Academy of Inventors. He is a Fellow of the U.K. Royal Society of Chemistry, the MRS, and the ACS. Marks has published 1260 peer-reviewed articles and holds 260 issued U.S. patents. He holds Honorary Doctorate Degrees from Hong Kong University of Science and Technology, the University of South Carolina, the Ohio State University, and the Technical University of Munich.

Functional Nanofibers

Seeram Ramakrishna

National University of Singapore, Singapore

E-mail: seeram@nus.edu.sg

Product innovations as well as manufacturing innovations are closely related to the materials innovations. Together they enable better living – clean air, water and energy, health and wellbeing, smart living and transportation, safety and security, and sustainability. Functional nanomaterials are emerging domains of materials innovation. Electrospinning has been developed as a viable manufacturing method for producing a range of nanofibers and nanoparticles of polymers, metals, ceramics, carbon, and their combinations. It has also evolved into a 3D printing or additive manufacturing method. Functional nanomaterials enable air filtration, water purification, clean energy generation and storage, controlled drug delivery, tissue engineering, regenerative medicine, food packaging, high performance apparel, electronic-skin, and light-weight, damage tolerant materials for transportation, electricity transmission, buildings and construction. They also enable self-cleaning, super-hydrophilic, super-hydrophobic, and anti-microbial surfaces. This lecture provides an overview, and discusses emerging opportunities.

BIOGRAPHY

Academician, Professor Seeram Ramakrishna, *FREng*, is the Director of Center for Nanofibers and Nanotechnology at the National University of Singapore (NUS). He is an Academician of UK Royal Academy of Engineering (*FREng*); Singapore Academy of Engineering; Indian National Academy of Engineering; and ASEAN Academy of Engineering & Technology. He is an elected Fellow of International Union of Societies of Biomaterials Science and Engineering (*FBSE*); Institution of Engineers Singapore; ISTE, India; Institution of Mechanical Engineers and Institute of Materials, Minerals & Mining, UK; and American Association of the Advancement of Science; ASM International; American Society for Mechanical Engineers; American Institute for Medical & Biological Engineering, USA. He co-authored five books and ~ 1,000 journal papers which attracted ~ 67,000 citations and ~121 H-index. He is a Highly Cited Researcher in Materials Science. Thomson Reuters (Clarivate Analytics) identified him among the World's Most Influential Scientific Minds. He is a member of World Economic Forum's Committee on Future of Production.

Fibrillar Structures, Transport Pathways and Stretchable Electronics

Elsa Reichmanis, Guoyan Zhang, Nils Persson, Michael McBride
School of Chemical and Biomolecular Engineering, Georgia Institute of Technology, Atlanta, GA 30332, USA
E-mail: ereichmanis@chbe.gatech.edu

Simultaneous enhancement of the electrical performance, mechanical stretchability, and optical transparency of conjugated polymers may significantly broaden the spectrum of realizable applications for these materials in future intelligent optoelectronics, i.e., wearable devices, electronic skin, stretchable displays, and a vast array of biomedical sensors. The successful introduction of these devices however, relies on the design and development of sustainable, robust and reliable materials and processes. Studies have shown that not only does device performance depend critically on semiconductor alignment at many length scales, materials' mesostructure can be manipulated in solution prior to device fabrication. Observations surrounding the behavior of these materials suggest that requisite macroscopic long-range order required for high performance devices may be achieved through process optimization coupled with quantitative analysis of processed semiconducting thin film images. This approach allows development of robust materials structure-process-property relationships that can then be used to guide subsequent materials selection and process approaches. This presentation will explore how we can derive critical insights into the structure-process-property space using well-coupled modeling and experimental approaches, and how those can impact the design and development of all-printed, stretchable electronic devices.

BIOGRAPHY

Elsa Reichmanis is Pete Silas Chair in Chemical Engineering and Brook Byers Professor of Sustainability in the School of Chemical and Biomolecular Engineering of the Georgia Institute of Technology. Prior to joining Georgia Tech, she was Bell Labs Fellow and Director of the Materials Research Department, Bell Labs, Murray Hill, NJ. She received her Ph.D. and BS degrees in chemistry from Syracuse University. She is an elected member of the National Academy of Engineering and has received several awards for her work. She has also been active in professional societies; she served as 2003 President of the ACS, and has participated in many National Research Council activities. Her research, at the interface of chemical engineering, chemistry, materials science, optics, and electronics, spans from fundamental concept to technology development and implementation. Her interests include the chemistry, properties and application of materials technologies for photonic and electronic applications, with particular focus on polymeric and nanostructured materials for advanced technologies. Currently, efforts aim to identify fundamental parameters that will enable sub-nanometer scale dimensional control of organic, polymer and/or hybrid semiconductor materials.

2.Abstracts of Keynote/Invited Lecture

Crystalline/Single Crystal Semiconductor Optoelectronic Fibers and Devices

John Badding, Pier J.A. Sazio,[5] Xiaoyu Ji,[2,4] Steven Aro,[1,4] Rongrui He,[1,4] Justin Sparks,[1,4] Subhasis Chaudhuri,[1,4] Todd Day,[1,4] Anna Peacock,[5] Noel Healy,[5] Venkatraman Gopalan[2,4]

[1]*Department of Chemistry, Pennyslvania State University, University Park, PA 16802 USA;*

[2]*Department of Materials Science and Engineering, Pennyslvania State University, University Park, PA 16802 USA;*

[3]*Department of Physics, Pennyslvania State University, University Park, PA 16802 USA;*

[4]*Materials Research Institute, Pennyslvania State University, University Park, PA 16802 USA;*

[5]*Optoelectronics Research Centre, University of Southampton, Highfield, Southampton SO171BJ UK*

E-mail: jbadding@chem.psu.edu

Polycrystalline or single crystal forms of technologically relevant elemental and compound semiconductors typically have optoelectronic properties much superior to the amorphous materials usually drawn into fibers. High Pressure Chemical Vapor Deposition (HPCVD) can scalably integrate precisely structured forms of crystalline semiconductors into appropriate optical fiber templates over many meters of length as well as into many other nanostructured and microstructured templates.[1] We have fabricated single crystal, small core (1-2 μm) crystalline silicon and germanium optical fibers with low optical losses by laser heating.[2] These elemental semiconductor fiber cores can serve as the building blocks for optoelectronic devices, such as meter-long *pin* junctions and GHz bandwidth in-fiber detectors.[3] There has been much interest in flexible, permeable solar or electronic fabrics woven from these fibers that incorporate efficient silicon and compound semiconductor junctions.

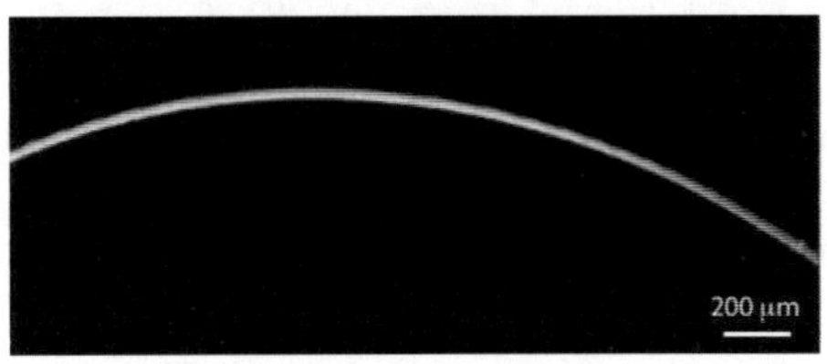

Figure 1. Zinc Selenide fiber cores can serve as the basis for a new generation of compound semiconductor optical fiber devices.

We later extended HPCVD to demonstrate the first small core crystalline compound semiconductor optical fibers (Figure 1) with low optical loss. We doped these ZnSe fibers with iron and chromium to fabricate continuous wave tunable infrared (2-5 μm) fiber lasers that have the potential to exceed the output power of lasers based on bulk gain media (~10 W) by 1 to 2 orders of magnitude. Many other types of light emitting devices based on crystalline compound semiconductor fibers should be possible. Finally, synthesis of diamond nanothreads inside optical fiber pores represents a promising direction for fibers of unprecedented mechanical strength to weight ratio and novel electronic properties associated with mixed sp^2/sp^3 carbon hybridization.[4]

References

[1] Sazio, P.J.A., Amezcua-Correa, A., Finlayson, C.E., Hayes, J.R., Scheidemantel, T.J., Baril, N.F., Jackson, B.R., Won, D.J., Zhang, F., Margine, E.R., Gopalan, V., Crespi, V.H., & Badding, J.V., Microstructured optical fibers as high-pressure microfluidic reactors. *Science* **311,** 1583-1586 (2006).

[2] Healy, N., Mailis, S., Bulgakova, N.M., Sazio, P.J.A., Day, T.D., Sparks, J.R., Cheng, H.Y., Badding, J.V., & Peacock, A.C., Extreme electronic bandgap modification in laser-crystallized silicon optical fibres. *Nat Mater* (2014).

[3] He, R.R., Sazio, P.J.A., Peacock, A.C., Healy, N., Sparks, J.R., Krishnamurthi, M., Gopalan, V., &

Badding, J.V., Integration of gigahertz-bandwidth semiconductor devices inside microstructured optical fibres. *Nat Photonics* **6,** 174-179 (2012).

[4] Fitzgibbons, T.C., Guthrie, M., Xu, E.S., Crespi, V.H., Davidowski, S.K., Cody, G.D., Alem, N., & Badding, J.V., Benzene-derived carbon nanothreads. *Nat. Mater.* **14,** 43–47 (2015).

BIOGRAPHY

John Badding is Professor of Chemistry, Physics and Materials Science and Engineering at Pennsylvania State University, USA. He has over 200 publications in the general area of materials chemistry. His research interests include new carbon materials, semiconductor optical fibers and photonics, and nanostructured semiconductor materials. His group has pioneered high pressure chemical vapor deposition for crystalline and single crystal elemental and compound semiconductor optical fibers. It has also discovered a new low-dimensional carbon nanomaterial, "diamond nanothreads". Badding leads interest research group 3 (IRG3) of the NSF-funded Penn State Materials Research and Engineering Center (PSU-MRSEC), "High Pressure Enabled Electronic Metalattices". He is the recipient of a David and Lucile Packard Fellowship and a Penn State Faculty Scholar Medal. He holds a Ph.D. from the University of California, Berkeley and did post-doctoral work as a Carnegie Fellow at the Carnegie Institution of Washington.

Fibrous Branched-Clusters: The Next Generation of Tissue Regeneration Templates

Benjamin A. Minden-Birkenmaier, Gretchen S. Selders, and Gary L. Bowlin

Department of Biomedical Engineering, Herff College of Engineering, University of Memphis, Memphis, TN, 38152, U.S.A

E-mail: glbowlin@memphis.edu

Three-dimensional structural templates such as electrospun, hydrogels, and cryogels have been fabricated to mimic the extracellular matrix of a variety of tissues with limited success, often limited in regard to seeded cell infiltration and tailorability of the cellular microenvironments. Thus, there remains a need for a flexible 3-D template system which could be combined with cell suspensions to promote viable tissue regeneration, and ultimately allow cells to freely reorganize and modify their microenvironment. In this study, traditional electrospun templates were processed into electrospun fibrous branched-clusters (Figure 1) using a custom-designed razor blade template blender adapted from a procedure by Thieme *et al.*[1] The fibrous branched-clusters were then separated into three different size categories (small, medium, and large) via centrifugation, with average branch lengths of 65, 326, and 349 μm, respectively. Human dermal fibroblasts ($5x10^5$ cell/ml) were then combined in culture media with various amounts (0.5 mg small/0.5 mg medium, 1 mg large, 1 mg small/1 mg medium, or 2 mg large) of fibrous clusters, and either centrifuged down into pellets or allowed to settle under gravity. After 1 or 3 weeks of static culture in 15 mL tubes, the fibrous cluster and cells had integrated to form three-dimensional tissue-mimicking constructs. Pycnometry was used to measure construct density, and found the densities (1.2-1.6 g/mL) were comparable to those of native soft tissues. Samples were then formalin fixed, cryosectioned, and stained with DAPI for visualization of cell nuclei. This staining revealed uniform cell distributions throughout, indicating that nutrient and waste diffusion was adequate to support cell viability from cluster edge to core (1.3-3.2 mm in diameter). Currently, the expression of Ki67 protein is being evaluated through immunohistochemistry to visualize ongoing fibroblast proliferation throughout the constructs. In conclusion, the results of this study demonstrate the fabrication of fibrous branched-clusters of various sizes, and the ability of these branched-clusters to serve as building blocks for cells to create continuous template constructs achieving homogenous cell distributions not yet realized from statically-seeded, traditional electrospun templates.

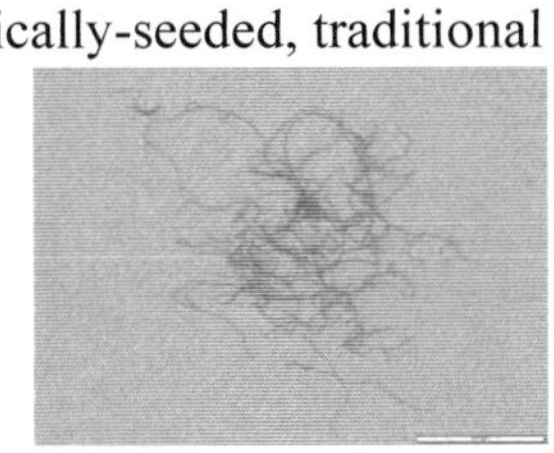

Figure 1. Micrograph of a fabricated medium-sized fibrous branched-cluster (scale bar = 100 microns).

References

[1] Thieme M, Agarwal S, Wendorff J H, Greiner A, *Polymers for Advanced Technologies*, **2011**, 1335-1344.

BIOGRAPHY

Dr. Bowlin is the Herbert Herff Chair of Excellence at The University of Memphis in the Department of Biomedical Engineering. He received his Ph.D. in Biomedical Engineering from the University of Akron in 1996. His laboratory has published extensively in the area of electrospinning for tissue regeneration templates with over 125 peer-reviewed manuscripts. Google Scholar data shows his group's publications have been cited over 15,000 times (H-index of 53). He has also been granted 12 U.S. Patents and over 35 foreign patents and is a Fellow of the National Academy of Inventors.

Development of New Optical Functional Materials based on Composition Design and Structure Analysis

Gemei Cai, Hangkong Li, Liumei Su, Jing Zhang, and Zhanpeng Jin

School of Materials Science and Engineering, Education Ministry Key Laboratory of Non-ferrous Materials Science and Engineering, Central South University, Changsha, 410083, Hunan, China

Email: caigemei@csu.edu.cn

It is known that phase diagrams are important to obtain composition or temperature dependent phase relationship, which is beneficial to rapid design of materials. While, crystal structure largely determines the function of the material. From these points of view, study on the phase diagram and determination of crystal structure has constructive meaning in the progress of developing new materials. Based on composition design and structure analysis, several oxides including borate, titanate, tungstate and phosphate were successfully fabricated in the past few years. These new oxides exhibit promising application prospects in the lighting and display fields due to the excellent optical properties.

Preparation and Piezoelectric Properties Investigation of Electrospun Poly(vinylidene fluoride) Fibrous Membranes

Sheng Chen, Runfang Fu, Yingchun Gu, Jie Jiang, Jingjing Li

Department of Textile Engineering, College of Light Industry, Textile and Food Engineering, Sichuan University, Chengdu, 610065, China

E-mail: chensheng@scu.edu.cn

Polyvinylidene Fluoride (PVDF) is a kind of piezo-polymer materials with extraordinary properties including easy moldability, good toughness, flexibility, and particularly excellent electrical properties such as ferro-, piezo-, and pyro-electricity[1]. Compared with other inorganic piezo-materials such as piezo-ceramics, PVDF materials have the characteristics of good plasticity and good toughness[2]. In this research, PVDF fibrous membranes and its fibrous composite were prepared by electrospinning and were deposited on a rotating drum at rotation speed of 2000 rpm. The composites were obtained by adding cellulose nanocrystals (CNCs) and polyaniline-modified CNCs (PCN), which worked as functional nanofillers in the PVDF matrix. In this research, the effects of the nanofiller contents on the piezoelectricity of electrospun PVDF fibrous membranes were also analyzed.

The results demonstrate that both the crystallinities and the contents of β phase crystals of fibrous composites increase at first and then decrease with increasing contents of CNCs or PCN. The piezoelectricities of the composites reach to the highest level when the content of nanoparticle is 2wt%, which is in accordance with the effects of these two nanoparticles on crystallinity and content of β phase. It is worthy to notice that the piezoelectric voltage of PVDF/CNCs-2wt% fibrous composite was about 50V. In contrast, the piezoelectric voltage of PVDF/PCN-2wt% fibrous composite was about 80V, which is 60% higher than that of PVDF/CNCs-2wt% fibrous composite. PVDF/PCN-2wt% fibrous composite was made into flexible piezoelectric power generators in two different kinds of forms which could be used in different areas. The power generators were prepared by using two pieces of conductive fabrics or copper foils to form sandwich-liked structure. For the power generators whose electrodes are conductive fabrics, the surface areas of the power generators and frequencies of applied forces on the generator play important roles in their piezoelectric voltages. Increasing surface areas and force frequencies are efficient ways to enhance their piezoelectric voltages. As to the power generator which was sandwiched by copper foils, it is able to light 20 blue and 25 white LED bulbs and also to charge capacitors with different capacity. The highest electric power that power generator can store in capacitor is 2.62μJ.

In conclusion, two efficient methods to improve piezoelectricity of electrospun PVDF fibrous membrane were studied in this research. The as-prepared PVDF fibrous composites are expected to be used in the field of flexible electronic device.

References

[1] Dagdeviren C, Joe P, Tuzman O L, et al, *Extreme Mechanics Letters*, **2016**, 9, 269-281.

[2] Martins P, Lopes A C, Lanceros-Mendez S, *Progress in Polymer Science*, **2014**, 39, 683-706.

BIOGRAPHY

Sheng Chen is an associate professor at Sichuan University. He received his bachelor's degree in chemical fibers from Sichuan University in 2002 and his doctor's degree in materials science from the same university in 2008. In 2012 he joined the group of Professor Gang Sun at the University of California at Davis as a postdoctoral fellow. In 2013 he returned to Sichuan University where he works as an associate professor. His current research interests are in the fields of functional polymer materials and nanocomposites, especially conducting and electroactive polymers.

Biomimetic Gelatin/Bacterial Cellulose Nanofiber Hybrid Scaffolds and Their Improved Cellular Response

Shiyan Chen, Baoxiu Wang, Huaping Wang

State Key Laboratory for Modification of Chemical Fibers and Polymer Materials, College of Materials Science and Engineering, Donghua University, Shanghai 201620, P. R. China

E-mail: chensy@dhu.edu.cn

A big challenge in tissue engineering is the design and fabrication of 3D biomimetic cellular environment to mimic the natural extracellular matrix (ECM). The present work reports the fabrication of 3D biomimetic gelatin/bacterial cellulose (BC) nanofiber composite scaffolds using two kinds of BC nanofibers. The scaffolds showed microporous structure with nanofibers which could mimic the natural ECM and facilitate nutrient, metabolic waste transport. The hybrid scaffolds showed excellent compressive mechanical property (an eleven-fold enhancement in compression modulus), super-elastic property which could bear a compressive strain as high as 80% and recovers its original shape afterwards. The BC nanofibers in the scaffolds improved the roughness and stiffness of scaffold surface, promoting cell behaviors in terms of cell adhesion, proliferation using pig iliac endothelium cells (PIECs) as cell model. These biomimetic hybrid scaffolds with improved cell adhesion, and proliferation would be highly promising for tissue engineering application, e.g. bone, cartilage and urethral regeneration.

BIOGRAPHY

Shiyan Chen is a professor of Donghua University. She received her MS degree in Shandong University in 2002, and PhD from Institute of Chemistry, Chinese Academy of Sciences in 2006. Then she joined in Donghua University. She focuses on bacterial cellulose preparation, modification and functional materials.

Inorganic Biomaterials

Yu Chen

State Key Lab of High Performance Ceramics and Superfine Microstructure, Shanghai Institute of Ceramics, Chinese Academy of Sciences, P. R. China

E-mail: chenyu@mail.sic.ac.cn

Nanomedicine provides the new strategies for early diagnosis and efficient therapy of various serious diseases. Biomaterials with unique structure, composition and performance are the basis for the extensive biomedical applications. Compared to traditional organic biomaterials, inorganic biomaterials are featured with high biocompatibility, multifunctionality, high chemical/thermal stability, etc. Therefore, they show the high performance and great promise for versatile applications in biomedicine. This presentation will focus on several important inorganic biomaterial nanoplatforms, including zero-dimensional mesoporous silica/organosilica nanoparticles, two-dimensional nanosheets (oxides, TMDCs, MXenes, etc.), as well as two-dimensional hybrid implants. We will introduce their design/fabrication principles, biological effects and versatile biomedical applications, including control drug releasing, molecular imaging, diagnostic imaging, gene therapy, ultrasound therapy, catalytic nanomedicine, tumor theranostics, etc.

References

[1] Y. Chen,* J. L. Shi,* et al., *Nat. Commun.*, **2017**, In press.

[2] Y. Chen,* J. L. Shi,* *Adv. Mater.*, **2016**, 28, 3235-3272. (Review)

[3] Y. Chen,* et al., *Adv. Mater.*, **2017**, 29, 1604105. (Review)

BIOGRAPHY

Prof. Yu Chen received his bachelor degree at Nanjing Tech University and PhD degree at Shanghai Institute of Ceramics, Chinese Academy of Sciences (SICCAS). He is now the full professor and Ph.D supervisor in SICCAS. His research focuses on the design and application of inorganic biomaterials, includes the design, synthesis and biomedical applications of zero-dimensional organic-inorganic hybrid mesoporous materials, two-dimensional biomaterials and three-dimensional composite implants, including mesoporous material for drug delivery, planar nanosheets for theranostic nanomedicine, molecular probes for molecular imaging (MRI, CT, PA and US), ultrasound therapy, cardiac therapy, immunotherapy, non-virus gene delivery vehicles and *in-situ* localized tumor therapy. He has published more than 110 SCI papers in nanomedicine with more than 5700 citations (h-index: 40), such as Nature Communications, Chem. Soc. Rev., Acc. Chem. Res., Adv. Mater., J. Am. Chem. Soc., Angew. Chem. Int. Ed., Nano Letters, ACS Nano, Biomaterials, etc.

Synthesis of Inorganic-organic Nanocomposites for the Imaging and Therapy of Tumor

Nuo Yu, Zhouqi Meng, Zhaojie Wang, Meifang Zhu, Zhigang Chen*

State Key Laboratory for Modification of Chemical Fibers and Polymer Materials,

College of Materials Science and Engineering, Donghua University, Shanghai 201620, China

E-mail: zgchen@dhu.edu.cn

Near-infrared (NIR) laser-induced photothermal ablation therapy (PAT) has great potential to revolutionize conventional therapeutic approaches for cancers, and a prerequisite is to obtain biocompatible and efficient photothermal nanoagents. To address this problem, we have used hydrophilic polymer as surface ligands to prepare several semiconductor nanoagents, including CuS [1] and $W_{18}O_{49}$ nanomaterials [2]. Their aqueous dispersions exhibit intense NIR absorbance and excellent photothermal effects. Importantly, if we inject aqueous dispersion into the tumor in mice, cancer cells *in vivo* can be efficiently ablated under the irradiation of NIR laser with a safe intensity for ~10 min. To further improve the therapeutic effects, the combination of NIR-PAT and chemotherapy has been proposed for the synergic therapy. Recently, by combing semiconductor nanoagents and thermosensitive polymer nanogels, we have constructed smart nanocapsules (G-CuS-DOX) that can be switched by NIR-laser[3]. The nanocapsules exhibit the controllable and efficient photothermal/ chemotherapy effect compared with single PAT or chemotherapy effect for the tumor. Furthermore, ideal theranostic nanoplatform for tumor should be one nanoparticle that has single semiconductor or metal component but contains all multi-model imaging and therapy abilities. By tuning vacancy concentrations via the morphology control or doping technology, we have developed FeS_2 (size: ~350 nm) [4] and Nb-doped TiO_2 [5] nanoparticle as "all-in-one" multifunctional nanoagents for the imaging guided PAT of tumor. Therefore, these inorganic-organic nanocomposites have great potential in the imaging and/or therapy of tumor.

References

[1] (a) Q. W. Tian, M. H. Tang, Y. G. Sun, R. J. Zou, Z. G. Chen, M. F. Zhu, S. P. Yang, J. L. Wang, J. H. Wang and J. Q. Hu. *Adv. Mater.*, **2011**, *23*, 3542; (b) R. Zhong, C. Peng, L. Chen, N. Yu, Z. Liu, M. Zhu, C. He and Z. Chen. *RSC Adv.*, **2016**, *6*, 40480.

[2] Z. G. Chen, Q. Wang, H. L. Wang, L. S. Zhang, G. S. Song, L. L. Song, J. Q. Hu, H. Z. Wang, J. S. Liu, M. F. Zhu and D. Y. Zhao. *Adv. Mater.*, **2013**, *25*, 2095.

[3] Z. Meng, F. Wei, R. Wang, M. Xia, Z. Chen, H. Wang and M. Zhu. *Adv. Mater.*, **2016**, *28*, 245.

[4] Z. Meng, F. Wei, W. Ma, N. Yu, P. Wei, Z. Wang, Y. Tang, Z. Chen, H. Wang and M. Zhu. *Adv. Funct. Mater.*, **2016**, *26*, 8231.

[5] N. Yu, Y. Hu, X. Wang, G. Liu, Z. Wang, Z. Liu, Q. Tian, M. Zhu, X. Shi and Z. Chen. *Nanoscale*, **2017**, *9*, 9148.

BIOGRAPHY

Zhigang Chen received his B.S. and M.S. degrees from Central China Normal University in 2002 and 2005, respectively. He received his Ph.D. degree from Fudan University in Jul 2008. Afterwards, he joined Donghua University as a lecture. From Oct 2008 to Dec 2009, he carried out postdoctoral research at Max Planck Institute for Colloids and Interfaces in Germany, supported by the Alexander von Humboldt Foundation. In 2010 and 2013, he was elevated to Associate Professor and Full Professor, respectively. His researches focus on the synthesis of photothermal nanomaterials and their application in biomedical and energy fields. Dr. Chen has authored over 110 peer-reviewed papers. He was listed as one of 'Highly Cited Researchers' (Materials) by Thomson Reuters in 2016.

Fluid Instability during Thermal Drawing Processing for Fibers

Daosheng Deng

Department of Aeronautics and Astronautics, Fudan University, Shanghai, China

E-mail: dsdeng@fudan.edu.cn

Instability of a thin liquid film, such as dewetting arising from Van der Waals force, has been well studied, and is typically characterized by formation of many droplets. Interestingly, a thin liquid film subjected to an applied stretching during a process of thermal drawing is evolved into an array of filaments, i.e., continuity is preserved along the direction of stretching while breakup occurs exclusively in the plane of cross section. Here, to understand this anisotropic instability, we build a physical model by considering both Van der Waals force and the effect of stretching. By using the linear instability analysis method and then performing a numerical calculation, we find that the growth rate of perturbations at the cross section is larger than that along the direction of stretching, resulting in the anisotropic instability of the stretching film. These results may provide theoretical guidance to achieve more diverse structures for nanotechnology.

BIOGRAPHY

Daosheng Deng is a professor at the Department of Aeronautics and Astronautics at Fudan University in Shanghai, China. He received the B.S. and M.S. degrees in Physics from Fudan University, China, in 2000 and 2003, respectively, and the Ph.D. degree in Materials Science from MIT, USA, in 2010. Then he continued his postdoctoral research at MIT and in Harvard University. His research group currently focuses on complex fluids and its technology applications at two perspectives: One perspective is the basic research to explore physical phenomena and the underlying mechanisms (including droplets, bubbles, vapor steam, electro-osmotic flow, electrolyte, fluid instability, etc.); the other perspective is the relevant technological applications for the sustainable environment and energy and among others (such as nanotechnology, desalination, solar steam, water purification and treatment, etc.). His research results have been published in *Nature*, *Phys. Rev. Lett.* and *Nano Lett.* and featured by *MIT news, PhysOrg, ScienceDaily,* and *Technology Review*.

Formation of Self-Assembled Nanofibers for Sensing and Imaging of Polyprotein

Dan Ding

College of Life Sciences, Nankai University, Tianjin, 30071, China

E-mail: dingd@nankai.edu.cn

Understanding specific polyprotein-peptide interactions could offer a deep insight into the development of therapeutics for many human diseases. In this work, we designed and synthesized a far-red/near-infrared (FR/NIR) fluorescence light-up probe (DBT-2EEGWRESAI) by simply integrating two tax-interacting protein-1 (TIP-1)-specific peptide ligands (EEGWRESAI) with one 4,7-di(thiophen-2-yl)-2,1,3-benzothiadiazole (DBT) unit. We firstly demonstrated that DBT is an environment-sensitive fluorophore with FR/NIR fluorescence due to its strong charge transfer character in the excited state. Thanks to the environmental sensitivity of DBT, the probe DBT-2EEGWRESAI is very weakly fluorescent in aqueous solution but lights up its fluorescence when the probe specifically binds to polyprotein (ULD-TIP-1 tetramer). It is found that DBT-2EEGWRESAI/ULD-TIP-1 tetramer could self-assemble to nanofiber network, respectively, which lead to the probe fluorescence turn-on through providing DBT with hydrophobic microenvironment. By virtue of the self-assembly-induced FR/NIR fluorescence turn-on, DBT-2EEGWRESAI shows unique merits in detecting and visualizing specific polyprotein-peptide interactions in both solution and live bacteria.

References

[1] H. Wang, J. Liu, A. Han, N. Xiao, Z. Xue, G. Wang, J. Long, D. Kong, B. Liu, Z. Yang, D. Ding, *ACS Nano*, **2014**, *8*, 1475-1484.

BIOGRAPHY

Dan Ding obtained his Bsc. in Chemistry (2005) and DPhil. (2010) in Polymer Chemistry and Physics from Nanjing University. After a postdoctoral training in the Department of Chemical & Biomolecular Engineering in National University of Singapore, he joined Nankai University in March 2013, where he is currently a professor in the College of Life Sciences and State Key Laboratory of Medicinal Chemical Biology. In 2013, he was selected in the Tianjin 1,000 Young Talents Plan. In 2014, he was selected in the 100 Young Academic Leaders Plan in Nankai University. In 2016, he was obtained the China National Science Fund for Excellent Young Scientists. His current research interest focuses on the design and preparation of nanomedicines and new molecular imaging probes as well as exploration of their biomedical applications.

DNA Nanotechnology-enabled Engineering at the Biosensing Interface

Chunhai Fan

Division of Physical Biology, and Bioimaging Center, Shanghai Synchrotron Radiation Facility, Shanghai Institute of Applied Physics, Chinese Academy of Sciences, Shanghai 201800, China

E-mail: fchh@sinap.ac.cn

Proteins and nucleic acids are dynamically organized in cells to realize their physiological functions with spatial and temporal orderliness. This type of elegant supermolecular assembly has inspired researchers to create molecular/biomolecular structures with dynamic organization outside of the cells. In particular, DNA nanotechnology has proven to possess extraordinary flexibility and convenience for "bottom-up" construction of exquisite nanostructures with high controllability and precision, which holds great promise in a wide range of applications, e.g., nanofabrication and molecular electronics, in-vivo and in-vitro sensing and drug delivery.

In this talk, I will present several examples of using tetrahedral DNA nanostructures (TDNs) for dynamic organization of biomolecules in vitro. TDNs are three-dimensional (3D) DNA architecture with high mechanical rigidity and structural stability, which are suitable for organization of higher-ordered nanocomplexes and nanodevices. As one of the examples,we employed single-particle tracking to visualize the internalization of TDNs, and dissect the cell entry pathways of these virus-like nanoparticles. In the second example, I hope to use the way that we employed TDNs to dynamically organize the biosensing interface, and realize macroscopic applications on diagnostics.

References

[1] J. Li, A. A. Green, H. Yan, C. Fan, *Nature Chem.* DOI:10.1038/nchem.20405, in press (**2017**).

[2] H. Zhang, J. Chao, D. Pan, H. Liu, Y. Qiang, K. Liu, C. Cui, J. Chen, Q. Huang, J. Hu, L. Wang, W. Huang, Y. Shi, C. Fan, *Nature Commun.* 8, 14738 (**2017**).

[3] M. Liu, Q. Li, L. Liang, J. Li, K. Wang, J. Li, M. Lv, N. Chen, H. Song, J. Lee, J. Shi, L. Wang, R. Lal, C. Fan, *Nature Commun.* 8, 15646 (**2017**).

[4] M. Lin, P. Song, G. Zhou, X. Zuo, A. Aldalbahi, J. Shi and C. Fan, *Nature Protocols* 11, 1244 (**2016**).

[5] M. Lin, J. Wang, G. Zhou, J. Wang, N. Wu, J. Lu, J. Gao, X. Chen, J. Shi, X. Zuo and C. Fan, *Angew. Chem. Int. Ed.* 54, 2151 (**2015**).

BIOGRAPHY

Chunhai Fan obtained his B.S. and Ph.D. from the Department of Biochemistry at Nanjing University in 1996 and 2000. After his postdoctoral research at University of California, Santa Barbara (UCSB), he joined the faculty at Shanghai Institute of Applied Physics (SINAP), Chinese Academy of Sciences (CAS) in 2004. He is now Professor and Chief of the Division of Physical Biology at SINAP and the Bioimaging Center of Shanghai Synchrotron Radiation Facility (SSRF). He is also an adjunct professor of the Biodesign Institute at Arizona State University. He is a fellow of Royal Society of Chemistry (FRSC) and an elected fellow of the International Society of Electrochemistry (ISE). He serves as an Associate Editor of ACS Applied Materials & Interfaces and serves on the editorial board of over ten international journals. Dr. Fan has published ~300 papers in peer-reviewed journals, and he was recognized as High Cited Researchers in 2014-2106 by Thomson Reuters.

Multifunctional Carbon Nanotube Fibers and Sheets that Sense, Actuate, and Store Energy

Shaoli Fang[1], Jiangtao Di[1], Ray H. Baughman[1], Zunfeng Liu[2,3], Mei Zhang[4]

[1]*NanoTech Institute, the University of Texas at Dallas, Richardson, USA;*

[2]*Jiangnan Graphene Research Institute, Changzhou, China;*

[3]*College of Electronic Information and Optics Engineering, Nankai University, Tianjin, China;*

[4]*High-Performance Materials Institute, Florida State University, Tallahassee, USA*

E-mail: sfang@utdalas.edu

Twist-spun carbon nanotube (CNT) yarns derived from CVD-grown spinnable multiwalled nanotube forest are of great interest for such diverse applications as artificial muscles, supercapacitors, batteries, intelligent textiles and structural composites. These twist-spun, torque stabilized nanotube fibers can be produced with strength over 700 MPa. Using ITAP (Incandescent Tension Anneal Process, i.e., thermally annealing twist-spun CNT yarns at a temperature of about 2000 °C while these yarns are under high tensile loads), we increased CNT yarn modulus and yarn strength by factors of up to 12 and 2.6, respectively. Using our biscrolling technology, which traps multifunctional powders inside nanotube yarns, we fabricated superconducting yarns, textile batteries and supercapacitors, and superelastic electronic yarns. We also demonstrated a bi-sheath fiber strain sensor for giant tensile and torsional displacements, where buckled structure was used to maximize the contact area between adjacent buckles using carbon nanotube sheets, which contained two layers of buckled CNT sheath and a buckled rubber interlayer coaxially coated on a elastic support fiber.

BIOGRAPHY

Shaoli Fang is Research Professor at the NanoTech Institute of the University of Texas at Dallas, received his Ph.D. in Materials Science from Institute for Molecular Science in Japan (1994). Before joining the NanoTech Institute of UTD in 2004, he has served various management and technical positions in semiconductor, telecommunication, and nanotechnology industries. Shaoli has over 80 publications, 2 Issued US Patents, over 10 pending U.S. patent filings while at UTD (430 page PCT filing being nationalized in the U.S.). He has received Nano 50 Awards from Nanotech Briefs Magazine for Carbon Nanotube Sheets and Yarns (2006), the Scientific American Magazine 50 recognition (2006), and Gold Winner of the R&D 100 Market Disruptor Product Recognition Award (2015).

Multifunctional Photonic Fibers: Sensing through Structure and Material

Xian Feng

Laser Institute of Engineering, Beijing University of Technology, Beijing 100124, China

E-mail: xfeng2015@bjut.edu.cn

The microstructured photonic crystal fiber [1] and the emerging multi-material multifunctional fiber [2,3] represent the trend of the next generation of photonic fibers. To construct the fiber with structures and materials, novel functionalities can be realized, compared to the conventional optical fiber. Using the structure approach, the performance of fiber waveguide can be widely tailored through building microstructure or nanostructure in the fiber. By introducing various non-glass, non-dielectric materials, e.g., semiconductors, metals, and so on, into dielectric glass or polymer matrix, multiple functionalities beyond the traditional optics capabilities can be achieved in a single integrated fiber.

I present here with our progress of multi-functional photonic fibers through structure and material approaches. First, an air-suspended-core (ASC) photonic crystal fiber (see Fig.1a) is employed as high-sensitivity sensors for monitoring bio-chemical substances in the aqueous environment around the fiber core. Due to the strong evanescent field around the small core, ppm-level of chemical impurities can be detected. The fluorescence signal or Raman signal has been utilized for characterizing the species and the concentration of various impurities. Nano-liter of aqueous solution is sufficient for the sampling in such a holey fiber microfluidic device. Second, a glass-clad semiconductor-core fiber (see Fig.1b) has been demonstrated as a photo-detector with fast response of 1MHz. Semiconductor germanium is chosen as the core material. In the future, a combination of structure and material would provide even complex and desirable sensing functions in photonic fibers.

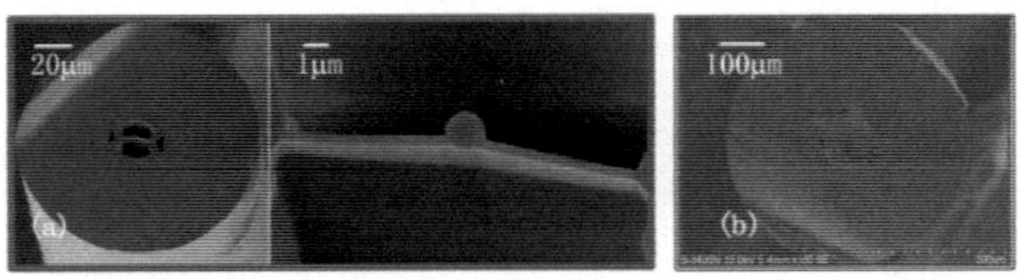

Figure 1. SEM photographs of (a) ASC fiber with 1μm core. (b) Ge fiber with 125 μm core.

References

[1] P. St. J. Russell, *Science*, **2003**, 299, 358-362.

[2] A. F. Abouraddy, et al., *Nature Materials,* **2007**, 6, 336-347.

[3] G. Tao, et al., *I. J. Appl. Glass Sci.,* **2012**, 3, 349-368.

BIOGRAPHY

Xian Feng is currently a full professor of Laser Institute of Engineering, Beijing University of Technology. In 1999, he was awarded Doctor degree of Engineering from Shanghai Institute of Optics and Fine Mechanics, Chinese Academy of Sciences. In 1999-2001, he worked in Kyoto University, Japan and Rutgers University, USA as postdoctoral researcher. In 2001, he joined Optoelectronics Research Centre, University of Southampton, UK. He has then been working in the pioneering team in ORC for designing and fabricating high-performance non-silica glass photonic crystal fibers until 2015. He has so-far published more than 140 peer-reviewed papers and his H-index is 25.

Graphene Fibers: Scalable Preparation and Energy-Related Applications

Chao Gao

Department of Polymer Science and Engineering, Zhejiang University, Hangzhou, 310027, China

E-mail: chaogao@zju.edu.cn

Graphene fibers, assembled from graphene nanoplatelets, belong to the new type of carbonaceous fibers with high expectations in terms of mechanical and functional performances. Our group has been focusing on this topic for decades, aiming to extend the charming attributes of graphene into macroscopic fibers. Nowadays, the strongest graphene fibers could reach a tensile strength up to 2 GPa, an ultrahigh stiffness of 282-400 GPa, and an outstanding electrical conductivity of $\sim 10^6$ S/m. Meanwhile, the productivity of high-quality graphene fibers is significantly improved, which are able to be continuously fabricated in kilometres and/or kilograms, making it possible for large scale applications. Combining with the low density of carbon-based materials, the lightweight, strong, conductive and flexible graphene fibers show great promise in various high-tech areas such as aerospace industry, vehicles, cables, electronic and energy-storage devices and so on. In this presentation, the scalable fabrication of graphene fibers and yarns, our full scale synergetic defect engineering protocol and doping strategy for high-performance graphene fibers, as well as their applications in wearable supercapacitors and fibrous electrical heating elements will be introduced.

BIOGRAPHY

Prof. Chao Gao joined the Department of Polymer Science and Engineering, Zhejiang University, and was promoted as a full Professor in 2008. In 2001, he was a faculty member in SJTU, and Associate Professor in 2002. Since Nov. 2003, he worked with Prof. Sir Harry Kroto as a visiting scholar and postdoc research fellow in the University of Sussex, UK, and then moved to Prof. Axel H. E. Müller's group at Bayreuth University, Germany in July 2005 as an Alexander von Humboldt research fellow.

His research interests include macro-assembly of graphene and its applications. He co-edited a book on hyperbranched polymers (published by John, Wiley & Sons) and published more than 150 papers with citation of more than 9400 times and H-index 50. His research of graphene fiber knot has been selected by Nature as "Images of the Year" in 2011. He was awarded or funded with 2013 Ten Pieces of News in Science and Technology in China, National Science Fund for Distinguished Young Scholars, the least dense solid Guinness World Records, "Gold Kangaroo" World Innovation Award, Academician of Asia-Pacific Academy of Materials (APAM), Zhejiang Youth Technology Award, and National Excellent Doctoral Dissertation of China.

Plasmonic Fiber-optic Biomedical Sensors

Tuan Guo

Institute of Photonics Technology, Jinan University, China

E-mail: tuanguo@jnu.edu.cn

Surface Plasmon resonance (SPR) optical fiber sensors can be used as a cost-effective and relatively simple-to-implement alternative to well established bulky prism configurations for in-situ high sensitivity biochemical and electrochemical measurements. The miniaturized size and remote operation ability offer them a multitude of opportunities for single-point sensing in hard-to-reach spaces, even possibly in vivo. Grating-assisted and polarization control are two key properties of fiber-optic SPR sensors to achieve unprecedented sensitivities and limits of detection. The biosensor configuration presented here utilizes a nano-scale metal-coated tilted fiber Bragg grating (TFBG) imprinted in a commercial single mode fiber core with no structural modifications. Such sensor provides an additional resonant mechanism of high-density narrow cladding mode spectral combs that overlap with the broader absorption of the surface Plasmon for high accuracy interrogation. In this paper, we briefly review our recent developments of plasmonic tilted fiber grating sensors, including the surface affinity studies of biomolecules for real life problems, electrochemical actives of electroactive biofilms for clean energy resources.

BIOGRAPHY

Tuan Guo is a Professor in the Institute of Photonics Technology, Jinan University, Guangzhou, China. He received the Ph.D. in Optics from Nankai University in 2007. Thereafter he worked as a Postdoctoral Fellow with the Department of Electronics at Carleton University (Canada) and the Photonics Research Centre at The Hong Kong Polytechnic University. He joined the Jinan University as an Associate Professor in 2011 and promoted to a full Professor in 2014. He has authored and coauthored more than 120 papers in the peer-reviewed international journals (included 4 invited review papers) and presented over 20 invited talks at international and national conferences. He holds 15 patents and pending patents. His research activities include optical fiber sensors, fiber lasers, fiber gratings, plasmonics, biophotonics, optofluidics. He was a Co-Chair of the "IEEE Photonic Technology in Instrumentation and Measurement" technical committee, an Associated Editor for *Journal of Sensors* from 2010 to 2014 and a Guest Editor for a Special Issue on "Recent Advances in Fiber Bragg Grating Sensing" in *MDPI Sensors* in 2016. He is a Senior Member of IEEE and Member OSA.

Bioengineering of Hydrogels for Smart Wound Care

Bhuvanesh Gupta[1], **Amlan Gupta**[2], **Surabhi Singh**[1], **Sadiya Anjum**[1], **Jincy Joy**[1]

[1]*Bioengineering Laboratory, Department of Textile Technology, Indian Institute of Technology, New Delhi-110016, India;*

[2]*Department of Pathology, Sikkim Manipal Institute of Medical Sciences, Gangtok, Sikkim, India*

E-mail: bgupta@textile.iitd.ernet.in

Hydrogels and hydrocolloids have been an advantage to human healthcare due to their very innovative and beneficial life support features. Hydrogels have excellent biocompatibility, hydrophilicity, processability and sometimes they are bioactive in nature as well. A series of hydrogels based on polyvinyl alcohol, chitosan, pectin and dextran have shown enormous potential in wound care system. These hydrogels may be combined with appropriate ingredients to develop a system which would offer optimum healing to the wound. Herbal bioactive agents have been used for this application. However, the development of such materials requires a close collaboration between medicine, microbiology and materials chemistry. The development of polymers with bioactive coating is an important area of research focused on solving the problem of contamination by infection in wound care systems. We have observed that a bioactive component may be incorporated within the hydrogel matrix to make it infection resistant. Herbal drugs and essential oils have been incorporated into either natural hydrogels like chitosan, pectin and dextran by blending approach. A wide range of natural bioactive agents such as aloe vera, curcumin, sandal wood oil, clove oil and honey are available to develop excellent materials for wound care. These dressings have been evaluated in wound healing using mouse as the animal model. Excellent healing with minimum scar by hydrogel dressings has been observed.

References

[1] S. Anjum, A. Arora, MS Alam, and B. Gupta, *Int J Pharm,* **2016**, *92, 508.*

[2] S. Anjum, A. Gupta, D. Sharma, A. Kapil, A. Sharma, and B. Gupta, *Mat. Sci. Eng. C.* **2016**, 64, 157.

BIOGRAPHY

Dr. Bhuvanesh Gupta is the Professor of Polymers at Indian Institute of Technology, New Delhi and is the President of three international bodies; Asian Polymer Association, Society of Biomaterials Artificial Organs India (Delhi), and Indo-Italian Forum on Biomaterials & Tissue Engineering. Dr. Gupta worked for eight years in France, Sweden and Switzerland under different capacities in several laboratories. Dr. Gupta initiated research career in the field of polymer functionalisation, biomaterials and tissue engineering and worked in collaboration with University of Uppsala, Sweden and INSERM Paris, France.

Protein Resistant and Antithrombogenic Properties of Amphiphilic Block Copolymer Coatings Prepared by RAFT Polymerization

Kazutoshi Haraguchi[1], **Tetsuso Takada**[1], **Kazumoi Kubota**[1], **Toshihide Mizuno**[2], **Toru Takehisa**[1]

[1]*Department of Applied Molecular Chemistry, College of Industrial Technology, Nihon University, Narashino, Chiba. 275-8575, Japan;*

[2]*Department of Artificial Organs, Research Institute, National Cerebral and Cardiovascular Center, Suita, Osaka, 565-8565, Japan.*

E-mail: haraguchi.kazutoshi@nihon-u.ac.jp

We have developed nanocomposite hydrogels (NC gels) and soft nanocomposite materials which have biocompatibilities in terms of cell cultivation, implantation, and protein resistant properties.[1] Based on these studies, we designed novel amphiphilic block copolymers composed of hydrophobic (poly(2-methoxyethyl acrylate): M) and hydrophilic (poly(*N,N*-dimethylacrylamide): D) segments. Two types of amphiphilic block copolymers, triblock MDM and 4-arm block $(MD)_4$ copolymers, were synthesized by living radical polymerization.[2] These copolymers show good adhesion to various types of substrates, and their coatings showed high protein repellency, a suspension cell culture capability, and antithrombogenicity.[2] In a Chandler-loop method with whole blood, thromboresistance and adsorption resistance to blood-derived proteins were observed.[3] Further, the high hemocompatibility was confirmed with human whole blood by thrombelastography and platelet function analyses.[3]

References

[1] K. Haraguchi, *Polym. J.*, **2011**, *43*, 223-241.

[2] K. Haraguchi, K. Kubota, T. Takada, S. Mahara, *Biomacromolecules*, **2014**, *15*, 1992-2003.

[3] K. Haraguchi, T. Takehisa, T. Mizuno, K. Kubota, *ACS Biomater. Sci. Eng.*, **2015**, *1*, 352-362.

BIOGRAPHY

Kazutoshi Harguchi is a Professor in Department of Applied Molecular Chemistry, College of Industrial Technology, Nihon University. He graduated from Kyushu University, and received his Doctor degree from Kyushu University in March 1978. He joined Kawamura Institute of Chemical Research (KICR) and he worked in Liverpool University, UK as a post-doctoral fellow in 1979-1981. He became a Director of KICR in 2001, and a General Manager (Senior Managing Director) in 2006. He became a Professor of Nihon University, and a Visiting Professor of University of Tokyo from April, 2014. His major research fields are polymer physics, nanocomposites, biomaterials, and polymer hydrogels. He received the Award of the Society of Fiber Science and Technology, Japan in 2003, and the Award of the Society of Polymer Science, Japan in 2010. Since 2016, he is a vice president of the Society of Polymer Science, Japan.

Molybdenum Disulfide-Based Nanoplatform for Cancer Theranostics

Liang Chen, **Chuanglong He***

College of Chemistry, Chemical Engineering and Biotechnology;

State Key Laboratory for Modification of Chemical Fibers and Polymer Materials,

Donghua University, Shanghai, 201620, China

E-mail: hcl@dhu.edu.cn

As a two-dimensional (2D) graphene-like layered material, molybdenum disulfide (MoS_2) has attracted increasing attention in the biomedical field by virtue of their outstanding photothermal conversion efficacy, good biocompatibility, and potential degradability.[1] However, most of reported MoS_2-based nanoagent for cancer therapy was prepared by "top-down" exfoliated strategy, which is always time-consuming and produces layered MoS_2 with uncontrollable size and morphology. On other hand, the fabrication of safe and stable theranostics is also highly rewarding to realize simultaneous cancer diagnosis and treatment.

Herein, bovine serum albumin−gadolinium (BSA−Gd) complexes and MoS_2 nanoflakes were successfully married as cancer theranostics (MoS_2−Gd−BSA) for dual-modality magnetic resonance (MR)/photoacoustic (PA) imaging-guided photothermal therapy (PTT).[2] In our design, BSA−Gd complexes were selected owing to their stability and excellent MR contrast capability. The preparation of theranostic agent MoS_2-Gd-BSA was schematically illustrated in **Figure 1**. The as-prepared MoS_2−Gd−BSA possessed a good photostability and photothermal effect. The cytotoxicity assessment and hemolysis assay suggested the excellent biocompatibility of MoS_2−Gd−BSA. Meanwhile, MoS_2−Gd−BSA could not only achieve in vivo MR/PA dual-modality imaging of xenograft tumors, but also effectively kill cancer cells in vitro and ablate the xenograft tumors in vivo upon 808 nm laser illumination. Thus, our results substantiated the potential of MoS_2−Gd−BSA for cancer theranostics.

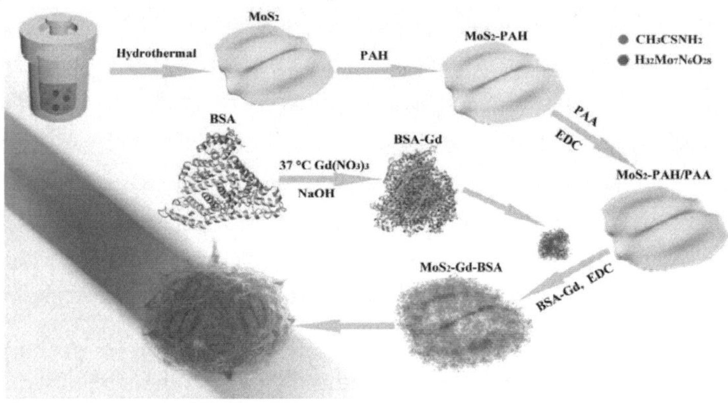

Figure 1. Schematic illustration of the preparation of BSA−Gd complexes functionalized MoS_2 nanoflakes as a theranostic nanoplatform.

References

[1] Hao, J. L., et al. In Vivo Long-Term Biodistribution, Excretion, and Toxicology of PEGylated Transition-Metal Dichalcogenides MS_2 (M = Mo, W, Ti) Nanosheets. *Adv. Sci.* **2017**, 4, 1600160.

[2] Chen, L., et al. Marriage of Albumin−Gadolinium Complexes and MoS_2 Nanoflakes as Cancer Theranostics for Dual-Modality Magnetic Resonance/Photoacoustic Imaging and Photothermal Therapy. *ACS Appl. Mater. Interfaces.* **2017**, 9, 17786−17798.

A Novel Hydrophilic Anti-Frogging Prepared by Cellulose Nanofibrils Manufactured via Formic Acid Hydrolysis plus the Followed High Pressure Homogenization

Dannong He[1]*, Bin Li[2], Fang Zhang[1], Lin Lin[1]

[1]National Engineering Research Center for Nanotechnology, shanghai, China;

[2]CAS Key Laboratory of Bio-based Material, Qingdao Institute of Bioenergy and Bioprocess Technology, Chinese Academy of Sciences, Qingdao, China

E-mail: hdn_nercn@163.com

Cellulose nanofibrils (CNFs) isolated from cellulosic materials (e.g. wood, bamboo, herbaceous plants, etc.) have received increasing attention recently due to their unique properties such as high length-to-width ratio, high specific surface area, low density, high elastic modulus, low thermal expansion, renewability, and biodegradability. Therefore, CNFs can be used as building blocks for the manufacture of many sustainable nanomaterials.

In this work, CNFs were firstly prepared by a sustainable process of formic acid (FA) hydrolysis combined with high pressure homogenization. After FA hydrolysis of bleached hardwood pulp, the fibrous cellulosic solid residue was fibrillated to CNFs with relatively low-pressure homogenization in the solvent of dimethylacetamide (DMAC). The obtained CNFs had high aspect ratio of 100-200 with the diameter in 5-25 nm and the length in 500-1200 nm, and they could be well dispersed in DMAC, dimethylsulfoxide (DMSO), and imethylformmide (DMF) because of the introduced carboxyl or hydroxy groups on the surface of the products. In addition, FA could be readily recovered and reused with the recovery rate higher than 90%.

Anti-frogging coating was prepared by spraying technology using the obtained CNFs dispersion in DMSO. The resultant coating exhibited high transparency, hydrophilic and contact angle < 12° with sudium glass. CNFs coating can withstand 70°C steam repeated 30 times and below 18° for 30 min to room temperature without fog. The CNFs coating shows the advantage of environmentally friendly non-toxic compared to the market hydrophobic anti-fog spray. This novel anti-frogging method has high potential application in the fields of the car window, bathroom mirror, etc.

References

[1] Angelika-Ioanna G., Argyro B., Maria K., Poonam N., Athanasios A. K., *Innovative Food Science and Emerging Technologies*, **2016**, 33:416-421.

[2] Abdul Khalil H. P. S., Davoudpour Y., Nazrul Islam Md. Mustapha, A., Sudesh K., Dungani R., Lawaid M., *Carbohydrate Polymers*, **2014**, 99: 649-665.

[3] Nechyporchuk O., Belgacem M.N., Bras J., *Industrial Crops and Products*, **2016**, 93: 2-25.

[4] Missoum K., Sadocco P., Causio J., Belgacem M.N., Bras J., *Materials Science and Engineering C*, **2014**, 45: 477-483.

BIOGRAPHY

Prof. He is the director of National Engineering Research Center for Nanotechnology and Application; doctoral tutor of Shanghai Jiao tong University; the member of the third national expert group on nanoscience and technology guidance coordination committee. Prof. He has devoted to the nano-materials and multi-fields application technology research, scientific and technological management. Prof. He has issued about 2 books, 200 papers and 150 patents.

Insights into the Flow Processing of Silks

Chris Holland

Natural Materials Group, The Department of Materials Science and Engineering, The University of Sheffield, Mappin Street, Sheffield, S1 3JD. United Kingdom.

E-mail: christopher.holland@sheffield.ac.uk

Silks are biological polymers that have evolved to be processed by controlled protein denaturation, a process depending on the researchers' background, with similarities to amyloidogenesis for some and flow induced crystallisation for others. Understanding the fundamental impact processing has on the performance of a silk will be the focus of this presentation.

Processing silk in the unspun liquid state has been largely explored over the past 15 years through the use of rheology. In this talk our contributions to this area will be presented and the tools that have been developed to probe structural hierarchies in silk as it self-assembles. Discussing more recent work we will draw on how whole animal and feedstock behaviour have supported new perspectives onto silk hydration, the natural spinning process, improved resolubilisation strategies and silk protein applications. We will conclude there is more to silk than just a fibre and that Nature may in fact hold unique solutions to the current challenges facing the synthetic polymer industry, i.e. routes towards low embodied energy, sustainable wet processing of polymers.

BIOGRAPHY

Chris is based in The University of Sheffield in the Materials Science and Engineering Department where he leads the Natural Materials Group (www.naturalmaterialsgroup.com). He currently holds an EPSRC Early Career Fellowship, is an Associate Editor for ACS Biomaterials Science and Engineering and Chair of the IoM3 Natural Materials Association.

His research uses tools developed for the physical sciences to better understand how processing effects performance in natural materials. Using silk as a model system and studying it is spun he has been able to gain unique insights into this material's biodiversity, structure and evolution. Additionally, this work has made important links between natural and industrial fibre processing which has led to several patents and a fundamentally new way of designing, testing and fabricating bio-inspired materials which is now being realised as Co-ordinator of the H2020 FET Open project FLIPT (www.h2020flipt.eu).

Electrospun Nanofibres of High-Performance Polymer Prepared by Solid-state Polymerization of Molecularly Self-assembled Monomers

Jian Zhu[1,2], Yichun Ding[1], Seema Agarwal[2], Andreas Greiner[2], Hean Zhang[3], and Haoqing Hou[1]

[1]*Department of Chem. and Chem. Engi., Jiangxi Normal Univ., Nanchang, 330022, China;*

[2]*Macromol. Chem. II, Univ. of Bayreuth, Universitätsstrasse 30, 95440, Bayreuth, Germany;*

[3]*Department of Chemistry, Nanchang University, Nanchang, 330006, China*

Polybisbenzimidazobenzophenanthroline-dione (BBB) is a high-performance polymer which is characterized by very high mechanical strength in combination with exceptional thermal stability,[1-4] but it cannot be processed to electrospun fibres for any useful applications due to its insolubility and infusibility.[5-8] We overcome all obstacles in an electrospinning of BBB by a new bottom-up, facile, and scalable approach for the solid-state polymerization of self-assembled monomer precursors (Figure 1). Key to this new strategy is the incorporation of a high molecular weight sacrificial polymer to aid fibre formation. The resulting electrospun BBB fibres and belts thereof prepared according to this new approach are extremely strong show excellent thermal stability. We envisage that this procedure could be applied to other classes of non-processable high-performance polymers for the preparation of electrospun fibres for applications such as filtration, sound insulation, battery separation, electrodes, fire protection, and reaction engineering under demanding conditions.

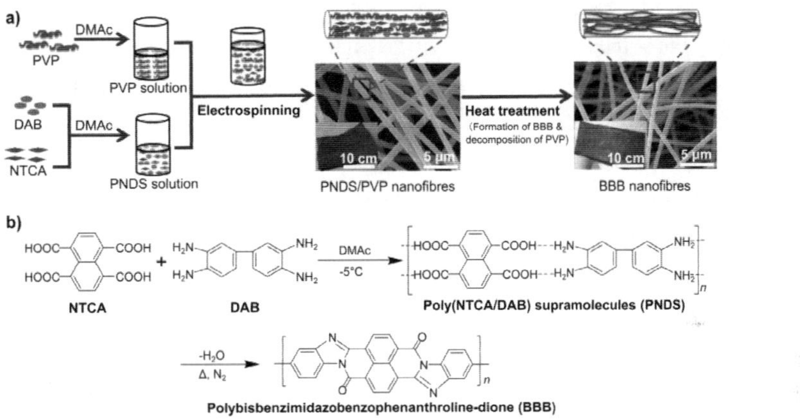

Figure 1. Schematic illustration of BBB nanofibre preparation: a schematic illustration of the preparation of BBB nanofibres by molecular self-assembly and electrospinning; **b** synthetic scheme and chemical structure of the BBB polymer.

References

[1] Gibbs W E, *Journal of Macromolecular Science: Part A-Chemistry*, **1968**, 2(7), 1291-1302.

[2] Arnold F E, Van Deusen R L, *Macromolecules,* **1969**, 2(5), 497-502.

[3] Van Deusen R L, *Journal of Polymer Science Part B: Polymer Letters*, **1966**, 4(3), 211-214.

[4] Berry G C, *Journal of Polymer Science: Polymer Physics Edition*, **1976,** 14(3), 451-478.

[5] Darrell H R, Iksoo C, *Nanotechnology,* **1996**, 7(3), 216.

[6] Agarwal S, Greiner A, Wendorff J H, *Progress in Polymer Science*, **2013**, 38(6),963-991.

[7] Dzenis Y, *Science*, **2004**, 304(5679), 1917-1919.

[8] Greiner A, Wendorff J H, *Angewandte Chemie International Edition*, **2007**, 46(30), 5670-5703.

BIOGRAPHY

Education:

Bachelor's Degree: Department of Chemistry, Jiangxi Normal University (China), 1982;

Master's Degree: School of Chemistry and Materials Science, University of Science and Technology of China, 1989;

Doctorate: Changchun Institute of Applied Chemistry of Chinese Academy of Sciences, 1998.

Postdoctoral Research:

Chemistry Department, Philipps University Marburg (Germany), 1999.01-2002.06;

Department of Polymer Sciences, the University of Akron (USA), 2002.07-2003.12;

Work Position:

Professor, College of Chem. and Chem. Engi., Jiangxi Normal Univ. (China), 2004.1-Now;

President, Jiangxi Advanced Nanofiber S&T Co., Ltd, 2016-Now.

Honor: Distinguished professor of Jiangxi Province (Jinggang Scholar); Chair professor of college of chemistry and chemical engineering, Jiangxi Normal University; Natural Science Prizes of Jiangxi Province (2005 and 2012) and Natural Science Award of Jilin Province (2015).

Research Interests: High-performance polymers; polymer nanocomposites, electrospinning, electrospinning industrialization, polymer nanofibers, carbon nanofibers, carbon nanotubes, supercapacitors, lithium battery separators, lithium battery anode materials and the like.

Multiscale Construction of Graphene Fibers with High-Performances

<u>PingAn Hu</u>, Jia Zhang, Xiaona Wang, Yunfeng Qiu

Research Centre for Micro/nanotechnology, Harbin Institute of Technology, Harbin, 150080, China.

E-mail: hupa@hit.edu.cn

High performance nanofibers are of significance for flexible and wearable microelectronics. Core–sheath fibers were massively fabricated from ultralong chemical vapor deposition (CVD) grown graphene bundles. They exhibited superior conductivity and excellent mechanical properties which exceeded the reduced graphene oxide (rGO) fibers. The intrinsic dynamic fracture procedure and mechanism of the core–sheath nanofibers were investigated. Furthermore, safe strain sensors based on as–prepared core–sheath CVD graphene fibers have been demonstrated as a proof–of–concept application. The performance of strain sensors has been greatly improved by using CVD graphene fiber. Organic–inorganic nacre–like composites excited scientists for decades and have been investigated for use as coatings and thin films. Realization of composite fibers with biomimetic–inspired layered structures offer novel mechanical properties, different deformation patterns, and new fields of application. Here we demonstrate that nacre–like fibers from nanoplatelets and polymers can be produced by shear–induced self–assembly. The combination of nanoscale brick–and–mortar and helical microscale structural organization gives rise to stretchability greater than 400% and a gravimetric toughness as high as 640 J g^{-1}. The unique toughness of the fibers originates from the multiscale deformation regime involving energy dissipation at nanoplate–polymer interfaces and macroscale decoiling. Moreover, luminescent CdTe nanowires incorporated into the fibers conveys the unique property of mechanically tunable circularly polarized luminescence associated with the helical nano– and meso–scale deformations in the material. The nacre–like fibers create a novel technological work space for investigating the optomechanics of biomimetic composites while their continuous spinning methodology makes scalable production realistic.

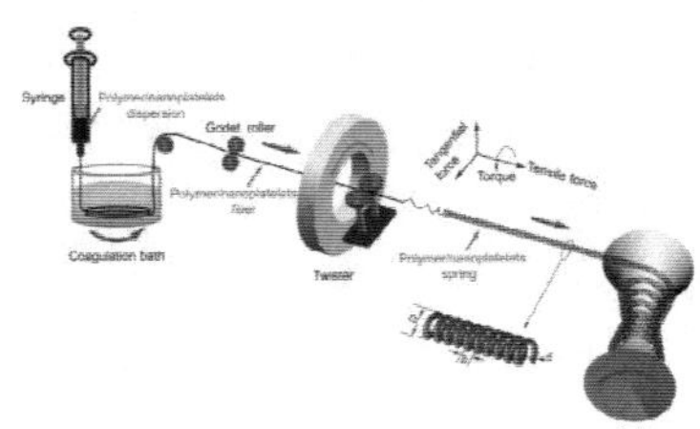

Figure 1. Schematic diagram of the two-step formation of nacre-like composite fiber.

References

[1] J Zhang, W Feng, H Zhang, ZL Wang, H A Calcaterra, B Yeom, P A Hu*, N A Kotov, *Nature Comm.*, **2016**, 7,10701.

[2] X N Wang, Y F Qiu, W Cao, PA Hu, *Chem. Mater.*, **2015**, 27 (20), 6969–6975.

BIOGRAPHY

Dr. PingAn Hu is a Professor at Research Centre for Micro/Nanotechnology in Harbin Institute of Technology (HIT). He obtained his Ph.D degree from Institute of Chemistry, Chinese Academy of Science, then he worked as a JSPS research fellow at Waseda University, Japan, and a Research Associate in University of Cambridge. He got New Century Excellent Talents in University of China in 2010 and Longjiang Special Professorship in 2011.

He published about 100 SCI papers in highly qualified international journal such as Nature Communications, Advanced Materials and Nanoletter. His research has been introduced and commented by some international journals such as Nature Photonics, New Scientist .

Constructing Highly Planar Conjugated Systems through Nonbonding Conformational Locks

Hui Huang*, **Tao Dong, Lei Yang, Lei Lv, Pan Ye**

College of Materials Sciences and Opto-electronic Engineering,
University of Chinese Academy of Sciences, Beijing, 100049, China

"Nonbonding conformational locks" has been widely employed as an important strategy to construct highly planar conjugated systems. Previously, we constructed a building block with S···O intramolecular nonbonding interactions and defined it as "conformational locks". Here, we designed and synthesized different building blocks with S···O and Se···O conformational locks. The single crystal X-ray diffraction studies supported the existence of the nonbonding conformational locks. Interestingly, TTVTT(OEt) molecule exhibits aggregation induced emission (AIE) phenomenon though it is highly planar, which was fully investigated with UV-vis, ^{1}H NMR, ESI-MS, and DFT calculations. Furthermore, the building block with Se···O conformation locks was used to build conjugated polymers for TFTs and OPVs. The results showed that introducing Se…O conformational locks can enhance the charge carrier transport, resulting in improved PCE.

Keywords: conjugated polymers; nonbonding conformations; organic electronics

BIOGRAPHY

Hui Huang is currently a Professor at College of Opto-electronic Engineering and Materials Science, University of Chinese Academy of Sciences (UCAS). He obtained his B.A. in Chemistry from Beijing Normal University, M.A. from Institute of Chemistry, Chinese Academy of Sciences, and Ph.D. from Dartmouth College under the supervision of Professor Russell P. Hughes in 2008. From 2008 to 2010 he carried out his postdoctoral training with Professor Tobin J. Marks and Professor Antonio Facchetti at Northwestern University. In 2010, he joined the Renewable Energy Group at Research Center of ConocoPhillips as a research scientist. He moved back to China to join UCAS as a member of "Hundred Talents Program" in 2013. He has published over 40 peer-reviewed papers and holds over 10 patents. He is named as an "Emerging Investigator" by the Journal of Materials Chemistry A of the Royal Society of Chemistry in 2017. His research interests include synthetic methodology and applications of organic/polymeric semiconductors for photovoltaics, thin-film transistors, photodetectors, and biosensors.

Influence of Treatment Conditions on Structure and Mechanical Properties of BPDA-PDA Polyimide Fibers

Xiangling Ji[1], **Wenke Yang**[1], **Fangfang Liu**[2], **Ensong Zhang**[1], **Xuepeng Qiu**[2]

[1] *State Key Laboratory of Polymer Physics and Chemistry, Changchun Institute of Applied Chemistry, Chinese Academy of Sciences, Changchun, 130022, China;*

[2] *Polymer Composites Engineering Laboratory, Changchun Institute of Applied Chemistry, Chinese Academy of Sciences, Changchun, 130022, China*

E-mail: xlji@ciac.ac.cn

During fabriction of polyimide (PI) fibers using two-step approach, the polyamic acid (PAA) fibers are spun first and then converted into PI fibers via thermal treatment. Definitely, the treatment conditions have great influence on the structure and mechanical properties of the obtained PI fibers. Herein, PAA fibers derived from 3,3',4,4'-biphenyltetra-carboxylic dianhydride (BPDA) and p-phenylenediamine (PDA) are thermally treated under different conditions, such as ramp heating, isothermal treatment, heating rate, atmosphere and force[1-2]. The evolution of structure and mechanical properties during thermal imidization is traced and characterized. Definitely, some suggestions for PI fiber fabrication are proposed.

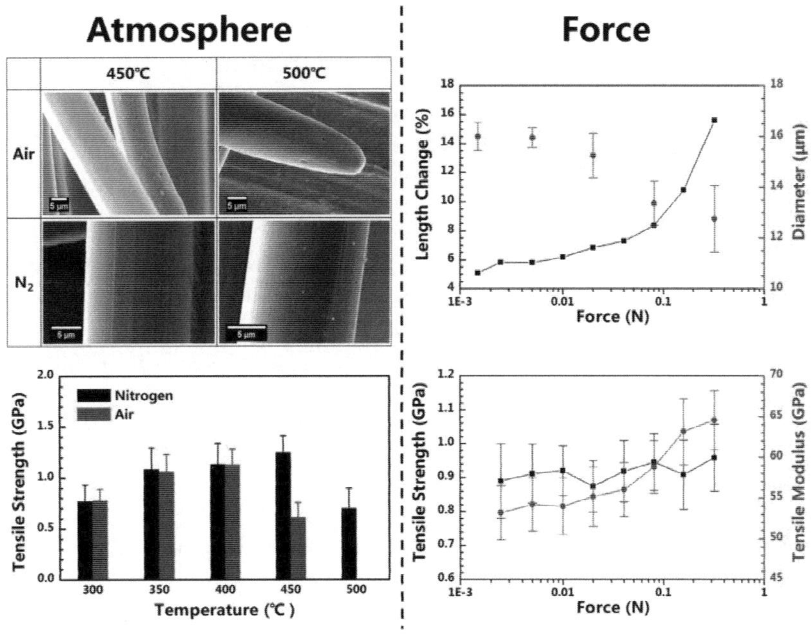

Figure 1. Influence of atmosphere and force on morphologies and mechanical properties of BPDA-PDA PI fibers.

References

[1] W. Yang, F. Liu, G. Li, E. Zhang, Y. Xue, Z. Dong, X. Qiu, X. Ji, *Chin. J. Polym. Sci.*, **2016**, *34*, 209-220.

[2] W. Yang, F. Liu, E. Zhang, X. Qiu, X. Ji, *Chem. J. Chin. Univ.*, **2017**, *38*, 150-158.

Precise Control of Membrane Structure for High-Performance Separation

Jian Jin

Suzhou Institute of Nano-Tech and Nano-Bionics, Chinese Academy of Sciences, Suzhou, 215123, China

E-mail: jjin2009@sinano.ac.cn

The key issue for separation membranes for practical application is to address the trade-off between high-precision and high-flux separation. From the viewpoints of increasing the porosity, reducing the critical breakthrough pressure and shortening the transport route, we put forward several strategies for achieving polymer separation membranes with high flux and simultaneously high selectivity. Through introducing non-facial and kink units into rigid polymer backbones, the effective control of the micropore structures of the membranes was realized, which effectively improved the membrane porosity. The achieved membranes exhibited highly improve performance for gas separation. Wastewater containing emulsified oil/water mixture is generated in many industrial processes, separation of which is always a tough work and challenging. We present here a facile methodology to fabricate superwetting membranes with hierarchical structures for separation of various oil/water emulsions with high-flux. Polymer membranes with micro/nano hierarchical structure on surface were fabricated by adjusting the phase inversion process during membrane preparation. The micro/nano hierarchical structure endows the membranes special wettability for water and oil, which could be used for fast separation of various oil-in-water and water-in-oil emulsions. To further increase the permeation flux, a series of ultrathin porous membranes with nano-scale thickness were successfully prepared in order to reduce the effective path for mass transfer. The separation with superhigh flux was thus achieved. Our results provided new inspirations for the fundamental and applied research of polymer-based separation membranes with high performance.

References

[1] S. Gao, J. Sun, P. Liu, F. Zhang, W. Zhang, S. Yuan, J. Li, Jian Jin*, *Adv. Mater.* **2016**, *28*, 5307-5314.
[2] Z. Wang, D. Wang, S. Zhang, L. Hu, J. Jin*, *Adv. Mater.* **2016**, *28*, 3399–3405
[3] W. Zhang, Y. Zhu, X. Liu, D. Wang, J. Li, L. Jiang, J. Jin*, *Angew. Chem. Int. Ed.* **2014**, *53*, 856-860.
[4] Z. Shi, W. Zhang, F. Zhang, X. Liu, D. Wang, J. Jin*, L. Jiang, *Adv. Mater.* **2013**, *25*, 2422-2427.

BIOGRAPHY

Jian Jin received her B. S. degree (1996), Ph.D. degree (2001) from Jilin University (China). She then worked as a JSPS postdoctoral fellow in the Research Center of Advanced Science and Technology in Tokyo University, Japan. From 2004 to 2009, she worked as a senior researcher in National Institute for Materials Science, Japan. In 2009, she joined Suzhou Institute of Nano-Tech and Nano-Bionics (SINANO), CAS as a group leader. Her interests are surface modification and control of materials wetting property, design and fabrication of advanced filtration membrane for gas, oil/water separation and purification.

Simultaneous Structure Formation and Dyeing of PET Fibers through Cold-Drawing in Ethanol – Utilization of Infusion Phenomena

Dong-Woo Go[1], Rina Khanum[1], Wataru Takarada[1], Arun Aneja[2], Takeshi Kikutani[1]

[1]*Department of Materials Science and Engineering, School of Materials and Chemical Technology, Tokyo Institute of Technology, Tokyo 152-8550, Japan;*

[2]*Department of Engineering, East Carolina University, Greenville, NC 27858, USA*

E-mail: kikutani.t.aa@m.titech.ac.jp

In our previous research, the batch-type drawing of undrawn poly(ethylene terephthalate) (PET) filament was carried out in ethanol at room temperature. The formation of multiple necking, reduction of drawing stress, infusion of ethanol into filament along with dye, and the occurrence of solvent-induced crystallization were observed[1-3]. Typical image of the neck-drawn PET fiber is shown in Fig.1. It should be noted that neither diffusion of ethanol into PET fiber nor solvent induced crystallization occurs if PET fiber is simply immersed in ethanol.

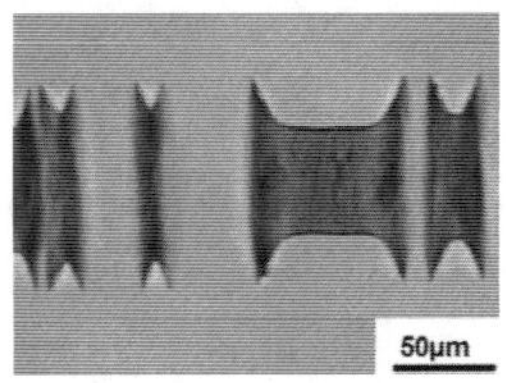

Figure 1. Photograph of PET fiber neck-draw in ethanol solution containing dye.

To explore the possibility of applying this unique behavior to industrial process, continuous drawing of PET fibers in ethanol was carried out using a set of rollers of different rotation speeds and an ethanol channel. In some experiments, the ethanol solution containing 0.5 wt% of disperse dye was used. In situ observation of the drawing process was carried out using a high-speed camera. In addition, a tension meter was applied for continuous monitoring of the tension in the drawing line.

Photographs of PET fibers captured from the continuous drawing line in ethanol of various feed roller rotation speeds are shown in Fig. 2. In these experiments, draw ratio of 2x was maintained.Formation of multiple necks even in the continuous drawing process was confirmed. The number of necks was found to increase with the increase of the speed of feed roller. Dyeing of drawn part during the drawing in ethanol dye solution was also confirmed.

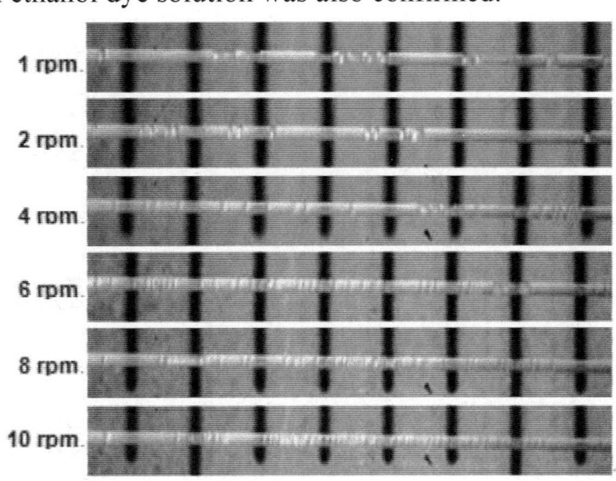

Figure 2. PET fiber drawing line in ethanol of various feed roller speeds. Draw ratio was fixed at 2x. Images were magnified 2x only in the vertical direction for better visibility of necking deformation.

Through the continuous monitoring of the drawing tension, it was found that the periodic occurrence of the formation of a group of multiple necks was accompanied by the oscillation of drawing tension. The installation of a drawing pin was discovered to be useful for the stabilization of the drawing process.

References

[1] R. Khanum, W. Takarada, A. Aneja, T. Kikutani, *Polymer*, **2015**, 59, 27-34.

[2] R. Khanum, D. W. Go, W. Takarada, A. Aneja, T. Kikutani, *Polymer*, **2015**, 70, 19-29 .

[3] R. Khanum, D. W. Go, W. Takarada, A. Aneja, T. Kikutani, *SEN'I GAKKAISHI*, **2015**, 71, 273-283.

BIOGRAPHY

Takeshi Kikutani is a Professor in the Department of Materials Science and Engineering, School of Materials and Chemical Technology, Tokyo Institute of Technology. He is currently the president of the Society of Fiber Science and Technology, Japan. He is a board member of the Polymer Processing Society (International), and served for the society as the president from 2009 to 2011. He also is a member of the Japan Society of Polymer Processing, and served as the president from 2012 to 2014. In addition, he is a member of the Governing Council of the Fiber Society (USA) and a board member of The Textile Machinery Society of Japan. His main research interest is the mechanism of structure development in the melt processing of fibers and films. He has been doing researches especially in the fields of high-speed melt spinning, bicomponent melt spinning, film stretching, fiber reinforced composites etc. He received his PhD degree from Tokyo Institute of Technology in 1982, and joined the Institute as a faculty member immediately after the graduation. He was a visiting scientist at the Department of Polymer Engineering, University of Akron, USA in 1986-1987.

Eco-Friendly Microporous Organic Polymers Based on Hyper-Crosslinked Pitch and Coal Tar: Preparation and Application

Lei Li, Wenqing Li, Hui Gao, Lei Ding

College of Materials, Xiamen University, Xiamen, 361005, China

E-mail: lilei@xmu.edu.cn

Microporous organic polymers (MOPs) have attracted enormous scientific attention due to high surface area, inherent light weight and large degree of flexibility in the design of structures, properties and functions. MOPs are one of important members of porous materials, which can be produced by knitting aromatic small molecules with rigid external crosslinkers. However, the preparation for high performance MOPs with novel and low cost building blocks still remains a difficult challenge.

Pitch is a low-cost and abundant product obtained as the distillation residue of coal tar, wood tar, or petroleum. It primarily consists of a wide variety of phenols, polycyclic aromatic hydrocarbons (PAHs) and heterocyclic compounds which can be used as monomers to prepare MOPs. Besides, it is well known that most PAHs are genotoxic and carcinogenic. Therefore, the suitable means for strengthening innocent treatment of pitch is essential for global sustainable development and environmental issues.

Here, we have employed a serial of pitch and coal tar as building blocks to prepare MOPs with high surface area through a one-step mild hypercrosslinking reaction. The reaction is performed in chloroform, which serves as both solvent and external crosslinker (scheme 1). Obtained HCPs have a maximum BET surface area of 1337 m^2 g^{-1} and exhibit high CO_2 and H_2 uptake capacity (17.74 wt% at 1.0 bar/273 K and 1.83 wt% at 1.13 bar/77 K, respectively). The utilization of sustainable and inexpensive building blocks will open a way to the scale-up preparation of MOPs, and push forward the innocent treatment of pitch.

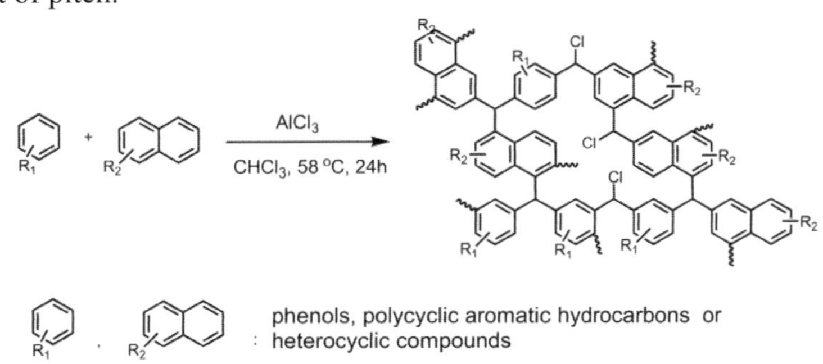

Scheme 1. Simplified schematic of the preparation of pitch and coal tar-based MOPs.

References

[1] W. Q. Li, A. J. Zhang, H. Gao, M. J. Chen, A. H. Liu, H. Bai and L. Li, *Chem. Commun.*, **2016**, *52*, 2780-2783.

[2] H. Gao, L. Ding, H. Bai and L. Li, *Chemsuschem*, **2017**, *10*, 618-623.

[3] H. Gao, L. Ding, H. Bai, A. H. Liu, S. Li and L. Li, *J. Mater. Chem. A*, **2016**, *4*, 16490-16498.

BIOGRAPHY

Institute: College of Materials, Xiamen University, Xiamen, 361005, China

E-mail add: lilei@xmu.edu.cn

Self-Assembled Monolayers of Functional Molecules for Solid-state Electronic Devices

Tao Li

School of Chemistry and Chemical Engineering, Shanghai Jiao Tong University, Shanghai 200240, China

E-mail: litao1983@sjtu.edu.cn

Molecular electronics describes a field that seeks to implement electronic components made of molecular building blocks. The development of synthetic molecules endowed with desirable functionalities, and the incorporation of these molecules into reliable solid-state electronic devices, have been two major challenges since the dawn of molecular electronics.

Compared with the junctions built on a very limited number of molecules, their self-assembled monolayers (SAMs) offer a more reproducible way for the fabrication of molecular electronic devices, and are more amenable to mass production and integration. However, direct vapor deposition of metals onto SAMs is very invasive and damaging, normally resulting in a high percentage of short-circuits. To address this problem, we introduce a new method using solution-processed ultrathin graphene films as soft top contacts for the non-destructive fabrication of molecular junctions. The device yield is over 90%, exhibiting good operational stability and long shelf life. By using this new testbed, we design a series of photochromic and redox-active small molecules and conjugated polymers, and investigated their electronic properties in solid-state devices. Transistor-like devices are successfully made from a single molecular monolayer.

References

[1] Z. Wang, H. Dong, T. Li, R. Hviid, Y. Zou, Wei, Z.; X. Fu, E. Wang, Y. Zhen,; K. Nørgaard, B. Laursen, and W.Hu, *Nat. Commun.* **2015**, *6*, 7478

[2] D. Ji, T. Li, and H. Fuchs, *Adv. Electron. Mater.* **2017**, *3*, 1600348.

BIOGRAPHY

Dr. Tao Li is an Associate Professor in School of Chemistry and Chemical Engineering, Shanghai Jiao Tong University. He received his Ph.D. from Institute of Chemistry, Chinese Academy of Sciences in 2010. Then he joined University of Copenhagen as a postdoc fellow. In 2013, he was appointed as an Assistant Professor in Department of Chemistry, University of Copenhagen. He returned to China and joined Shanghai Jiao Tong University as an Associate Professor in 2015. His research mainly focuses on molecular electronics, especially on the fabrication of organic field-effect transistors and solid-state molecular scale devices.

High Performances of Plant Fiber Reinforced Composites

Yan Li

School of Aerospace Engineering and Applied Mechanics, Tongji University, Shanghai, 200092, China

E-mail: liyan@tongji.edu.cn

Plant fiber reinforced composites (PFRCs) have raised great attentions among materials scientists and engineers in the recent decades due to their light weight, environmental friendly, interesting mechanical and physical properties, et. al. Realizing high performances of PFRCs are the necessity for expanding their application potentials.

On the basis of the shear-lag model, a double interface model by considering the unique multi-scaled structures was proposed and successfully calculated the interfacial behaviors of PFRCs so that the multi-scaled interfaces could be designed [1]. Improving the mechanical properties of PFRCs were attempted by hybridizing with glass and carbon fibers and by incorporating carbon nanotubes. Multi-scaled failure modes were obtained and the mechanical properties were remarkably improved.

The acoustic properties of PFRCs were also investigated theoretically and experimentally based on the above multi-scaled methodology [2-3]. It was found that both plant fibers and their composites had superior capability for noise reduction. The sound absorption mechanisms of plant fibers were proposed and the multi-scaled and hollow lumen structures provided major contributions.

Demonstration parts were manufactured and applied in the national key projects. It can be concluded that the development of structure-function integrated PFRCs appears to be more important and feasible than that of mechanically "high-performance" biocomposites.

References

[1] Li Q, Li Y, Zhang Z S, Zhou L M. *CCCM-2*, Zhenjiang, China, **2015**.

[2] Yang W D, Li Y. *Science China-Technological Sciences*, **2012**, *55(8)*, 2278-2283.

[3] Zheng Z Y, Li Y, Yang W D. *Journal of Reinforced Plastics and Composites*, **2013**, *32(20)*, 1561-1568.

BIOGRAPHY

Prof. Yan Li is currently the dean of the School of Aerospace Engineering and Applied Mechanics, Tongji University. She received her BEng and MEng from Beijing University of Aeronautics and Astronautics, and her PhD from the University of Sydney. She has been working in the University of California, San Diego (UCSD) from 2001 to 2003 and joined Tongji University in 2003. She obtained Distinguished Young Scholars Award from the National Natural Science Foundation of China (NSFC) (2016). Prof. Li has over 80 publications in the peer-reviewed journals with more than 1000 citations.

Prof. Li's major research interests include plant fibre reinforced polymer composites, durability of advanced composite materials and structures, nanomaterials and nanotechnology, et.al.

Nitrogen-containing Conjugated Microporous Polymers and Their Derived Carbons for Gas Capture and Energy Storage and Conversion

Yaozu Liao*, Haige Wang, Zhonghua Cheng, Jiahuan Li

State Key Laboratory for Modification of Chemical Fibers and Polymer Materials, College of Materials Science and Engineering, Donghua University, Shanghai 201620, P. R. China

*Email: yzliao@dhu.edu.cn

Porous materials for the selective uptake of gases or micropollutants have attracted great interest in recent years. For instance, efficient CO_2 uptake has received worldwide interest, as it is a potential solution for the reduction of greenhouse gas emissions. As another example, volatile iodine uptake is of special interest, since the long-lived radioactive iodine isotopes (e.g. ^{129}I or ^{131}I) need to be removed from exhaust fumes of nuclear power plants regularly.

Conjugated microporous polymers (CMPs) combing π-conjugation of the organic backbone with permanent micropores and high surface areas, have emerged as a new class of porous materials for gas Capture. Compared to traditional CMPs, which are mainly based on aromatic hydrocarbons, nitrogen-containing CMPs have advantages owing to their polarity, basicity and coordination ability to metal ions. In this presentation, several nitrogen-containing CMPs prepared *via* Buchwald-Hartwig, Yamamoto and chemical oxidative polymerization cross-coupling reactions will be presented and the effect of nitrogen-functionalities such as amine, pyridine, and nitrile groups on CO_2 and iodine uptake efficiencies will be discussed. Moreover, the application of the nitrogen-containing CMPs as electrodes and carbon-precursors for electrochemical energy storage as well as hydrogen evolution will be presented.

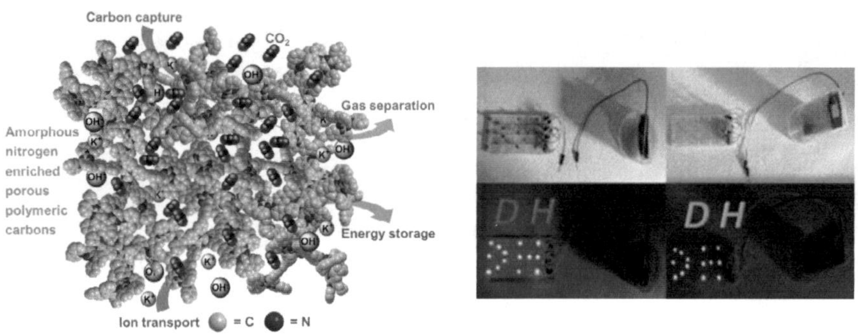

Figure 1. Nitrogen-containing CMPs used for gas capture and energy storage.

References

[1] Y. Liao, et al, *Chem. Commun.* **2014**, 50, 8002.

[2] Y. Liao, et al, *Macromolecules* **2016**, 49, 6322.

[3] Y. Liao, et al, *Chem. Mater.* **2017**, 29, 4885-4893.

Unique Energy Harvesting Performances of Electrospun Nanofiber Webs

Tong Lin[1], Jian Fang[1], Chenhong Lang[2], Hao Shao[1]

[1]*Institute for Frontier Materials, Deakin University, Geelong, VIC 3216, Australia;*

[2]*College of Textiles, Donghua University, Shanghai, 201620, China*

E-mail: tong.lin@deakin.edu.au

Kinetic energy is the energy carried by objectives with mechanical motions. Large kinetic energies from water fall and strong wind have been used to generate electric power. Small kinetic energies such as those from human body movement, blood flow, gentle wind, vibrations and sound are available everywhere, and their accumulation can be sufficient to power microelectronic devices. However, small kinetic energies have remained little developed until very recently.

Recently, our and other research groups' studies have indicated that piezoelectric nanofibers prepared by electrospinning show great potential in mechanical-to-electric energy conversion[1]. Electrospinning uses high electric field to stretch a viscous polymer solution into nanofibers. When a piezoelectric polymer is electrospun, the nanofibers obtained have high piezoelectric response without extra stretching and poling treatment which are the essential processes for making piezoelectric dense films. Piezoelectric electrospun nanofiber webs could be used to develop mechanical energy harvesters[2] and acoustoelectric generators[3,4]. This talk gives the background and state-of-the-art of the piezoelectric nanofiber research undertaken mainly in the Institute for Frontier Materials, Deakin University.

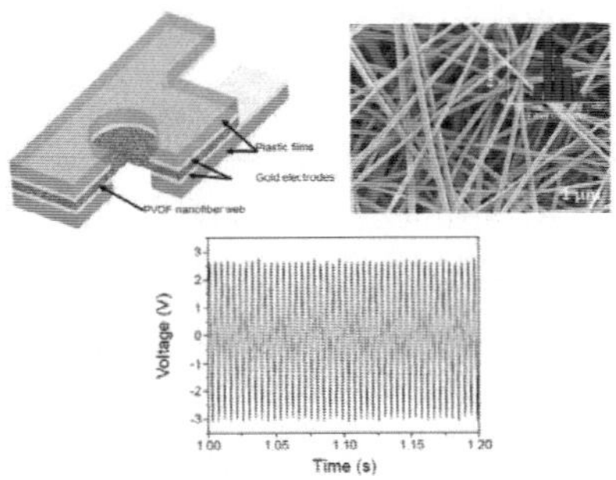

Figure 1. Nanofiber device for conversion of acoustic waves into electricity (left up), morphology of the nanofibers under SEM and histogram of the fiber diameter distribution (right up), and voltage output of the device under sound.

References

[1] J. Fang, X. Wang, T. Lin, *J. Mater. Chem.* **2011**, 11088.

[2] J. Fang, H. Niu, H. Wang, X. Wang, T. Lin, *Energy Enviro. Sci.* **2013**, 2196.

[3] C. Lang, J. Fang, H. Shao, X. Ding, T. Lin, *Nat. Commun.* **2016**, 11108.

[4] C. Lang, J. Fang, H. Shao, H. Wang, G. Yan, X. Ding, T. Lin, *Nano Energy*, **2017**, 146.

BIOGRAPHY

Professor Tong Lin received his PhD degree in physical chemistry from Chinese Academy of Sciences (CAS) in 1998. After several years of postdoctoral research, he joined Deakin University in 2003, and has served as Professor and Personal Chair since 2013. He has been an active researcher in the field of electrospinning, functional fibers and polymers. He contributes to the development of needleless electrospinning for large-scale nanofiber production and novel applications of nanofibrous materials in diverse areas. He has also been awarded the Australian Research Council (ARC) Future Fellow and Fellow of the Royal Society of Chemistry (RSC, UK).

Fibril Network Bulk Heterojunction Blends for High Efficiency Organic Photovoltaics

Feng Liu

School of Physics and Astronomy, Shanghai Jiao Tong University, Shanghai, 200240, China

E-mail: fengliu82@sjtu.edu.cn

Organic photovoltaics have emerged as one of the most promising renewable energy source that nowadays showed power conversion efficiency larger than 14%. To enable high performance organic solar cells, the bulk heterojunction device, which employs donor-acceptor mixture, are used. The ability to control and optimize the donor-acceptor active layer morphology is critical. Conjugated polymers, which featured with an aromatic backbone and aliphatic side-chain possess self-assembly ability to form nano fibrils, which entangled to establish a network structure to transport carriers. Solvent choice, processing additives, coating rate and coating temperatures are often used to control the fibril network morphology. New material design, of enabling better material crystallinity and donor-acceptor phase separation is another parameter space control that is critical in improving solar cell device performance. The current work discusses the general fibril network morphology and its relevance with device power conversion efficiency, to show how tiny structure change in pushing the efficiency in OPV research.

References

[1] Gao, K.; Miao, J.; Xiao, L.; et al. *Adv. Mater.* **2016**, *28* (23), 4727–4733.

[2] Liu, F.; Gu, Y.; Wang, C.; et al. *Adv. Mater.* **2012**, *24* (29), 3947–3951.

BIOGRAPHY

Dr. Feng Liu obtained Ph.D in polymer science and engineering from University of Massachusetts in 2014 under the supervision of Professor Thomas P. Russell, focusing on understanding the non-equilibrium structure of blended thin films. He is currently a distinguished research fellow in Department of Physics and Astronomy in Shanghai Jiao Tong University. His expertise includes polymer science, synchrotron x-ray and neutron techniques and thin film electronic devices. He has coauthored more than 150 research papers and book chapters. His research interests include soft materials, energy science, biomaterials and surface/interface sciences. In this talk, Feng will focus on the morphology characterization of bulk heterojunction (BHJ) photovoltaic blends to study the fibril network structure. Basics of synchrotron structure characterization methods, especially grazing incidence x-ray wide-angle scattering (GIWAXS) and resonant soft x-ray scattering (RSoXS) will be discussed. Structure-property relationship will be summarized.

From Small Molecule to Supramolecular Polymer to Supramolecular Nanotube Yarn

Minghua Liu, **Yaqing, Tianyu Wang, Ling Wang**

Institute of Chemistry, The Chinese Academy of Sciences, Beijing, 100190, China

E-mail: liumh@iccas.ac.cn

Over the past decades, there has been a huge amount of research into supramolecular gels, in which organic small molecules self-assemble into a wide variety of entangled nanostructures to form a 3D network and immobilize the solvents. Different from the covalent linked polymers to be fabricated into macroscale materials, for supramolecular systems, it remains a great challenge to use non-covalent interaction to assemble the small blocks into robust materials with structural ordering over macro-scale. Although many polymer fibers have been spun into yarns, supramolecular nanofiber are hardly spun into macroscopic yarns so far. We report the first macroscopic supramolecular nanotube yarns directly spun from the diluted aqueous solution. We have found that an L-histidine terminated bolaamphiphile could self-assemble into ultra-long supramolecular nanotubes in slightly alkaline aqueous solution. By an automatic spinning process, such solution can be spun into bundled nanotube yarns with lengths in meterscale and diameters of 1.5–5 μ m at a large yield. Although the yarns are built totally via non-covalent interaction, they showed nominal tensile strength of 45–60 MPa with Young's modulus coming to 6.8–9.9 GPa, which could be comparable to many covalent-linked polymers. These results provided the new horizon of hierarchical self-assembly and the continuous spinning of microscopic yarns from the solution of self-assembled small amphiphilic molecules opens a new strategy for building the nanoscale assemblies into ordered macroscopic supramolecular materials.

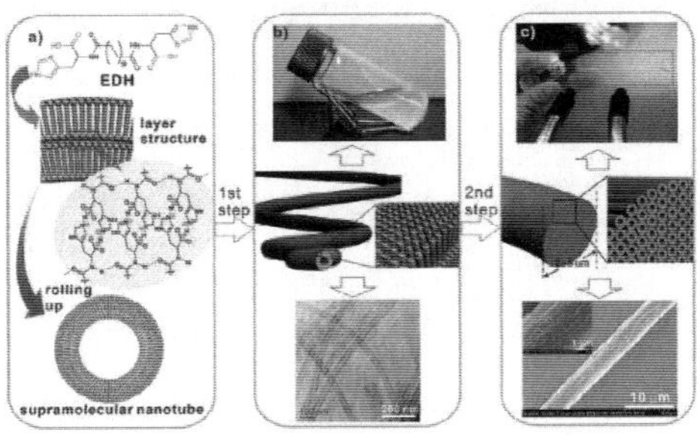

Figure 1. Schematic illustrations on the heierachiral self-assembly of small molecule into nanotube and aged into nanofiber, which can be further spun into macroscopic nanotube yarns.

References

[1] Y. Liu, T. Wang, Y. Huan, Z. Li, G. He, M. Liu, *Adv. Mater.* **2013**, *25*, 5875-5879.

[2] L. Zhang, X. Wang, T. Wang, M. Liu, *Small* **2015**, *11*, 1025-1038.

[3] T. Wang, Z. Shen, M. Liu, *ACTA POLYMERICA SINICA* **2017**, 50-62.

BIOGRAPHY

Minghua Liu graduated from Nanjing University in 1986 and received his Ph.D. in 1994 from Saitama University, Japan. After doing postdoctoral research in RIKEN (1994-1997) and Tokyo University of Agriculture and Industry (1997-1998), he joined the Institute of Photographic Chemistry, in 1998 and then Institute of Chemistry, CAS, from 1999 till now. From the June of 2013, he has served as the Director of National Center for Nanoscience and Technology, China. His research interests cover the colloid and interface sciences, supramolecular chemistry, self-assembled nanostructures and the gel-based soft nanomaterials. He served as the international advisory board of Soft Matter, Langmuir, Advanced Science, Nanoscale Horizons, and Chirality. He has published more than 300 papers in international peer reviewed journals.

New Insight into the Gelation of Na-alginate Aqueous Solution

Congxian Zhao, Chao Zhang, Hongliang Kang, Ruigang Liu*

Sate Key Laboratory of Polymer Physics and Chemistry, Beijing National Laboratory of Molecular Sciences, Institute of Chemistry, Chinese Academy of Sciences, Beijing 100190, China

E-mail: rgliu@iccas.ac.cn

Sodium alginate (SA) hydrogels have a wide range of applications including tissue engineering, drug delivery and formulations for preventing gastric reflux. The dynamics of sodium ions during the gelation process of SA solution is critical for clarification of the gelation mechanism. In this work, nuclear magnetic resonance (NMR) relaxometry and pulsed-field-gradient (PFG) NMR diffusometry were used to investigate the dynamics of the sodium ions during the gelation of SA alginate. We find that sodium ions are in two different states with the addition of divalent calcium ions, corresponding to the bound-state and free-state. The sodium ions in free-state are those released from the alginate chains via the replacement of calcium ions during the crosslinking. The relative content of sodium ions in bond state decreased with the increase in the content of calcium ions in the system. The relaxation time T_2 of sodium ions in bound- and free-states shifts to shorter and longer relaxation time with the increase in concentration of calcium ion, which indicates the closer package of SA chains and the larger space for the diffusion of free sodium ions. This work clarifies the gelation mechanism of SA at molecular level.

3D Printed High Strength Hydrogels as Tissue Engineering Scaffolds

Wenguang Liu[1], **Fei Gao**[1], **Xinyun Zhai**[1,2], **Changshun Ruan**[2]

[1]*School of Materials Science and Engineering, Tianjin University, Tianjin 300350, China;*

[2]*Research Center for Human Tissue and Organs Degeneration, Shenzhen Institutes of Advanced Technology, Chinese Academy of Sciences, Shenzhen 518055, China*

E-mail: wgliu@tju.edu.cn

Repair of bone and cartilage defects with suitable scaffolds finely fitting the original surrounding bone tissues is one of important options for patients' rehabilitation. Hydrogels are promising scaffolds for repairing bone and cartilage defects; however, the poor mechanical properties intrinsic to conventional hydrogels severely restrain their application as loading support for bone and cartilage regeneration. Recently, we invented a hydrogen bonding-reinforced supramolecular polymer hydrogel (SPH) which exhibited high tensile/compressive strengths and toughness [1]. Further, this SPH combined with bioactive inorganic nanoparticles was developed as a bioink, which was 3D printed to bone and cartilage scaffolds. Implantation of the 3D printed scaffold in the tibia defect and articular cartilage defect of rats could efficiently elicit the new bone and cartilage formation, respectively [2]. 3D printing of hydrogen bonding supramolecular polymer hydrogels with growth factors and cells will open up a new avenue to construct load-bearing tissue engineering scaffolds for precision and individualized repair of degenerated tissues.

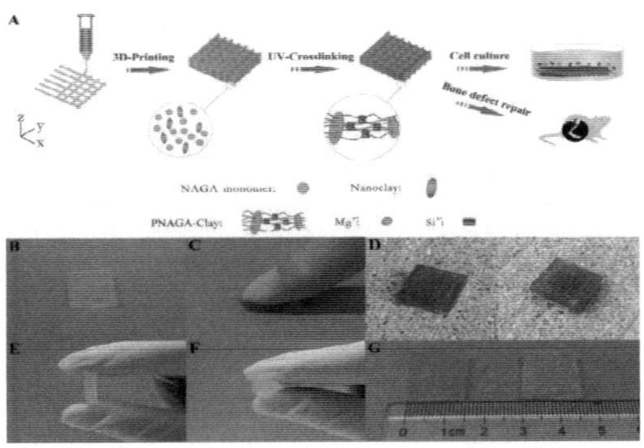

Figure 1. Procedure of 3D-printing scaffold (A) and photographs of the scaffolds showing its ability to resist finger compression (B and C), car wheel pressing (D, left and right denote before and after pressing) and hand folding (E and F). The scaffold is very stable even after immersing in water for a long time (G).

References

[1] Xiyang Dai, Yinyu Zhang, Lina Gao, Tao Bai, Yuanlu Cui, Wenguang Liu. *Adv Mater* **2015**, 27, 3566-3571.

[2] Xinyun Zhai,Yufei Ma, Chunyong Hou, Fei Gao, Yinyu Zhang, Changshun Ruan, Haobo Pan, William Weijia Lu, Wenguang Liu. *ACS Biomater Sci Eng* **2017**, 3, 1109-1118.

BIOGRAPHY

Dr. Wenguang Liu is a full Professor of School of Materials Science and Engineering at Tianjin University. He earned his PhD in Biomedical Engineering in 1999 from Tianjin University. Dr. Liu was a visiting scholar at The University of Hong Kong from July 2003 to January 2004. He did postdoctoral research at the Department of Cellular and Molecular Medicine, University of Ottawa (Canada) from March 2005 to November 2006. His current research interests are functional polymers, hydrogels, regenerative medicine and tissue engineering. Dr. Liu is the recipient of 2013 National Natural Science Funds for Distinguished Young Scholar.

Soft Functionalization of Silk Fibroin Materials and Bio-Flexible Devices

Xiangyang Liu

Department of Physics, FOS, National University of Singapore, Singapore

Research Institute for Biomimetics and Soft Matter, Xiamen University, Xiamen, PR China

E-mail: phyliuxy@nus.edu.sg; liuxy@xmu.edu.cn

Bombyx mori silk fibers have been utilized as a textile material for over 7000 years. Nevertheless, the application of silk fibroin materials to fabricating flexible photoelectric devices has become popular since the beginning of this century. As an excellent flexible biomaterial, silk fibroin materials offer exquisite mechanical, optical, and electrical properties which are advantageous toward the development of next-generation biocompatible electronic devices. In turning SF materials into bio-soft devices, to acquire some particular performance is very crucial. In this concern, to re-engineer the hierarchical structure of soft materials and to functionalize the materials are the two common approaches to achieve the functions. The functionality of soft materials is originated from the synergy of structures among different levels. In contrast to the conventional hard materials that various functions can be acquired by doping or re-constructing at atomic/molecular levels, re-constructing the hierarchical structure of soft/SF materials at the mesoscale is the major way to acquire different performance. Furthermore, in functionalize SF materials, a systematic approach, so-called Mesoscopic Material Assembly (MMA) has been put forward, which is to add and bind some specific nanomaterials or molecule to the networks so as to acquire some additional functions without jeopardizing the original performance. In this talk, I will cover the principles and strategies of mesoscopic structural re-engineering and functionalization of SF materials, which allows in the design and integration of high-performance bio-integrated devices for future applications in consumer, biomedical diagnosis, and human–machine interfaces.

References

[1] NB Lin, XY Liu*, *Chem. Soc. Rev.,* **2015**, 44, 7881-7915.

[2] AT Nguyen, QL Huang, ZYang, NB Lin, GQ Xu and XY Liu*, *Small* **2015**, 11, 1039-1054.

[3] NB Lin, LW Cao, QL Huang, CY Wang, Y Wang, J Zhou, and XY Liu*, *Adv. Funct. Mater.*, **2016**, 26(48): 8885-8902.

[4] ZW Chen, HH Zhang, ZF Lin, YH Lin*, JH Esch, and XY Liu*, *Adv. Funct. Mater.*, **2016,** 26(48): 8978-8990.

[5] N Lin, F Hu, Y L Sun, CX Wu, HY Xu* and XY Liu*, *Adv. Funct. Mat.*, **2014**, 24, 5284-5290.

[6] NC Tansil, Y Li, CP Teng, SY Zhang, KY Win, X Chen, XY Liu, and MY Han, *Adv. Mat.,* **2011**, 23, 1463–1466.

[7] YY Diao, XY Liu*, GY William Toh, L Shi and J Zi, *Adv. Func. Mat.,* **2013**, 23, 5373-5380.

[8] NB Lin, XY Liu*, YY Diao, HY Xu, CY Chen, XH Ouyang, HZ Yang, W Ji, *Adv. Funct. Mat.,* **2012**, 22, 361-368.

[9] N Du, Z Yang., XY Liu*, Y Li, HY Xu, *Adv. Funct. Mat,.* **2011**, 21, 772-778.

[10] RC Liu, QQ Deng, Z Yang, DW Yang, MY Han, XY Liu*, *Adv. Funct. Mat,.* **2016**, 26(30), 5534-5541.

[11] Y Song, ZF Lin, LQ Kong, Y Xing, NB Lin, ZS Zhang, BH Chen and XY Liu*, *Adv. Funct. Mater.,* **2017**, DOI: 10.1002/adfm.201700628.

BIOGRAPHY

Professor Liu Xiang Yang received his Ph.D. degree with the cum laude title from the Radboud University Nijmegen (Netherlands), in 1993. After two years postdoc research, he joined the Unilever Research Port Sunlight Lab. as a Senior Scientist. Dr. Liu joined the Department of Physics, the National University of Singapore in 1999, as a full Professor. He is also currently the State Distinguished Professor of The 1000 Talent Program, Chair Professor of China MOE Changjiang Scholars Program.

Nanotechnology for Innovative Cancer Therapy

Zhuang Liu

Institute of Functional Nano & Soft Materials Laboratory (FUNSOM), Soochow University Suzhou, Jiangsu, 215123, China
E-mail: zliu@suda.edu.cn

Nanomaterials especially functional inorganic nanomaterials exhibit a range of unique inherent physical and chemical properties useful in biomedicine. Starting from 2009, our group has been working on the development of functional nanomaterials for applications in multimodal biomedical imaging, drug delivery, as well as photo-therapies of cancer. In the first part of this talk, I will introduce our recent efforts on the development of nanotechnology to enhance the efficacy of cancer radiation therapy (RT), one of three mainstream cancer treatment methods used in the clinic. We have uncovered that certain types of inorganic nanomaterials could absorb X-ray, and thereby concentrate a great deal of ionization energy locally in the tumor to enhance the efficacy of RT. With the help of nanotechnology, we could also change the tumor microenvironment (e.g. tumor hypoxia) to overcome hypoxia-associated RT resistance by tumors. In the second part of this talk, I will then discuss our recent studies in the area of nano-immunotherapy. We have tried to combine photothermal therapy with immunotherapy using nano-agents which not only exhibit strong near-infrared (NIR) absorbance for efficient photothermal conversion, but also serve as nano-adjuvants to promote the immunological responses. Stimulated by the tumor-associated antigens released after photothermal tumor ablation, the triggered immunological responses if in combination with checkpoint blockade therapy to suppress the activity of regulatory T cells could result in effective inhibition of tumor cells remaining in the body, promising for treatment of cancer metastasis.

BIOGRAPHY

Dr. Zhuang Liu is a professor at Soochow University in China. He received his BS degree from Peking University in 2004 and PhD degree from Stanford University in 2008. In 2009, Dr. Liu joined Institute Functional Nano & Soft Materials (FUNSOM) at Soochow University. He is now working in the field of nanomedicine to develop various functional nanomaterials for cancer theranostics. Dr. Liu has authored over 200 peer-reviewed papers, which have received a total citation of > 28,000 times and given him an H-index at 79 (based on Web-of-Science search). He was listed as one of 'Highly Cited Researchers' (Materials, Chemistry) by Thomson Reuters in 2015 and 2016. Now he is serving as an associate editor for 'Biomaterials', and as the editorial board members for a number of other international journals. He has been invited to be the Fellow of the Royal Society of Chemistry (FRSC) in 2015. The awards he received include 'NSFC Distinguished Young Scholar Award (Fund)', Biomaterials Science Lectureship (RSC), Chinese Chemical Society Young Chemist Award, etc.

Hierarchically Buckled Sheath-Core Fibers for Superelastic Electronics, Sensors, and Torsional Muscles

Zunfeng Liu[1,2,3,*], **Ningyi Yuan**[2], **Shougen Yin**[4], **Jianning Ding**[2]

[1]*Alan G. MacDiarmid NanoTech Inst., Univ. of Texas at Dallas, Richardson, TX 75080, USA;*

[2]*Jiangsu Collaborative Innovation Center of Photovoltaic Science and Engineering, Changzhou University, Changzhou 213164, China;*

[3]*Jiangnan Graphene Research Institute, Changzhou 213149, China;*

[4]*Institute of Materials Physics, Tianjin University of Technology, Tianjin 300384, China*

E-mail: liuzunfeng@nankai.edu.cn

Superelastic conducting fibers providing improved properties and new functionalities are needed for applications ranging from electronic interconnects to sensors and electrically driven artificial muscles. Presently described carbon-nanotube/rubber sheath-core fibers function as novel conductors that can be reversibly stretched 1320%, and provide a ratio of percent elongation in fiber length to percent resistance change of up to 421, which is 97 times higher than for previously reported elastomeric fiber conductors. Since extreme twist does not degrade conductance, the elastic range can be increased to 2470% by tightly coiling the fiber. This performance results from short-period and long-period sheath buckling that reversibly occurs out-of-phase in axial and belt directions. These sheath-core structures enable strain sensors generating 860% capacitance change as the fiber is elongated 950%, as well as electrically powered torsional muscles that operate by a new mechanism. The strain dependences of resistance and capacitance, as well as characteristics of torsional and tensile actuation are accurately predicted [1].

References

[1] Z.F. Liu et al. *Science*, **2015**, *349*, 400-404.

BIOGRAPHY

Zunfeng Liu is Professor in Nankai University. He was the assistant research fellow in UTDallas, Nanotech Institute. His research focuses on the stretchable electronics and artificial muscles. He led and participated in many international major projects, including the 2009 VENI grant, etc. He published above 50 papers such as Science, Adv. Mater, and Adv. Funct. Mater. etc., with the citations more than 5000 times. He applied more than 10 patents which one has been authorized. He is the editor board member of Nanoscience and Nanotechnology, HOAJ Biology, the academic association member of IEEE, ACS, Graphene flagship, and NVVM.

Synthesis, Characterization and Application of Two-Dimensional Materials

Jun Lou

Department of Materials Science and NanoEngineering, Rice University, Houston, 77005, USA

E-mail: jlou@rice.edu

Two-dimensional (2D) materials, such as Graphene, h-BN and MoS_2, are promising candidates in a number of advanced functional and structural applications, owing to their exceptional electrical, thermal and mechanical properties. In this talk, we first discuss the synthesis of high quality 2D materials and related heterostructures, with our current understanding of some underlying mechanisms. Next, we will report the electrical, opto-electrical and mechanical characterizations of the synthesized 2D materials, with an emphasis on the effect of defects on relevant properties. Finally, we will demonstrate some potential applications of selected 2D materials including using h-BN as high-performance oxidation resistant and anti-corrosion coatings, and 2D catalyst for hydrogen evolution reactions.

BIOGRAPHY

Jun Lou obtained B.E. and M.S. degrees in Materials Science and Engineering from Tsinghua University and Ohio State University, respectively, and his Ph.D. degree from the Department of Mechanical and Aerospace Engineering and Princeton Materials Institute at Princeton University. After a brief postdoc at Brown University he joined Rice University, and directs the Nanomaterials, Nanomechanics and Nanodevices Lab (N^3L). He is currently a full professor and the associate chair of the Department of Materials Science and NanoEngineering. Lou is an AFOSR Young Investigator and a recipient of Charles Duncan Award for Outstanding Academic Achievement at Rice. He is the Editor-in-Chief of Materials Today, the Elsevier flagship journal (Impact factor: 21.7) covering original research and reviews in the broader materials science community. Lou has extensive experience in the synthesis and design of 2D materials beyond Graphene and other nanomaterials; mechanical and multi-physics characterization, and fabrication of advanced material systems and devices. His group has made pioneering contributions in the area of nanomechanics including discovering of the "cold welding" phenomena in Au nanowires and performing the first experimental measurement of the fracture toughness of graphene. He currently serves as the site director for the NSF industry university collaborative research center (IUCRC) of Atomically Thin Multifunctional Coatings (ATOMIC) that has 13 member companies, exploring potential applications of 2D materials in different industries with commercial partners.

Nano-functionalized Silk for Smart Wearable Devices

Zhisong Lu, **Cuiping Mao, Huihui Zhang, Mei Meng, Chang Ming Li**
Institute for Clean Energy & Advanced Materials, Faculty of Materials & Energy, Southwest University, Chongqing, 400715, China
E-mail: zslu@swu.edu.cn

With rapid advances of miniaturization and wireless technologies, electronics has been associated with traditional textiles to develop smart wearable devices for daily health monitoring and fitness tracking.[1] It is of great demand to fabricate flexible materials that possess both wearability and practical functions for wearable devices.

Silk from *Bombyx mori* cocoons, a natural protein fiber consisting of 18 amino acids, is considered as an ideal support for wearable electronics because of its softness, high hygroscopicity and superior skin affinity[1], which fulfill the basic requirements of wearable devices on comfort and flexibility. However, practical applications of silk as building components of wearable devices are still rare due to its lack of specific functions such as sensing and conductivity.

In this presentation, I will first introduce properties and applications of silk, a natural protein fiber with great flexibility, environmental friendliness and excellent mechanical strength.[2] Our works on surface modification of silk with functional nanomaterials for flexible electronics will then be presented. Thirdly, fabrication of a silk-based sensor for physiological measurements will be discussed. Finally, I will show our recent achievements on the preparation of conductive silk textiles and their applications as energy elements.[3-9]

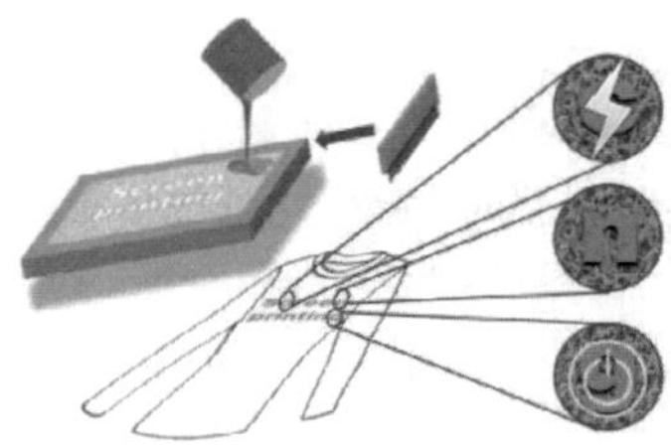

Figure 1. Fabrication of textile-based wearable devices with printing technology.

References

[1] Cherenack K, van Pieterson L. *J. Appl. Phys.*, **2012**, 112, 091301.
[2] Chen X, Qi Y Y, Wang L L, Yin Z, Yin G L, Zou X H, Ouyang H W. *Biomaterials*, **2008**, 29, 3683.
[3] Lu ZS, Meng M, Jiang Y K, Xie J L. *Colloids Surfaces A*, **2014**, 447, 1.
[4] Lu ZS, Mao CP, Meng M, Liu SG, Tian Y, Yu L, Sun B, Li C M. *J. Colloid Interface Sci.*, **2014**, 435, 8.
[5] Lu ZS, Xiao J, Wang Y, Meng M. *Journal of Colloid and Interface Science*, **2015**, 452, 8.
[6] Lu ZS, Mao CP, Zhang H H. *Journal of Materials Chemistry C*, **2015**, 3(17), 4265.
[7] Lu ZS, Zhang HH, Mao CP, Li CM. *Applied Energy*, **2016**, 164, 57.
[8] Zhang HH, Qiao Y., Lu ZS. *ACS Appl. Mater. Interfaces*, **2016**, 8 (47), 32317.
[9] Lu ZS, Zhou MY, Zhang HH. *Flex. Print. Electron.*, **2017**, 2, 025001.

BIOGRAPHY

Dr. Zhisong Lu received his PhD degree in School of Chemical & Biomedical Engineering from Nanyang Technological University, Singapore in 2011, and is currently a professor of Institute for Clean Energy & Advanced Materials (ICEAM) in Southwest University, P. R. China. Dr. Lu's research expertise is in the field of functional nanomaterials and smart textiles. With an h-index of 27 and total citation of 2500, he has co-authored more than 110 papers in refereed journals and has filed 9 patents. He also serves as an advisory board member of *Fundamental Toxicological Sciences*.

Toward Multi-functional Poly(3,4-ethylenedioxythiophene)-based Conducting Biointerfaces through Polymerization Techniques and Molecular Design

Shyh-Chyang Luo

Department of Materials Science and Engineering, National Taiwan University

No. 1, Sec. 4, Roosevelt Road, Taipei, 10617 Taiwan

E-mail: shyhchyang@ntu.edu.tw

Poly(3,4-ethylenedioxythiophene) or PEDOT has been demonstrated in various biomedical applications, ranging from biosensing to medical bionics. Being a conductive biointerface, PEDOTs can be served as an electrode for either electrochemical biosensing or to provide electrical stimuli on attached biomolecules. The functional groups on PEDOTs play an important role for the modification of surface property. In this presentation, I would like to introduce our strategies to manipulate the surface properties of PEDOT thin films for multi-functional and conducting biointerfaces. As a sensing platform, the control of functional density is a key for tuning the bio-recognition event. Besides, the surface properties, such as morphology and charges, are also critical to the detection sensitivity. For example, the control of surface charge can enhance the interaction between charged biomolecules, such as dopamine, and PEDOT surfaces through coulomb force, which leads to a higher sensitivity of these biomolecules. By fine-tuning the process of electropolymerization, it is possible to create PEDOT thin films of various surface morphology, hydrophobicity, and patterns. Moreover, the immobilization of redox active functional groups on the surfaces allows electro-responsive smart surfaces, which can be used as a controlled release platform.

BIOGRAPHY

Shyh-Chyang Luo received his B.S. in Chemistry (1992 - 1996) and M.S. in Materials Science and Engineering (1996 - 1998) from National Taiwan University. After he finished his military service in Taiwan, he then went to United States in 2001 to pursue his Ph.D. degree. He received his Ph.D. degree from the Department of Materials Science and Engineering, University of Florida in 2005.

Although the focus of his Ph.D. dissertation was mainly on light-emitting polymers and light-emitting devices, he found he is more interested in the field of bioengineering applications. So, after he received his Ph.D. degree, he decided to do his postdoctoral research at Institute of Bioengineering and Nanotechnology in Singapore to explore the field of conducting polymers for electrochemical biosensors (2006 - 2009). After that, he joined RIKEN in Japan as a research scientist. During his stay in Japan, he successfully received Grant-in-Aid for Young Scientist (KAKENHI) from Japan Society for Promotion of Science (JSPS) as a principal investigator to conduct independent research focused on the capture of rare cells. In 2013 August, he received an offer from National Cheng Kung University in Taiwan to join the Department of Materials Science and Engineering. In 2015, he moved back his hometown to contribute to his alma mater-- National Taiwan University.

Dr. Luo has published more than 30 papers in the field of conducting polymers. Because of his achievement in the field, in 2013, he was invited to be as a guest editor of a top journal in polymer science, *Polymer Reviews*, to organize a special issue "*conducting polymers as biomaterials and biointerfaces*". His current research interests include organic conducting materials, biointerfaces, electrochemical biosensors, bioelectronics, and stimuli-responsive materials for bioengineering applications.

Self-Powered Systems with Fiber-Shaped Devices

Wenjie Mai
Department of Physics, Jinan University, Guangzhou, 510632, China
E-mail: wenjiemai@gmail.com

Flexible and wearable devices has become more and more important nowadays. We have investigated the self-powered systems for wearable device applications, espeically using fiber-shaped devices, and we have obtained some important progresses in this field.

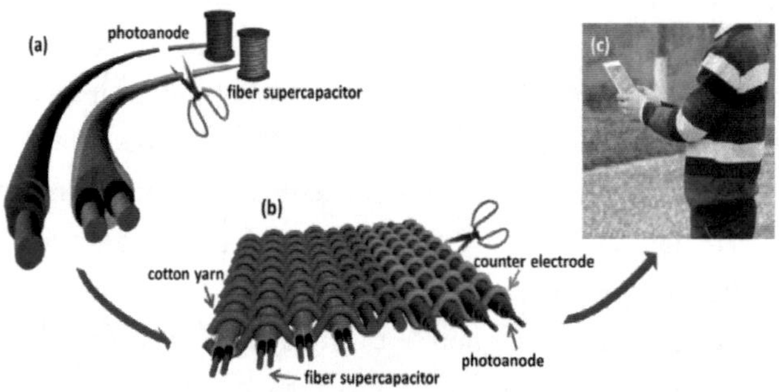

Figure 1. Scheme of tailorable textitle devices.

References

[1] P. Sun, R. Lin, Z.L. Wang, M.J. Qiu, Z.S. Chai, B.D. Zhang, H. Meng, S.Z. Tan, C.X. Zhao*, W.J. Mai*, *Nano Energy* **2017,** 31, 432-440.

[2] Z.S. Chai, N.N. Zhang, P. Sun, Y. Huang, C.X. Zhao, H.J. Fan, X. Fan*, W.J. Mai*, *ACS Nano*, **2016**, 10, 9201-9207.

BIOGRAPHY

Dr. Mai received his B.S. degree in Physics from Peking University in 2002 and Ph.D. degree in Materials Science and Engineering from Georgia Institute of Technology in 2009 under guidance of Prof. Zhong Lin Wang. Currently, he is full professor and department head in Jinan University. His research focuses on the science and engineering of nanomaterials and nanodevices used in energy harvesting, conversion, storage, and utilization. He has authored ~100 SCI papers published in prestigious peer-reviewed journals, including Science, Advanced Materials, Nano Letters, ACS Nano, Nano Energy, and these papers have been cited for more than 3000 times.

Polymer-functionalized Graphene Nanoribbons with Excellent Liquid-phase Dispersibility

Yiyong Mai[1], Yinjuan Huang[1], Deyue Yan[1], Klaus Müllen[2], Xinliang Feng[3]

[1]*School of Chemistry and Chemical Engineering, Shanghai Jiao Tong University, 800 Dongchuan Road, Shanghai 200240, China;*

[2]*Max Planck Institute for Polymer Research, Ackermannweg 10, 55128 Mainz, Germany;*

[3]*Department of Chemistry and Food Chemistry, Technische Universitat Dresden, Mommsenstrasse 4, 01062 Dresden, Germany*

E-mail: mai@sjtu.edu.cn

Structurally well-defined graphene nanoribbons (GNRs) have attracted great interest for next-generation semiconductor materials.[1,2] The functionalization of GNRs with polymers, which can widely broaden GNR-related studies on physiochemical properties and potential applications, has remained unexplored. Very recently, we achieved the bottom-up solution synthesis of structurally defined GNRs grafted with flexible poly(ethylene oxide) (PEO) chains.[3-5] The GNR backbones possessed a width of 1.7 nm and mean lengths of 15–60 nm, enabling near-infrared absorption and a low bandgap of 1.3 eV. Remarkably, the PEO coils rendered the GNRs superb dispersibility in common organic solvents, with a record concentration of ~1 mg mL^{-1} (for GNR backbone) that is greatly higher than that (<0.01 mg mL^{-1}) of reported GNRs. Scanning probe microscopy revealed raft-like self-assembled monolayers of uniform GNRs on graphite substrates. Thin-film based field-effect transistors (FETs) of the GNRs exhibited high carrier mobility of ~0.3 $cm^2\ V^{-1}\ s^{-1}$, indicating a promising application of the polymer-functionalized GNRs in electronic devices.[3]

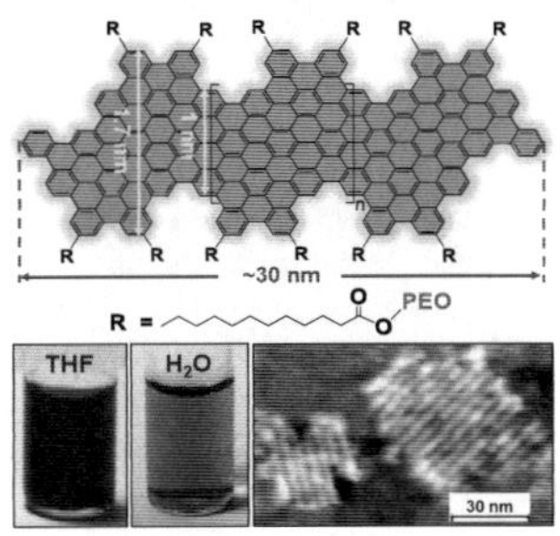

Figure 1. PEO-functionalized graphene nanoribbons and their dispersibility in common solvents.

References

[1] Cai, J.; Ruffieux, P.; Jaafar, R.; Bieri, M.; Braun, T.; Blankenburg, S.; Muoth, M.; Seitsonen, A. P.; Saleh, M.; Feng, X.; Müllen, K.; Fasel, R. *Nature* **2010**, 466, 470.

[2] Ruffieux, P.; Wang, S.; Yang, B.; Sanchez-Sanchez, C.; Liu, J.; Dienel, T.; Talirz, L.; Shinde, P.; Pignedoli, C. A.; Passerone, D.; Dumslaff, T.; Feng, X.; Müllen, K.; Fasel, R. *Nature* **2016**, *531,* 489.

[3] Huang, Y.; Mai, Y.; Yan, D.; Müllen, K.; Feng, X.; et al. *J. Am. Chem. Soc.* **2016**, 138, 10136.

[4] Huang, Y.; Yuan, R.; Xu, F.; Mai, Y.; Feng, X.; Yan, D. *Polym. Chem.* **2016**, 7, 1234.

[5] Huang, Y.; Mai, Y.; Yang, X.; Beser, U.; Liu, J.; Zhang, F.; Yan, D.; Müllen, K.; Feng, X. *J. Am. Chem. Soc.* **2015**, 137, 11602.

BIOGRAPHY

Dr. Yiyong Mai received his Ph.D. from Shanghai Jiao Tong University in April 2007 under the supervision of Professor Deyue Yan. He joined the group of Professor Adi Eisenberg at McGill University as a Postdoctoral Fellow in September 2008, where he was promoted as a Research Associate in March 2012. He has been a Professor in School of Chemistry and Chemical Engineering at Shanghai Jiao Tong University since March 2013. His present research interest involves synthesis and self-assembly of polymers of various architectures as well as self-assembled functional materials for optoelectronic or energy storage and conversion applications.

Multiaxial Mechanical Behavior at Single Fiber Scale: From Micro-Mechanisms to

Single Fiber Compression Tests

Alba Marcellan, Sébastien Joannès, Anthony Bunsell, Lucien Laiarinandrasana, Roland Piques

Laboratoire Sciences et Ingénierie de la Matière Molle, CNRS, ESPCI Paris, PSL Research University, Université Pierre et Marie Curie, Sorbonne-Universités, 10 rue Vauquelin, France

E-mail: alba.marcellan@espci.fr

High performance fibers, as Polyamide 66 or Aramid fibers, are designed to face material industrial challenges that demand high resistance and lightness. Very often, fibers are exploited as threads, made of a bundle of twisted individual fibers or woven in fabric. In that context, this implies that the single fiber undergoes multi-axial mechanical loadings but also complex fiber/fiber contacts. Identifying the mechanical responses at the single fiber scale along the fiber axis and in the transverse direction and understanding the micro-mechanisms involved are crucial[1-2].

Recently, we developed Single Fiber Transverse Compression Tests (SFTCT) to access to the mechanical anisotropy[3]. Mechanical characterization at the single fiber scale is challenging especially when the fiber diameter is as small as tens of micrometers. Our approach consisted of coupling morphological observations and mechanical experiments with SFTCT analysis (see Fig. 1) by comparing analytical solutions and finite element modeling. New insights on the analysis of the transverse direction response are highlighted as the determination of the transverse elastic limit and a skin/core property mismatch.

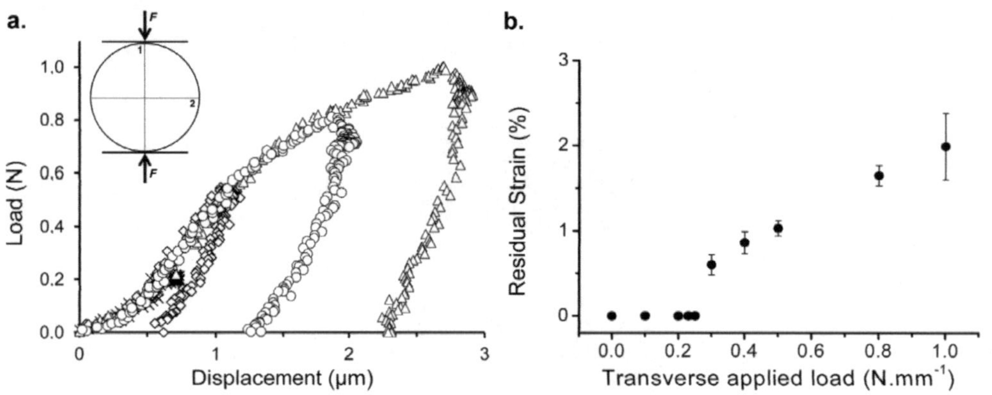

Figure 1. From Single Fiber Transverse Compression Test (SFTCT) to the determination of the transverse elastic limit. a, First transverse loading/unloading cycles for different maximum loads. b, Transverse residual strains after a transverse compression test at different maximum loads.

References

[1] Marcellan, A; Bunsell, AR; Laiarinandrasana, L; Piques R, *Polymer*, **2006**, *47*, 367–378.

[2] Marcellan, A; Colomban, Ph; Bunsell, A., *Journal of Raman Spectroscopy*, **2004**, 35, 308–315.

[3] Wollbrett-Blitz, J ; Joannès, S ; Bruant, R ; Le Clerc, Ch ; Romero De La Osa, M ; Bunsell, A ; Marcellan, A, *Journal of Polymer Science Part B-Polymer Physics*, **2016**, *54*, 374–384.

BIOGRAPHY

Career and References dates

After my Ph.D. in Mechanics and Materials at the École des Mines in Paris, I began working in a Polymer Chemistry and Soft Matter research environment at ESPCI. My research activities are mainly focused on the design and the understanding of the reinforcement mechanisms in polymer "model" systems, essentially polymer fibers and gels.

By using the concepts of polymer physics, with my colleagues, I have developed a new method for gluing gels and biological tissues. Beyond the promising results for surgery, this concept opens up a new paradigm for adhesion: gluing without polymers, but using solid nanoparticles. This work was rewarded by the Prize of the French Polymers Group in 2014.

Since 2016, I am invited as associate professor at the Global Station for Soft Matter (GSS). Serving as an international research hub, GSS is expected to provide innovative platform of soft matter research and education.

Prizes / Fellowships

2017 Fellowship of the Institut Universitaire de France (5 years)

2016 Fellowship of the Global Station for Soft Matter, international hub for research and education (5 years)

2014 Award of the Groupe Français des Polymères/ Société Française de Physique, Société Chimique de France
This prize rewards every two years a young researcher (<40 years) for his/her contribution in Polymer Science
Personal Grant from UPMC for research and doctoral supervision (4 years)

2010 Personal Grant from UPMC for research and doctoral supervision (4 years)

2007 Fellowship of the International Center for Materials Research (ICMR), "Polymeric and Self Assembled Gels" at the Materials Research Laboratory (MRS), University of California at Santa Barbara

Recent major publications – Graphical abstract

•"Multiaxial mechanical behavior of aramid fibers and identification of skin/core structure from single fiber transverse compression testing"; Wollbrett-Blitz, J ; Joannès, S ; Bruant, R ; Le Clerc, Ch ; Romero De La Osa, M ; Bunsell, A ; Marcellan*, A / Journal of Polymer Science Part B-Polymer Physics (2016)

•"Nanoparticle solutions as adhesives for gels and biological tissues" ; Rose, S; Prevoteau, A; Elziere, P; Hourdet, D ; Marcellan*, A ; Leibler*, L. / Nature (2014) – **News and Views** – From **Altmetric**: in the **98 percentile** of the most cited articles of similar age •"Thermoresponsive Toughening with Crack Bifurcation in Phase-Separated Hydrogels under Isochoric Conditions" ; Guo, H ; Sanson, N ; Hourdet, D ; Marcellan*, A / Advanced Materials (2016)

J. Polymer Science Polymer Physics (2016)

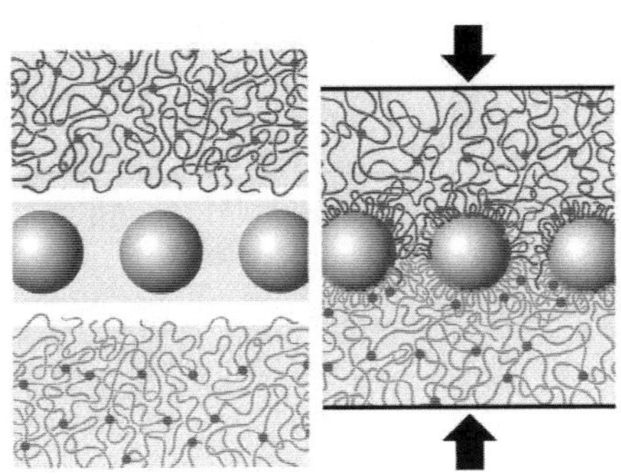

Nature (2014)

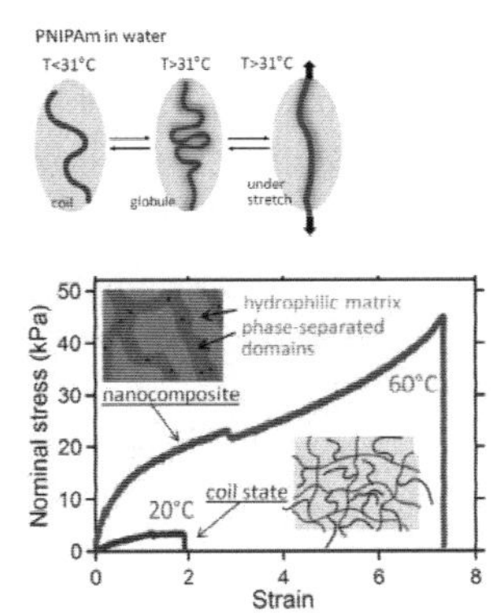

Advanced Materials (2016)

Structural Biopolymers – Using Nature's Building Blocks as an Inspiration for Advanced Manufacturing

Benedetto Marelli

Department of Civil and Environmental Engineering, Massachusetts Institute of Technology, Cambridge, MA, 20139, USA

E-mail: bmarelli@mit.edu

Structural biopolymers are materials engineered by Nature as building blocks of living matter. These materials have unique and compelling properties that allow for their assembly and degradation with minimal energy requirements as well as their performance at the biotic/abiotic interface. By combining basic material principles and biomedical science with advanced fabrication techniques and additive manufacturing, it is possible to define new strategies to drive the assembly of structural biopolymers in advanced materials with unconventional forms and functions such as edible coating for perishable food, inkjet prints of silk fibroin that change in color in the presence of bacteria, three dimensional monoliths that can be heated by exposure to infrared light and flexible keratin-made photonic crystals[1,2].

References

[1] B. Marelli, N. Patel, T. Duggan, G. Perotto, E. Shirman, C. Li, D.L. Kaplan and F.G.Omenetto, *PNAS*, **2017**, *114*, 451-456.

[2] B. Marelli, M. Brenckle, D.L. Kaplan and F.G.Omenetto, *Sci. Rep.*, **2016,** *6*, 25263.

BIOGRAPHY

Benedetto Marelli is the Paul M. Cook Career Development Assistant Professor in the Department of Civil and Environmental Engineering at the Massachusetts Institute of Technology. He received a B.Eng. and a M.Sc. in Biomedical Engineering from Politecnico di Milano in 2005 and 2008 under the supervision of Prof. M.C. Tanzi and Prof. Silvia Farè. Benedetto pursued his Doctoral studies in the Department of Materials Science and Engineering at McGill University. In 2012, he completed his PhD under the supervision of Professors Nazhat and Barralet with a dissertation on the biomineralization of tissue-equivalent collagenous constructs and their use as rapidly-implantable osteogenic materials. As a Postdoctoral Scholar in the SilkLab at Tufts University, Benedetto has worked with Professors Omenetto and Kaplan to study and direct the assembly and the polymorphism of structural proteins, particularly silk fibroin and keratin, with applications in food security, food safety, regenerative medicine and optoelectronics.

New Methods for Making Filament Nanofibers and Their Use for Functional Materials

Sergiy Minko

Nanostructured Materials Lab, University of Georgia, Athens, GA 30602, USA

E-mail: sminko@uga.edu

Over the past decade, interest in polymeric nano- and microfibers has rapidly grown for a number of applications including bone and tissue regeneration, drug delivery, biosensors, fuels cells, composite materials and fibrous polymer films. In particular, tissue-engineering technologies utilize scaffolds made of resorbable nanofibers with adequate porosity and surface area to mimic natural extracellular environment. Water and air purification systems, personal care products, energy storage devices, and membranes utilize the high surface area of nanofibers to enhance performance, barrier and filtration properties.

Several new methods for drawing of micro and nanofibers made of synthetic or natural polymers and fabrication of nanofibers using renewable resources are discussed in this presentation. Brush-, touch- and magnetospinning are recently developed fiber-spinning methods that are simple, user-friendly, cost-efficient, and independent of the solution dielectric properties, do not need high voltages, high RPM or large amounts of organic solvents compared with other fiber formation methods. The discussed methods are adapted for the fabrication of single-polymer, composite, or porous fibers that can be collected as 2D and 3D aligned meshes, scaffolds or yarns for a wide variety of applications.[1-3]

The essential difference of the brush-spinning method from other commonly used methods of nanofiber fabrication is in the mechanical control and manipulation of nanofibers that it offers. In contrast to all other methods, in brush-spinning single filament nanofibers are drawn by the mechanical force that determines not only fiber stretching but also guides fibers onto the spool or frame providing better control over fiber alignment. The brush-spinning method offers a very fast and practical method to produce scaffolds for cells with controlled mesh size. Our experiments demonstrated the capability of the spinning methods for fabrication of biomimetic scaffolds on different scales from macroscopic shape and dimensions to microscopic fiber dimensions and alignment into various meshes that are relevant to mesh-like structures in human tissues.

It was found that the formation of nanofibers using these spinning methods is quite different from the electro-spinning method due to differences in dynamics of fiber drawing and solvent evaporation. The difference is manifested in a much higher crystallinity of the spun fibers for the fibers of the same diameter obtained at the same linear speed of drawing. As discussed in the presentation, this difference has essential impact on fiber properties, including fibers for tissue-engineering scaffolds when the fiber properties affect the cell response and proliferation.

References

[1] Tokarev, A.; Trotsenko, O.; Griffiths, I. M.; Stone, H. A.; Minko, S. *Adv. Mater.* **2015**, *27,* 3560–3565.

[2] Tokarev, A.; Trotsenko, O.; Asheghali, D.; Griffiths, I. M.; Stone, H. A.; Minko, S. *Angew. Chem. Int. Ed.* **2015**, *54*, 13613-13616.

[3] Tokarev, A.; Asheghali, D.; Trotsenko, O.; Gruzd, A.; Lin, X.; Griffiths, I. M.; Stone, H. A.; Minko, S., *Adv. Mater.* **2015**, *27*, 6526-6532.

BIOGRAPHY

Sergiy Minko is the Georgia Power Professor in Fiber and Polymer Science with joint appointments in the Department of Textiles and Chemistry Departments, the University of Georgia. Sergiy Minko's research concentrates on nanostructured materials, functional fibers and textiles, responsive materials, biomaterials, single molecule research, colloids, nanoparticles, metal clusters, thin films, gels, coatings, polymers, polymer interfaces and polymer composites.

In 2003-2013, he occupied the position of Egon Matijevic Chair Professor in the Department of Chemistry and Biomolecular Science at Clarkson University, Potsdam, NY, USA. Before joining Clarkson in 2003, he held a senior research position at the Leibniz Institute of Polymer Research in Dresden, Germany. He completed his academic education in the Ukraine, gaining an MS in Chemical Engineering and a PhD in Chemistry from Lviv Polytechnic National University. He also holds a DSc in Macromolecular Chemistry from the Institute of Physical Chemistry, National Academy of Sciences and Ministry of Education, Ukraine.

In 1997-1998, Sergiy Minko was an Alexander von Humboldt Research Fellow at the University of Ulm, Germany. He has been a Visiting Professor at Vienna University of Technology in Austria, at the Institut Charles Sadron, CNRS, Strasbourg, France, the Max Planck Institute for Polymer Research, Mainz, Germany, and Massachusetts Institute of Technology (MIT), Cambridge, MA, USA.

Electrospun Nanofiber and Nanoyarn for Tissue Engineering

Xiumei Mo

Department of Bioengineering, College of Chemistry, Chemical Engineering & Biotechnology, Donghua University, Shanghai, 201620, China

E-mail: xmm@dhu.edu.cn

Electrospinning fabrication technique most commonly produces relatively 2D mats and the construction 3D structure nanofibers with higher porosity is still a major challenge. In this study, two methods were used to fabricate the 3D nanofiber scaffolds.

A dynamic electrospinning method were developed to fabricate the nanoyarn scaffold, by this way electrospun poly(L-lactide-co-ε-caprolactone)/collagen (P(LLA-CL)/Col) nanofibers were deposited and twisted into yarns in a water vortex before collecting on a rotating mandrel to form a nanoyarn scaffold. The nanoyarn scaffold contained 3D aligned microstructures with larger interconnected pores and higher porosity comparing with nanofiber scaffold. The biomimetic nanoyarn has a positive influence on cell proliferation and morphology. Pig iliac endothelial cells (PIECs) and MC3T3-E1 pre-osteoblastic cells cultured on the nanoyarn scaffolds showed significantly higher proliferation rates than that on traditional electrospun nanofiber scaffolds. After implanting the nanoyarn scaffold seeded with tendon derived stem cells into the nude mice, the fluorescence imaging indicated that the tendon derived stem cells had long-term survival, and the macroscopic evaluation, histology and immunohistochemistry examinations showed high-quality neo-tendon formation under mechanical stimulation in vivo.

Gelatin/PLA nanofiberous scaffold was fabricated by using combined electrospinning and freeze-drying. Gelatin/PLA nanofibers membranes were prepared by electrospinning; secondly, nanofibers membranes were cut into small pieces and were dispersed in tert-butanol by homogenizer; thirdly, the dispersions were frozen and freeze dried. Finally the scaffold was crosslinked by heating at 190°C or grafting HA on the nanofibers surface of scaffold by EDC/NHS. Thus obtained 3D nanofiber scaffold could promote cells growth and proliferation. In order to evaluate the repair capacity of 3D scaffold in vivo, an articular cartilage defect was created on rabbits and scaffolds were implanted into the defect. The in vivo study indicated that the cartilage repair capacity of scaffold without HA was limited, but scaffold modified with HA could enhance the repair of cartilage.

BIOGRAPHY

Dr. Xiumei Mo is a professor in Donghua University. She was granted 23 projects related with nanofiber fabrication for different tissue regeneration. Her researches are also related with electrospinning nanofiber and injectable hydrogel for tissue engineering. She has published more than 230 papers. ISI Web of Science showed that she ranking No.4 in the world on electrospinning nanofiber publication. She got the Science Technical Invention Awards from Shanghai Municipality in 2008, Science and Technology Progress Awards from State Department of People's Republic of China in 2009, Nature Science Awards from Shanghai Government in 2015. She is the committee members of China Biomaterials Society and the committee members Biomedical Engineering Society Biomaterials Branch.

Silk Biomaterials for Engineering Complex Tissues: Going Beyond the Fiber

Ana Leite Oliveira

Universidade Católica Portuguesa, CBQF - Centro de Biotecnologia e Química Fina – Laboratório Associado, Escola Superior de Biotecnologia, Rua Arquiteto Lobão Vital, Apartado 2511, 4202-401 Porto, Portugal

E-mail: aloliveira@porto.ucp.pt

Nowadays biomedical researchers area clearly recognizing that some of the widely used scaffolds for tissue engineering do not fairly recapitulate the cell microenvironment of the extra-cellular matrix (ECM), which is a dynamic and hierarchically organized structure regulating essential cellular functions such as morphogenesis, differentiation, proliferation, adhesion and migration[1]. As a consequence, new tools and biomaterials are being developed for tissue engineering to design advanced scaffolds that can establish a close dialogue with the ECM and eventually assemble more complex and larger functional tissues.

The orchestration between the chemical, biological and mechanical properties of a scaffold must result in instructive cues for cells to develop into a functional tissue *in vivo*. Among the several strategies that have been proposed to prepare polymeric scaffolds for tissue engineering (TE), fiber-based structures are particularly interesting as they present higher porosity, interconnectivity and surface area, which can facilitate cellular attachment and consequently improve scaffold cell colonization and new tissue formation[2]. Silk fibroin (SF), is a natural protein fiber which is spun by a variety of species including silkworms and spiders. Long standing FDA regulatory approval of silk as a suture, its abundance as raw fiber material and controlled proteolytic degradability *in vitro* and *in vivo* has established the ground for a range of tissue engineering applications, including bone, cartilage, tendon/ligament and skin regeneration.[2,3] In all of these approaches, the use of silk is associated with targeted functional microenvironments supporting tissue morphogenesis. This broad range of different applications is justified by its processing versatility and possibility of chemical modification, allowing for further biofunctionalization such as the immobilization of growth factors or cell binding domains able to modulate cell behavior. In this work are presented some examples of this enormous potential.

References

[1] Dvir, Tal, Langer, Robert *et al*, *Nature Nanotechnology*, **2011**, 6, 13–22.

[2] Ribeiro, Viviana P., Oliveira, Ana L. *et al.*, *Biomaterials*, **2017**, 123, 92-106.

[3] Thurber, Amy E., Fiorenzo G. Omenetto, David L. Kaplan, *Biomaterials*, **2015**, 71, 145-157.

BIOGRAPHY

Ana Leite Oliveira is the Head of the Biomaterials Group at the Centre for Biotechnology and Fine Chemistry (CBQF - Government Associated Laboratory), Portuguese Catholic University, Porto, Portugal. She is also Assistant Professor at the School of Biotechnology, Portuguese Catholic University, where she is responsible for the courses of Biomaterials taught to the Master of Biomedical Engineering and the Bachelors' Degree in Bioengineering. She obtained her PhD in 2008, in collaboration with the company Depuy Orthopaedics, Inc. (Johnson & Johnson), Warsaw, IN, USA. Her research interests are natural-based biomaterials, in particular silk for tissue engineering applications; biotextiles; sterilization of sensitive products. From her work have resulted already 5 international patents and more than 60 publications (articles in international journals with peer review and book chapters). She has organized and participated in different international conferences and workshops where she has presented more than 80 communications, mostly oral communications and invited lectures. She has been involved in the preparation and coordination as PI of various national and international projects funded by several National/European agencies and private companies. Parallel to her research activity she is a referee for several international journals in the biomedical field and since 2008, she has been assisting the European Commission as an Expert (Evaluator, Reviewer) for the Research Executive Agency (REA) in the 7th Framework Programme, COST actions (European Cooperation in Science and Technology) and Horizon 2020.

Morphological Controls Toward Intrinsically Stretchable Organic Electronics

Qibing Pei[a], Huier Gao[a], Jiajie Liang[a,b], Lu Li[a,c], Dustin Chen[a], Junpeng Li[a], Weili Hu[a], Fangchao Zhao[a], Shu-Yu Chou[a], Tibor Hajagos[a], Zhibin Yu[d]

[a]*Department of Materials Sciences and Engineering, University of California, Los Angeles, California 90095, USA;*

[b]*School of Materials Science and Engineering, The National Institute for Advanced Materials, Nankai University, Tianjin 300350, China;*

[c]*Institute for New Materials Technology, Chongqing University of Arts and Science, Chongqing, China;*

[d]*Department of Industrial and Manufacturing Engineering, High-Performance Materials Institute, Florida State University, USA*

E-mail: qpei@seas.ucla.edu

We present materials innovations for the fabrication of stretchable thin film electronic devices. Silver nanowire-polymer composite is demonstrated as a new generation of transparent electrode to replace ITO with added mechanical flexibility. Interpenetrating networks of conjugated polymer and polymer electrolyte blend are investigated as the emissive layer in stretchable OLEDs. A nanograin blend containing a PTB7 and PCBM is employed to demonstrate stretchable photovoltaic cells. Semiconducting carbon nanotubes in combination with an elastomeric dielectric allows for large strain deformation without substantial loss of carrier mobility. Peakforce quantitative nanomechanical AFM is used to analyze the stretchable morphologies, while polarized absorption and photoluminescent emission spectroscopy determines the molecular orientations. Silver nanowire polymer composite electrodes are developed as stretchable transparent conductors for the fabrication of the intrinsically stretchable devices.

References

[1] W. Hu, X. Niu, R. Zhao, and Q. Pei, *Appl. Phys. Lett.*, **2013**, 102, 083303.
[2] H. Gao, S. Chen, J. Liang, Q. Pei, *ACS Appl. Mater. & Inter.*, **2016**, 8(47), 32504-32511.
[3] D. McCoul, W. Hu, M. Gao, V. Mehta, Q. Pei, *Adv. Elect. Mater.*, **2016**, 1500407.
[3] J. Liang, L. Li, D. Chen, T. Hajagos, Z. Ren, S.Y. Chou, W. Hu, Q. Pei, *Nature Comm.*, **2015**, 6:7647.

BIOGRAPHY

Qibing Pei is Professor of Materials Science and Engineering at UCLA, specializing in synthetic polymers and composites for electronic, electromechanical, and photonic applications. He has over 170 peer-reviewed journal publications, and 42 issued US patents. His current research activities include synthesis of stretchable electronic materials, nanostructured polymer composites, dielectric elastomers and bistable electroactive polymers for muscle-like actuation, and fabrication of flexible and stretchable electronic devices. He received a B.S. degree from Nanjing University and a PhD from the Institute of Chemistry, Chinese Academy of Science. He was a postdoctoral fellow in Linköping University, Sweden, a senior chemist at UNIAX Corporation (now DuPont Display), Santa Barbara, and a senior research engineer at SRI International, Menlo Park, California. He is a Fellow of SPIE, member of ACS and MRS, Associate Editor of Smart Materials & Structures and Soft Robotics, and Advisory Board Member of Advanced Electronic Materials and Scientific Reports.

Smart Fiber Materials and Devices

Ye Zhang, Huisheng Peng
State Key Laboratory of Molecular Engineering of Polymers, Department of Macromolecular Science and Laboratory of Advanced Materials, Fudan University, Shanghai 200438, China.
E-mail: penghs@fudan.edu.cn.

A variety of smart fiber materials and devices has been developed to output various functionalities including energy harvesting, energy storage, sensing, lighting, displaying and the integration of the above functionalities. Compared with the conventional planar counterparts, the smart fibers can be woven into textiles with light weight, high integration and breathability.

References

[1] Peng, H. *et al.*, *Nat. Nanotech.* **2009**, 4, 738.
[2] Chen, P. *et al.*, *Nat. Nanotech.* **2015**, 10, 1077.
[3] Zhang, Z. *et al.*, *Nat. Photon.* **2015**, 9, 233.
[4] Deng, J. *et al.*, *Nat. Protoc.* **2017**, 12, 1349.
[5] Sun, H. *et al.*, *Nat. Rev. Mater.* **2017**, 2, 17023.

BIOGRAPHY

Huisheng Peng received his BEng in Polymer Materials from Donghua University in China in 1999, MSc in Polymer Science from Fudan University in China in 2003 and PhD in Chemical Engineering from Tulane University in USA in 2006. He worked at Los Alamos National Laboratory, US Department of Energy, from 2006-2008. Dr. Peng has been appointed as a Professor at the Department of Macromolecular Science and Laboratory of Advanced Materials at Fudan University since October 2008. His research focuses on the development of smart fiber materials and devices aiming at the application in wearable electronics.

RAFT, SNAPPS AND SUPERBUGS

Greg Qiao

Department of Chemical Engineering, The University of Melbourne, Victoria 3010 Australia

E-mail: gregghq@unimelb.edu.au

This talk will introduce our recently developed two areas of research based on mordent polymer chemistry. The first area is on our latest development of new methods to activate RAFT polymerization process. It will cover Photo-RAFT,[1] Fenton-RAFT,[2] Bio-RAFT and Sono-RAFT[3]. As shown in Fig 1a, monomers are polymerized under ultrasound wave in water in the present of a RAFT agent and the polydispersity is usually below 1.05. The polymerization can be turned on or off by simple switching the power of ultrasound device (Fig 1b). The second area will introduce our recently developed structurally nanoengineered antibacterial peptide polymers (SNAPPs) (Figure 1c) which can kill multi-drug resistant gram-negative superbugs without using any antibiotics and without causing toxicity to healthy cells.[4] Fig 1d & e demonstrates how the outer membrane of superbugs are "snapped" away by SNAPPs.

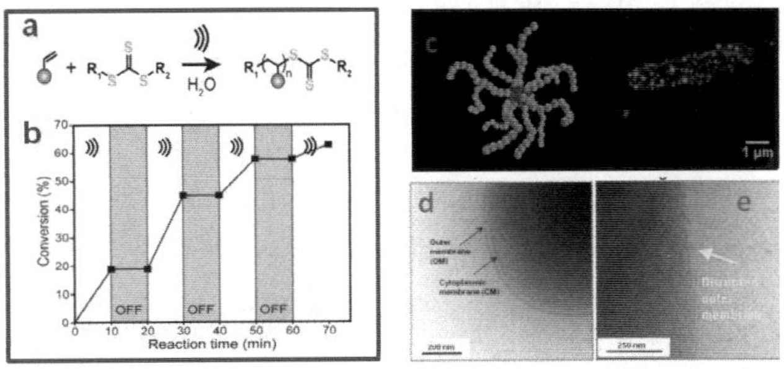

Figure 1. (Left) Schematic of sonochemically-induced RAFT polymerization (sono-RAFT) process (a) and monomer conversion with alternating on/off periods of applied ultrasound (b); (Right) peptide star polymers interacting with bacteria surface under fluorescent microscopy (c) and Transmission electron microscopy showing before and after the disrupted outer membrane of the Gram-negative bacteria (d and e).

References

[1] (a)McKenzie, T. G.; Fu, Q.; Wong, E. H. H.; Dunstan, D. E.; Qiao, G. G. *Macromolecules* **2015**, 48, 3864. (b) Fu, Q.; McKenzie, T. G.; Ren, J. M.; Tan, S.; Nam, E.; Qiao, G. G. *Scientific Reports*, **2016**, 6, doi:10.1038/srep20779

[2] Reyhani, A. McKenzie, T. G. Ranji-Burachaloo, H. Fu, Q. and Qiao, G. G. *Chem. Eur. J.* **2017**, 23, 7221.

[3] McKenzie,T.G.; Colombo,E.; Fu, Q.; Ashokkumar, M.; G.G. Qiao. *Angew. Chemi.*, **2017**, in press.

[4] Lam, S. J.; O'Brien-Simpson, N. M.; Pantarat, N.; Sulistio, A.; Wong, E. H. H.; Chen, Y. -Y.; Lenzo, J. C.; Holden, J. A.; Blencowe, A.; Reynolds, E. C.; Qiao, G. G. *Nature Microbiol.* **2016**, 1, 16162.

BIOGRAPHY

Professor Greg Qiao received his Ph.D. at the University of Queensland in 1996. He joined the University of Melbourne in 1996 and became a full Professor in 2009. From 2012-2015, he was an Australian Research Council's professorial Future Fellow. He is a Fellow of Royal Australia Chemical Institute (RACI) and Royal Society of Chemistry (RSC). Prof Qiao was the Chair of Polymer Division of the RACI from 2015-2016 and currently is a member of Australian Research Council (ARC) College of Experts. He is an Associate Editor for Australian Journal of Chemistry and is also on the International Advisory Boards of Macromolecular Bioscience, Macromolecular Materials & Engineering and Ullmann's Encyclopedia of Industrial Chemistry. Prof Qiao received ExxonMobil Award of Excellence for Chemical Engineering in 2015, RACI's Polymer Division Citation in 2011 and Freehills Award in 2010. He has published more than 230 journal papers and is a co-inventor of more than 20 patents.

Achieving High-powered Zn/air fuel cells through Doped Hierarchically Porous Carbons Derived from Quaternized Copolymers as Oxygen Electrocatalysts

Jinli Qiao

College of Environmental Science and Engineering, Donghua University, 2999 Ren'min North Road, Shanghai 201620, China

E-mail: qiaojl@dhu.edu.cn

Zn-air battery is considered as a promising energy storage solution for its low-pollution, safer property and provide impressively high specific (~1218 Wh kg^{-1}) and volumetric (~6136 Wh L^{-1}) energy densities. However, it still encounters some big technical challenges because of the inherently sluggish of cathodic oxygen reduction reaction (ORR), which usually requires a significant number of noble metals (such as Pt, Ru and Ir). A successful large-scale implementation of these technologies depends on the development of active, stable and selective ORR catalysts, preferably free of precious metals. Here, we report a series of Me/N/S-composited hierarchically porous carbon materials derived from quaternized copolymers with excellent ORR performance in both alkaline and acidic media. PUB, PAADDA and PICP with different chemical structures are used to pursue Fe/N/S-tailored ORR electrocatalysts. Surprisingly, they all display high discharge power density as applied to Zn-air fuel cells, and the HPCs-PAADDA catalyst thrillingly reaches 516.3 mW cm^{-2} when catalyst loading is optimized to 5.0 mg cm^{-2}.[1,2] The results elucidate that the polymer with long aliphatic chain is propitious to trap metals to create active sites and enwrap silica template to construct uniform pore structure. Only two kinds of nitrogen configuration (pyridinic-N and graphitic-N) are found with distinct structure in these HPCs, which happens to be active sites.[3]

References

[1] M. J. Wu, J. L. Qiao, K. X. Li, X. J. Zhou, Y. Y. Liu, J. J. Zhang, *Green Chem.* **2016**, *18*, 2699-2709

[2] M. J. Wu, Q. W. Tang, F. Dong, Y. Z. Wang, D. H. Li, Q. P. Guo, Y. Y. Liu, J. L. Qiao, *Phys. Chem. Chem. Phys.*, **2016**, *18*, 18665-18669

[3] Q. W. Tang, L. M. Wang, M. J. Wu, N. N. Xu, L. Jiang, J. L. Qiao*, *J. Power Source,* **2017,** *in press*

BIOGRAPHY

Dr. Jinli Qiao is a Professor, Supervisor, and Scientific Core Competency Leader of the College of Environmental Science and Engineering, Donghua Universtiy, China. She received her Ph.D in Electrochemistry from Yamaguchi University, Japan, then she joined the National Institute of Advanced Industrial Science and Technology (AIST) as research scientist. As the first/corresponding author, she has published over 100 research papers in peer-reviewed journals, 40 conferences and keynote/invited oral presentations, 4 co-authored/edited books/8 book chapters, and holds over 30 JP/China patents. Dr. Qiao has contributed actively to the progress of the profession and society. She is the vice-president/the board committee member of International Academy of Electrochemical Energy Science (IAOEES), co-founding editor of the Electrochemical Energy Reviewers (EER), the editorial board member of Green Energy & Environment (GEE), Frontiers in Energy Research, Current Graphene Science and the guest editor of Electrochim. Acta, Int. J. Hygrogen Energy and Applied Energy. She is the board committee member of the Solid State Ionics of China (CSSI), Electro-membrane Association of China (EMAC), and the member of the International Society of Electrochemistry (ISE), the Electrochemical Society (ECS), the American Chemical Society (ACS), Chemical Society of China (CSC), the Chemical Industrial and Engineering Society of China (IESC) and China Association of Hydrogen Energy (CAHE). Prof. Qiao's main research interests include PEM fuel cell catalysts/membranes, metal–air batteries, CO_2 electroreduction and supercapacitors.

Rheological and Enthalpic Response of Ultra-high Molecular Weight Polyethylene Melts

Tamito Igarashi[‖,†], **Kangsheng Liu, Dario Romano, Efren-Andablo Reyes, Ele de Boer, Sara Ronca, and Sanjay Rastogi**[†,‡]
[†]*Department of Materials, Loughborough University, Loughborough, LE11 3TU, UK;*
[‖]*Kureha Corporation, 3-3-2, Nihonbashi Hamacho, Chuo-ku, Tokyo 103-8552, Japan;*
[‡] *Department of Biobased Materials, Faculty of Humanities and Sciences, Maastricht University, Brightlands Chemelot Campus, Urmonderbaan 22, 6167RD Geleen, The Netherlands*
E-mail: sanjay.rastogi@maastrichtuniversity.nl

Recently, we investigated equilibration process of initially low entangled ultra-high molecular weight polyethylene (UHMWPE) melts. The observations are that the storage modulus and relaxation time of the melt increase with the increasing number of entanglements per chain. Resulting changes in the phase angle, as a function of frequency, can be shifted to a single curve. Horizontal and vertical shift factors applied to the phase angle reflect regular trend in the evolution of the storage modulus and the relaxation time, with the transformation of melt from non-equilibrium to equilibrium state. Different processes involved in the modulus build-up of the non-equilibrium melt are reflected in the development of the two shift factors. The studies are performed on UHMWPEs synthesized using a single-site post metallocene catalyst and compared to commercial UHMWPE synthesized using Ziegler-Natta catalyst. No rheological differences between the equilibrium melt states of UHMWPEs, synthesized under different conditions, could be obtained. However crystallization response from their melts shows major distinction. The distinction in the enthalpic response of the equilibrated polymer melts is attributed to differences in the topological constraints that rheology in the linear viscoelastic region does not differentiate. The paper draws parallels with learnings on differences in entangled state, and its influence on crystallization, in linear and cyclic polymers of similar molar masses.

References

[1] E.A. Andablo-Reyes, D. Auhl, E.L. de Boer, D. Romano, S. Rastogi. *Journal of Rheology* **2017**, 61 (3), 503-513.
[2] E.A. Andablo-Reyes, E. de Boer, D. Romano, S. Rastogi; *Journal of Rheology* **2014**, 58(6), 1981-1991
[3] K. Liu, E.L. de Boer, Y. Yao, D. Romano, S. Ronca, S, Rastogi; *Macromolecules* **2016**, 49 (19), 7497–7509

BIOGRAPHY

Sanjay Rastogi holds chair in polymer physics at Maastricht University (NL) and is honorary professor at Loughborough University (UK). His research interest is understanding of molecular architecture and its influence on mechanical properties. In his laboratory polymers are synthesized, and physics under quiescent and flow conditions is investigated to achieve the ultimate mechanical properties.

Implantable and Biodegradable Fibers for Optical Neural Interfaces

Xing Sheng
Department of Electronic Engineering, Tsinghua University
E-mail: xingsheng@tsinghua.edu.cn

Advanced optical fibers and photonic structures play important roles in neuroscience research, along with recent progresses of genetically encoded optical actuators and indicators. Most techniques for optical neural implants rely on fused silica or long-lasting polymeric fiber structures. In this paper, we present implantable and biodegradable optical fibers based on poly(L-lactic acid) (PLLA). PLLA fibers with dimensions similar to standard silica fibers are constructed using a simple thermal drawing process at around 200 °C. Formed PLLA fibers exhibit high mechanical flexibility and optical transparency, and their structural evolution and optical property changes are systematically studied during *in vitro* degradation. In addition, their biocompatibility with brain tissues are evaluated in living mice, and fully *in vivo* degradation is demonstrated. Finally, we implement PLLA fibers as a tool for intracranial light delivery and detection, realizing deep brain fluorescence sensing and optogenetic interrogation *in vivo*. The presented materials and device platform offers paths to fully biocompatible and bioresorbable photonic systems for biomedical uses.

BIOGRAPHY

Xing Sheng is currently an assistant professor in the Department of Electronic Engineering at Tsinghua University, China. From 2012 to 2015, he was a postdoctoral researcher at University of Illinois at Urbana-Champaign, working with Prof. John A. Rogers. He obtained his PhD degree at Massachusetts Institute of Technology in 2012 (advisor: Prof. Lionel C. Kimerling), and his B.Eng degree at Tsinghua University in 2007.

Xing Sheng's research includes fundamental and applied aspects of light and materials interactions and advanced optoelectronic devices and systems. His past experiences include design and fabricating novel electronic and photonic structures for energy and biomedical applications. He has authored more than 20 publications in journals like Nature Materials, Nature Communications, Advanced Materials, etc.

Negative Sorting of Circulating Tumor Cells Using a Microfluidic Platform Integrated with Streptavidin-Functionalized PLGA Nanofibers

Mengyuan Wang, Gangwei Xu, Di Yin, Yunchao Xiao, Xiaoyue Zhu,* Xiangyang Shi*

State Key Laboratory for Modification of Chemical Fibers and Polymer Materials, College of Chemistry, Chemical Engineering and Biotechnology, Donghua University, Shanghai, 201620, China

E-mail: xshi@dhu.edu.cn, zhu@dhu.edu.cn

Detection of circulating tumor cells (CTCs) plays an important role in early diagnosis of cancer and personalized therapy. However, CTCs are difficult to be separated from peripheral blood with conventional means due to their extreme rareness. In this study, a negative sorting method has been developed for isolation and identification of circulating tumor cells (CTCs) through a microfluidic platform integrated with streptavidin (SA)-functionalized electrospun polylactic-co-glycolic acid (PLGA) nanofibers. Through the specific biotin-streptavidin interation, the device is able to simultaneously capture biotin-CD45-labled white blood cells and enrich cancer cells or CTCs. We show that SA-functionalized PLGA nanofibers are hemocompatible and cytocompatible, and display a good binding affinity to biotin molecules as well as biotin-labeled WBCs. With these properties, the SA-functionalized PLGA nanofibers embedded within a microfluidic chip are able to selectively capture biotin-labeled WBCs, and model cancer cells or CTCs from patients can be enriched to the receiving tank through a dynamic microfluidic system and identified. We demonstrate that the WBC capture efficiency is as high as 97%, the recovery rate of cancer cells reaches up to 97.5%, and the average number of CTCs from randomly selected 15 patients is 5 per milliliter. These results suggest that the integrated negative sorting device is promising to be used for diagnosis of cancer and cancer metastasis.

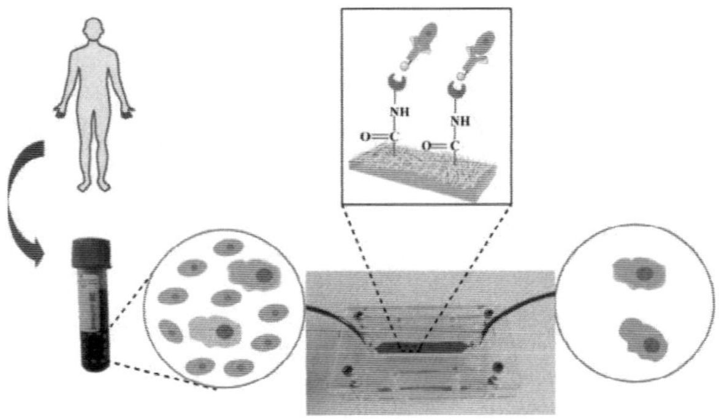

Figure 1. Schematic representation of the WBCs capture and CTCs enrichment through a microfluidic platform integrated with SA-functionalized electrospun PLGA nanofibers.

References

[1] Yili Zhao, Xiaoyue Zhu, Hui Liu, Yu Luo, Shige Wang, Mingwu Shen, Meifang Zhu, Xiangyang Shi. *J. Mater. Chem. B* **2014**, 2, 7384-7393.

[2] Gangwei Xu, Yulong Tan, Tiegang Xu, Di Yin, Mengyuan Wang, Mingwu Shen, Xiaofeng Chen, Xiangyang Shi, Xiaoyue Zhu. *Biomater. Sci.* **2017,** 5, 752-761.

[3] Yili Zhao, Zhangyu Fan, Mingwu Shen, Xiangyang Shi. *Adv. Mater. Interfaces* **2015**, 2, 1500256.

BIOGRAPHY

Dr. Xiangyang Shi graduated with B.Sc. (Chemistry, Henan Normal University,Xinxiang), M.E. (applied chemistry, Beijing Institute of Technology, Beijing), and Ph.D. (organic chemistry, Institute of Photographic Chemistry, the Chinese Academy of Sciences, Beijing) in 1992, 1995 and 1998, respectively. He worked as a Postdoctoral Fellow in Tsinghua University (1998-2000, Beijing) and Max-Planck Institute of Colloids and Interfaces (2000-2001, Potsdam, Germany). He then moved to California State University, Los Angeles (2001-2002) as a visiting scholar. From 2002- 2008, he worked as a Research Fellow, Research Associate II, Research Investigator, and Research Assistant Professor at the University of Michigan, Ann Arbor. He then became a professor of special appointment both in Donghua University and in Shanghai Institutions of High Learning (Eastern Scholar) since 2008. Since 2010 he is also appointed as an "Invited Chair in Nanotechnology" at University of Madeira, Portugal. Prof. Shi has authored or coauthored 258 SCI-indexed peer-reviewed journal articles and 236 technical conference abstracts or proceeding papers, along with 12invited book chapters and 63 approved patents. His research expertise includes dendrimer chemistry and organic/inorganic hybrid nanomaterials for nanomedicine applications, in particular, cancer diagnosis and therapy, and electrospun polymer nanofiber-based technology for applications in biosensing, pharmaceutical sciences, regenerative medicine and environmental sciences.

Nanostructure Control for Fabrication of CNF Filaments

Daniel Söderberg, **Nitesh Mittal**

Wallenberg Wood Science Center and Department of Mechanics

KTH Royal Institute of Technology, 100 44 Stockholm, Sweden

E-mail: dansod@kth.se

Spinning of cellulosic filaments is far from a novelty, viscose filaments could be spun already in thelate 19th century. Today there is a wide variety of filament materials originating from cellulose that are spun under industrial conditions using well-understood spinning processes. Furthermore, significant efforts have been made aimed at the fabrication of filaments from cellulose nanofibrils (CNF) during the last decade, and they typically apply conventional spinning processes and aim at achieving favorable properties by developing spinnable liquids that allow the use of these processes. However, given the nature of CNF, being very slender particles, the application of existing knowledge on macroscopic fiber suspension flows provides insights that offer new routes process routes that give improved mechanical performance[1], which also have been applied to proteins[2,3]. The ideal process would align CNF in the direction of the formed thread and by some means preserve this controlled structure during drying.

In order to achieve maximum nanostructure control in the flowing dispersion during spinning, there are a set of parameters that need to be tuned. Examples are the macroscopic time-scales connected to the flow and geometry, gelling by diffusion of salts or protons, as well as mechanisms on the nanoscale such as e.g. Brownian motion. If alignment can be achieved by tuning the flow-fields, Brownian motion would quickly force nanoscale rod like particles to return to an isotropic state assuming that the suspension is dilute. The time-scale for Brownian return-to-isotropy of a nanorod with a diameter of 10 nm and aspect ratio of 100 would be of the order 10^{-5} s, and an increase in concentration would slow down return-to-isotropy about one order of magnitude[4,5,6]. These time-scales assumes dilute to semi-dilute suspensions and, in principle, that the nanoscale rods are perfectly stiff. However, if the rods have some flexibility and friction-type forces are present at contact points, the rods may form what we can be described as a loose network[4]. Although the suspension is clearly below the percolation threshold, entanglement still allows flow-field based alignment but with a marked reduction in Brownian effects, allowing the fabrication of e.g. CNF-based filaments with exceptional properties.

The principles of selecting suitable parameters for achieving near perfect alignment will be illustrated as well as examples of mechanical and functional properties that can be achieved.

References

[1] K.M.O. Håkansson *et al.*, *Nature Communications*, **2014**, 5.
[2] N. Mittal *et al.*, *ACS Nano,* **2017**, 11, 5148–5159.
[3] A. Kamada *et al.*, *PNAS*, **2017**, 114, 1232–1237.
[4] H. Brenner. *Int. J. of Multiphase Flow,* **1974**, 1, 195–34.
[5] L. Tian *et al.*, *Aerosol Science and Technology*, **2016**, *50:5, 474*–486.
[6] Y.-G. *et al.*, *The Journal of Chemical Physics,* **2005**, 122, 244903.
[7] C. F. Schmid, D. J. Klingenberg, *Physical Review Letters,* **2000**, 84, 290.

BIOGRAPHY

Daniel received an MSc in Vehicle Engineering (1994) and PhD in Fluid Mechanics (1999) at Royal Institute of Technology (KTH), Stockholm. He was employed at the Swedish Pulp and Paper Research Institute, which later changed name to Innventia, from 1999 to 2013. From 2009 was Deputy Director of the business area Material Processes. During the period 2008-2014 he was also Adjunct Professor at KTH. Early 2014 he became as a full-time employee at KTH, and is, since March 2016, Head of the Department of Mechanics at KTH. His research is related to multiphase flows, specifically in connection to processes related to the manufacturing of forest-based materials. Daniel is currently Director of BiMaC Innovation (Bio-fibre Materials Center) and active within the Wallenberg Wood Science Center. He is also since 2013 chairman of the UK-based Pulp and Paper Fundamental Research Society. He is holder of several patents, including being the inventor of the Aq-vane technique that received the ATIP gold innovation award 2004, and which today is a commercially available technology.

Advanced Multi-Material Fibers

Fabien Sorin and **Yunpeng Qu**
Institute of Materials, Ecole Polytechnique Fédérale de Lausanne, 1015 Switzerland.
E-mail: Fabien.sorin@epfl.ch

The development of advanced fiber processing technologies is enabling the fabrication of fiber structures with different materials and hence increasingly complex functionalities. The thermal drawing process in particular, originally developed to make optical fibers, has experienced a series of breakthroughs that have expanded the range of cross-sectional architectures and materials that can be integrated in fibers[1]. Beyond the functional fibers it can generate, the thermal drawing process exhibit several intriguing materials scientific and processing aspects. Recent studies have identified interfacial tension and viscosity as playing a crucial role in the type of materials, structure and feature sizes that can be achieved in thermally drawn fibers. While this has advanced our understanding of, and opened new prospects for, the thermal drawing process, a large number of scientific questions and opportunities associated with the tailoring of interfacial tension and the rheological properties of materials remain unexplored. In this talk, we will present our recent work to deepen our understanding of the materials science behind thermal drawing patterning to engineering interfacial tension and viscosity. We will first show how we can tailor the interfacial tension of polymers to realize fibers and micro-channels with sub-micrometer surface textures[2]. We will then demonstrate how applying this understanding to electrically conducting polymer nano-composite can lead to intriguing fiber devices such as electromechanical one-dimensional distributed touch sensor[3]. Turning to semiconductor-based fibers, we will show how modifying the surface energy of semiconducting materials in solution can enable the fabrication of single-crystal nanowire-based optoelectronic fibers with unprecedented performance[4]. Finally, we will show how we can expand the map of materials compatible with the thermal drawing process, by looking at the required rheological and microstructural attributes at a deeper level. We will exploit this finding to show that some elastomeric materials can be drawn with flow properties similar to their thermoplastic counterparts, while maintaining their elastic properties at room temperature. This opens novel opportunities for fiber-based devices in the fields of stretchable optics and electronics as well as biological probes and smart textiles.

References

[1] M. Schmidt, A. Argyros and F. Sorin, *Advanced Optical Materials*, **2016**, 4, 13.
[2] T. Nguyen-Dang et. al..... Y. Qu and F. Sorin *Advanced Functional Materials*, **2017**, 27, 1605935.
[3] T. Nguyen-Dang, A. Page et. al..... Y. Qu and F. Sorin, *J. of Physics D: Appl. Phys.*, **2017,** 50, 144001.
[4] W. Yan, Y. Qu et. al....and F. Sorin, *Advanced Materials*, **2017**, 29, 1700681.

BIOGRAPHY

Prof. Fabien Sorin studied Physics at Ecole Polytechnique in France, and obtained his PhD from the department of Materials Science at the Massachusetts Institute of Technology (MIT), Cambridge, USA. After 3 years at the Research Laboratory of Electronics at MIT, he joined the company Saint-Gobain as a research scientist in Aubervillers, France. In 2013, he joined the Ecole Polytechnique Féderlae de Lausanne (EPFL, Switzerland) as an assistant professor in the Institute of Materials. His research interests lie in investigating innovative materials, fabrication methods, and nanoscale device architectures to integrate advanced photonic and electronic functionalities within one-dimensional fibers and two-dimensional flexible and stretchable substrates.

Development of a Novel Injectable Matrix for Cartilage Repair

Rui A. Sousa, Cristina Correia, Cristiana B. Cunha and David A. Learmonth
Stemmatters, Biotecnologia e Medicina Regenerativa SA, 4805-017, Guimarães, Portugal
E-mail: rasousa@stemmatters.com

Articular cartilage defects are present in the majority of patients undergoing arthroscopy, for which autologous chondrocyte implantation (ACI) remains the reference cell-based therapy. Still, ACI presents several performance drawbacks resulting in surgical complications. Improvements have been accomplished by using three-dimensional collagen matrices (MACI), but limitations have justified development of alternative matrices. An attractive approach is using a cell encapsulation matrix by which cells can be delivered and retained at the cartilage defect. Present work describes development of a novel tissue engineering strategy based on a proprietary encapsulation matrix. This study covers rational design optimization of its chemical structure, physicochemical characterisation, *in vitro* and *in vivo* performance evaluation, as well as on development of a regulatory strategy for clinical development and approval. A large library of hydrogels has been established exhibiting appealing physicochemical and biological performance. Screening enabled the identification of chondrogenic favourable formulations that improve phenotypic commitment and assure tissue adhesion. *In vitro* studies demonstrated high viability of the cells and progressive metabolic activity along culture. Intense cartilage matrix deposition was detected by up-regulation of chondrogenic markers by end of culture. *in vivo* studies demonstrated the establishment of lateral and subchondral integration with good tissue regeneration outcome. Proposed strategy enables minimally invasive repair of focal cartilage defects. Constructs demonstrated intense cartilage development *in vitro* through evaluation of specific markers. In addition, *in vivo* data demonstrated promising repair performance in focal cartilage lesions, suggesting a valuable approach for further therapeutic exploitation in humans. A preclinical development plan for evaluation of safety and efficacy (large animal model) is currently being executed and clinical development scheduled to start end of 2018.

BIOGRAPHY

Rui A. Sousa started to work in biomedical engineering in 1997, where he gained interest in the development of medical devices and novel advanced biomaterials. His work has focused in tissue engineering and regenerative medicine, both as researcher and entrepreneur. Rui co-founded Stemmatters and led the development of the overall business by transforming a single page business concept into a fully integrated biotechnology company. During this period, Stemmatters was able to build and reinforce considerably its intellectual property position in regenerative medicine, as well establish a relevant position regarding manufacturing of biologics. Rui developed key scientific and industrial experience in several business sectors with strong emphasis on product development and innovation. Academically, he holds a degree in Metallurgical and Materials Eng. (University of Porto), a PhD in Mechanical Engineering (Brunel University) and a Master in Business Administration (EGP-PBS). He is an author/co-author of +130 publications, including +70 papers in scientific journals with referee and book chapters. Rui A. Sousa is an inventor of at least 13 patent applications or granted patents.

Upconversion Luminescent Materials: Optical Tuning and Their Bioapplications

Qianqian Su

Institute of Nanochemistry and Nanobiology,School of Environmental and Chemical Engineering, Shanghai University, No. 99, Shangda Road, Shanghai, 200444, P.R.China

Email:chmsqq@shu.edu.cn

Upconversion luminescent materials that convert low energy excitation wavelength to high energy emission have attracted considerable attention due to the advantages of high penetration depth and high resistance to autofluorescence interference capability. We present a rational design and synthesis of lanthanide-doped upconversion nanoparticles displaying unprecedented optical properties. We also demonstrate the use of upconversion materials as probes for cell imaging and drug release monitoring application. In addition, we demonstrate molecular upconversion probe, which shows excellent photostability, high quantum yield and rapid excretion capability. The NIR-to-NIR upconversion technique also offers a promising detection method, which shows very high sensitivity and selectivity for the detection of Cu^{2+} ions in vitro and in vivo.

BIOGRAPHY

Qianqian Su obtained her BE degree in Materials Science and Engineering from Shandong University. She received her MS degree in Polymer Chemistry and Physics from Guangzhou Institute of Chemistry, Chinese Academy of Sciences and completed her PhD degree in the group of Professor Liu Xiaogang at the National University of Singapore. She worked as a postdoctoral researcher at Fudan University in the group of Professor Fuyou Li. Then she joined Institute of Nanochemistry and Nanobiology at Shanghai University in 2017. Her research interest focuses on the development of novel luminescent nanomaterials for bioapplications.

Dr. Su has authored 3 patents and 20 peer-reviewed scientific journals, including J. Am. Chem. Soc., Adv. Mater., Chem. Commun., Acc. Chem. Res., Chem. Soc. Rev., etc., some of which were selected as Highly Cited Papers in ESI.

Designing and Tuning the Optical Properties of Perylene Tetracarboxylic Diimides by POSS Side Chain and Bay Area Substitutions

Bin Sun[1*], Xinlin Zhang[1], Xin Wang[1], Ting Zeng[1], Yu Shao[1], Chien-Lung Wang[3], Wenbin Zhang[2], Meifang Zhu[1*], Stephen Z. D. Cheng[4]

[1]*State Key Lab for Modification of Chemical Fibers and Polymer Materials, College of Material Science and Engineering, Donghua University, Shanghai 201620, China;*

[2]*Key Laboratory of Polymer Chemistry & Physics of Ministry of Education, Center for Soft-Matter Science and Engineering, College of Chemistry and Molecular Engineering, Peking University, Beijing 100871, P. R. China;*

[3]*Department of Applied Chemistry, National Chiao Tung University, 1001 Ta Hsueh Road, Hsin-Chu, 30010, Taiwan;*

[4]*The Maurice Morton Institute and Department of Polymer Science,The University of Akron, Akron, Ohio 44325-3909, USA*

E-mail: sunbin@dhu.edu.cn, zmf@dhu.edu.cn

Perylene Tetracarboxylic Diimides (PDI) have found a wide variety of applications in gerenal dyes, fluorescencent dyes and chemosensors due to its stability, high fluorescence. The challengeing is how to adjust and controll its self-assembly behavior for its highly fluorescent under condensed state in its bulk low dimensional materials and/or its polymer composites. In our group, series of highly soluble and fluorescent core-twisted perylene diimide dyes (PDIs) substituted by Polyhedral Oligomeric Silsesquioxane (POSS) at the N-core and at the bay 1,7-position were synthesized and well characterized by NMR, MALDI-TOF mass spectrometry. Their self assembly and optical properties of this new functional dyes in solution and under solid crystal state were investigated and the realationships between optical properties and condensed structure have been tried to build up based on their crystal structure analysis in detail. Packing by using dimmer as motif due the effect of bulky POSS restraining the strong effect of $\pi-\pi$ to cause the molecules to be tightly packed in the aggregates is the main reason for their action. The highly fluorescent POSS-PDIs with adjustable packing mode has potentially use in smart fluorescent fiber.

This research is financially supported by Science and Technology Commission of Shanghai Municipality (16JC1400700), the Program for Changjiang Scholars and Innovative Research Team in University (T2011079, IRT1221), the Program of Talents of Discipline to University (111-2-04).

Keywords: Perylene Tetracarboxylic Diimides, POSS, Self-assembly, Crystal, Fluorescence

References

[1] Yu Shao, Guang-Zhong Yin,Xiangkui Ren,Xinlin Zhang,Jing Wang, Kai Guo, Xiaopeng Li, Chrys Wesdemiotis,Wen-Bin Zhang, Shuguang Yang, Meifang Zhu* and Bin Sun*. Engineering p–p interactions for enhanced photoluminescent properties: unique discrete dimeric packing of perylene diimides. *RSC Advances*, **2017**, 7, 6530-6537.

[2] Yu Shao, Xinlin Zhang, Kai Liang, Jing Wang, Yuejian Lin, Shuguang Yang, Wen-Bin Zhang, Meifang Zhu* and Bin Sun*. How does the interplay between bromine substitution at bay area and bulky substituents at imide position influence the photophysical properties of perylene diimides? *RSC Adv.*, **2017**, 7, 16155- 16162.

[3] Xin Wang, Ting Zeng, Mohamed Nourrein, Bo-Han Lai, Kaiwen Shen, Chien-Lung Wang*, Bin Sun*, Meifang Zhu*. Concentration-Dependent Self-Assembly Structures of An Amphiphilic Perylene Diimide with tri(Ethylene Glycol) Substituents at Bay Positions. *RSC Adv.*, **2017**, 7 (42): 26074-26081.

Chlorine Rechargeable Biocidal Polymers and Fibers for Biological Protection and Food Safety

Gang Sun[1], **Yang Si**[1], **Yue Ma**[1], **Dong Wang**[2]

[1]*Fiber and Polymer Science, University of California, Davis, CA 95616, USA;*

[2]*College of Materials Sciences and Engineering, Wuhan Textile University, Wuhan, China*

E-mail: gysun@ucdavis.edu

This research group has been working on development of polymeric and fibrous materials with rechargeable biocidal functions for personal biological protections and food safety for decades and has developed novel materials mainly in these two categories: Rechargeable halamine and Refreshable light-induced biocidal polymers and fibers. In this presentation, we will concentrate on preparation and applications of rechargeable halamine precursor polymers, fibers, and films. Olefin polymers can be effectively and economically modified by a radical graft polymerization reaction in a twin-screw extruder during an extrusion process, and thus modified polymers can be fabricated to films, fibers, and even nanofibers by using compressive film casting, a melt extrusion, and an electrospinning process. The processed products all demonstrated desired rechargeable biocidal functions against a broad spectrum of bacteria and viruses. The fibers and nanofibrous membranes could serve as ideal protective textile materials. In addition with the desired biological protective functions for biomedical and military personnel, these functionalized polymers and fibers can significantly reduce foodborne microbial contamination through contacts on solid surfaces, provide high efficient disinfection to filtered water, and prevent formation of biofilms on solid materials. The presentation will discuss the latest progresses.

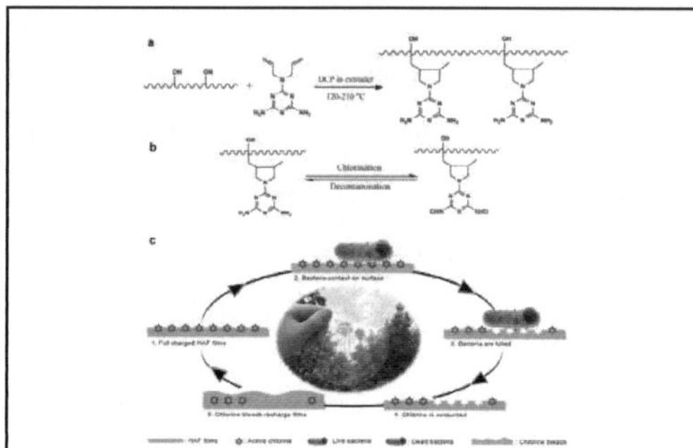

Figure 1. Refreshable biocidal polymerss.

References

[1] Jing Zhu, Qazaleh Bahramian, Philip Gibson, Heidi Schreuder-Gibson, Gang Sun, *J. Mater. Chem.*, **2012**,22, 8532.

[2] Yang Si, Andrea Cossu, Nitin Nitin, Yue Ma, Cunyi Zhao, Bor-sen Chiou, Trung Cao, Dong Wang, Gang Sun, *Macromol. Biosci.*, **2017**, 17, 1600304.

[3] Yang Si, Jiaying Li, Cunyi Zhao, Yue Deng, Yue Ma, Dong Wang, Gang Sun, *ACS Biomater. Sci. Eng.*, **2017**, 3 (5), 854–862.

BIOGRAPHY

Gang Sun is a professor of Fiber and Polymer Sciences in Division of Textiles and Clothing at University of California, Davis, and has been conducting research on functional textiles, textile chemistry, and nanotechnologies since 1995. Most his research activities and efforts have been devoted on development of novel antibacterial textiles and polymers for personal protection equipment, including medical, chemical, and military protective clothing and products. He is a recipient of the Olney Medal in 2016, the highest science award by American Association of Textile Chemist and Colorist (AATCC). He has published over 200 peer-reviewed journal articles, and is serving on editorial boards in several major textile professional journals.

Microstructural Design of Fibrous Electrode for Wearable Energy Storage

Gengzhi Sun, Peipei Shi, Li Li, Li Hua
Inititute of Advanced Materials, Nanjing Tech University, Nanjing 210009, China
E-mail: iamgzsun@njtech.edu.cn

Solid-state fiber-based supercapacitors have been considered promising energy storage devices for wearable electronics due to their light weight and amenability to be woven into textiles. Efforts have been made to fabricate high performance fibrous electrode by depositing pseudocapacitive materials on the outer surface of carbonaceous fiber using, *e.g.*, hydrothermal or electrochemical deposition to form a core-shell structure. Unfortunately, the performance of the fibrous electrodes is still far away from satisfactory and the critical issues lie in: 1) often only the shell is active while the bulk core of the fiber does not participate in the electrochemical charge-discharge processes, and 2) the shell active material is not efficiently utilized due to their poor conductivity, thus the loading capacity of active material is limited. Therefore, the areal/volumetric capacitances and rate capability of the hybrid fiber electrode are significantly compromised. Moreover, owing to the rigidness and brittleness of the shell active materials, the mechanical robustness of the hybrid fiber is poor. As a step forward, we focus our research on the microstructural design of fibrous electrode, aiming to realize the synergistic interaction between conductive filaments and active nanomaterials, and achieve high specific capacitance/energy density without compromising their high rate capability and cycling stability.

References

[1] P. P. Shi, L. Li, L. Hua, Q. Q. Qian, P. F. Wang, J. Y. Zhou, G. Z. Sun,* W. Huang, *ACS Nano*, **2017**, *11*, 444-452.

[2] L. Hua, Z. Y. Ma, P. P. Shi, L. Li, K. Rui, J. Y. Zhou, X. Huang, X. Liu, J. X. Zhu, G. Z. Sun,* W. Huang, *J. Mater. Chem. A*, **2017**, *5*, 2483-2487.

[3] G. Z. Sun, X. Zhang, R. Lin, B. Chen, L. Zheng, X. Huang, L. Huang, W. Huang, H. Zhang, P. Chen, *Adv. Electronic Mater.*, **2016**, *2*, 1600102.

[4] G. Z. Sun, X. Wang, P. Chen, *Mater. Today,* **2015**, *8,* 215-226.

[5] G. Z. Sun, X. Zhang, R. Lin, J. Yang, H. Zhang, P. Chen, *Angew. Chem. Int. Ed.*, **2015**, *54*, 4651-4656.

BIOGRAPHY

Gengzhi Sun obtained his Master degree in 2008 from Changchun Institute of Applied Chemistry (CIAC), Chinese Academy of Sciences (CAS), and Ph.D. in 2013 from Nanyang Technological University (NTU), Singapore. After three years postdoctoral study in NTU, he joined Institute of Advanced Materials, Nanjing Tech University, as a full professor. His research interests focus on micro-/nano-structural design of functional materials for flexible/wearable electronics.

Rare-earth Doped Upconversion Luminescent Nanomaterials for Sensing, Chemo-Photothermal Therapy and Bioapplications

Ruoyan Wei, Lining Sun*

Research Center of Nano Science and Technology, and School of Material Science and Engineering, Shanghai University, Shanghai 200444, China.

E-mail: lnsun@shu.edu.cn

Up to now, nanomaterials for chemical or physical therapy guided by near infrared (NIR) light imaging play an increasingly important role in the field of visual cancer therapy. As a matter of fact, both chemical and physical cancer therapies have their own limitations. Bioimaging–guided cancer therapy based on nanocomposites have attracted significant attention in recent years with higher therapeutic efficiency and reduced side effects [1]. In this case, a multifunctional cancer therapy nanocomposite was proposed and synthesized by linking pH–responsive prodrug onto gold nanocrystals that were grown *in situ* on the surface of upconversion nanoparticles (UCNPs). In the structure of prodrug, a hydrazone bond was utilized for the following pH–responsive drug release in the intracellular acidic microenvironment of cancer cells. The innovative assembly method is facile and mild to obtain the nanocomposites of UCNPs and gold, which show excellent photostability and biocompatibility. The final nanocomposites offer efficient treatment effects *in vitro* under irradiation of a 808 nm laser due to the synergy effect of chemotherapy and photothermal therapy. In addition, the final nanocomposites show excellent intracellular locating ability *via* upconversion luminescence imaging by Er^{3+} ion and magnetic resonance imaging by Gd^{3+} ion, which is potential as visual tracking agent in cancer treatment. Therefore, the presented bioimaging guided–multifunctional synergistic therapy nanocomposites are promising tools for imaging guided cancer therapy[2].

References

[1] X. Ge, Z. Song, L. N. Sun, Y. Yang, L. Shi, R. Si, W. Ren, X. Qiu, H. Wang, *Biomaterials*, **2016**, 108, 35.

[2] R. Wei, W. Xi, H. Wang, J. Liu, T. Mayr, L. Shi, L. N. Sun, *Nanoscale*, **2017**, 9, 12885.

BIOGRAPHY

Professor Lining Sun earned her Ph.D. degree in then end of 2007 from Changchun Institute of Applied Chemistry, Chinese Academy of Sciences. Then, from March 2008 to 2009, she has been as an Alexander-von-Humboldt Research Fellow in University of Regensburg, Germany. She joined Shanghai University as an Associate Professor in 2009. She has been a full professor since March 2016. Her research focuses on rare-earth doped upconversion luminescent nanomaterials for sensing and bioapplications.

From Optical Fibers to Advanced Functional Fibers

Guangming Tao

Huazhong University of Science and Technology, China

E-mail: Derek.Tao@msn.com

Optical fibers are the best example of fibers that have improved the quality of life throughout the world and have fundamentally altered the human conditions in both obvious and imperceptible ways. Over the past decade, a new class of fiber – multimaterial fiber – leverages the capabilities of traditional fiber fabrication, but aims at developing new fiber structures, functionalities, and applications that stem from altering the materials composition of the fiber. Materials with different optical, electronic, thermomechanical, and acoustic properties have now been incorporated into the fiber form factor. Furthermore, fibers that can be further weaved into textiles and fabrics. Fabrics made of highly functional fibers in optimized manners could realize new functionalities beyond staying warm and pretty. The insights emerging from both optical fiber and multimaterial fiber research, along with the textiles and fabrics industries are pointing to new ways that the manufacturing of advanced functional fibers and fabrics itself may be leveraged as a fabrication methodology. Such fibers and fabrics can integrate optoelectronics elements, nonlinear waveguides, electrodes, metamaterials, semiconductor junctions, and carbon-polymer composite into a simple or comprehensive fashion into a single fiber for various applications that stem from the functionality of materials themselves, the optimized architecture of these materials, and the inherent merits of fibers – extend length, flexibility, scalability and the low-cost mass production ability.

BIOGRAPHY

Dr. Tao received his B. Eng. (2006), M.Sc. (2009), and Ph.D. (2014) in optics/optoelectronic from Shandong University, Fudan University, and the University of Central Florida, respectively. He was a visiting scholar at Chinese Academy of Science (2007-2008), the Massachusetts Institute of Technology (2012), and Centre national de la recherche scientifique (2017). He is a Sr. Research Scientist at CREOL, The College of Optics & Photonics since 2014. He has been awarded several scholarships/honors from Optical Society of America (OSA), the International Society of Optical Engineering (SPIE), and the American Ceramic Society (ACerS), the Extraordinary Potential Prize winner of Chinese Government Award for Outstanding Self-financed Students Abroad (2013) and "1000 Young Talents" Award (2017). He has years of research experience in optical sciences and engineering in academia, industry, and government institutes with expertise in the areas of infrared material, infrared fiber, in-fiber nano-fabrication, and advanced functional fibers and textiles. He has also been the organizer/session chair of multiple international conferences. Dr. Tao is also the cofounder of Lambda Photonics LLC, a startup company which developed out of his Ph.D. work.

From “Active” Polymers to Smart Fibers via Post-modification Polymerization

Patrick Theato*, Shaojian Lin, Jiaojiao, Shang

Institute for Technical and Macromolecular Chemistry, University of Hamburg, Bundesstrasse 45, D-20146 Hamburg, Germany.

E-mail: theato@chemie.uni-hamburg.de

“Smart” polymeric fibers (SPFs) have received rapid development inspired by the progress of “smart” polymers recently.[1] Based on the characteristics of “smart” polymers, such SPFs can undergo reversible physical and chemical transitions upon external stimulation, including temperature, CO_2 and light etc. Compared to traditional bulk “smart” polymers, SPFs possess a higher surface area to volume ratio.[2] Moreover, the majority of SPFs can be fabricated with controlled diameters, pore sizes and shapes, using different methods such as melting spinning, wetting spinning or electrospinning.[3] Until now, numerous SPFs were spun directly from the corresponding polymers. In this case, the corresponding spinning process and parameters change with polymers, which are time-consuming. Moreover, some “smart” polymers are not suitable for spinning processes due to the limitations ranging from short polymer chains and unstable properties of some pendant groups during spinning processes. To address these issues, facile fabrication of SPFs via post-polymerization modification has received considerable attention recently.[4] The reason is that the post-polymerization modification strategy is a facile and efficient method to prepare “smart” polymers.[5] Inspired by the advantages of post-polymerization modifications, our group has addressed the successful synthesis of SPFs recently. First of all, CO_2-responsive nanofibers were facilely prepared from versatile photo-crosslinked poly(pentafluorphenyl acrylate) (PPFPA) nanofibers via “amine-active ester” chemistry due to high reactivity of pentafluorophenyl (PFP) esters toward amines. Additional, the versatile nature of PPFPA nanofibers was also proved by reaction with different amines to fabricate various functional nanofibers with azide groups, alkyne groups and ethenyl groups. In another study, versatile precursor poly(glycidyl methacrylate) (PGMA) nanofibers with pendant reactive epoxide groups were fabricated. In view of features of reactive epoxide moieties, PGMA nanofibers can be post-modified by reaction with hydroxyl, thiol and amine groups to obtain targeted “smart” nanofibers. Therefore, spirospyran functional nanofibers as well as sensitive CO_2 sensors were prepared according to this strategy. All in all, reactive precursor nanofibers represent a facile and versatile platform to fabricate various functional nanofibers via post-polymerization modification.

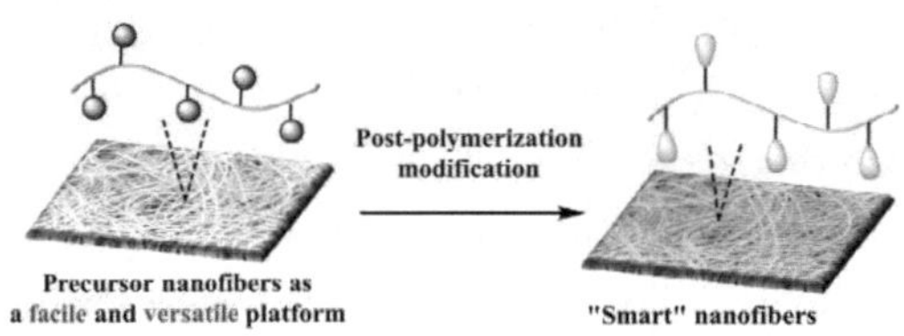

Scheme 1. Illustration of post-polymerization modificaton of precursor “active” nanofibers.

References

[1] Huang, C., Soenen, S. J., Rejman, J., Lucas, B., Braeckmans, K., Demeester, J., De Smedt, S. C. *Chem. Soc. Rev.*, **2011**, 40, 2417–2434.

[2] Thielke, M. W., Secker, C., Schlaad, H., Theato, P. *Macromol. Rapid Commun.*, **2016**, 37, 100–104.

[3] Zong, X., Kim, K., Fang, D., Ran, S., Hsiao, B. S., Chu, B. *Polymer (Guildf).*, **2002**, 43, 4403–4412.

[4] Sanyal, A., Sanyal, R., Kalaoglu-Altan, O. *Polym. Chem.*, **2015**, 6, 3372–3381.

[5] Das, A., Theato, P. *Chem. Rev.*, **2016**, 116, 1434–1495.

BIOGRAPHY

Prof. Dr. Patrick Theato studied chemistry at the University of Mainz and the University of Massachusetts, Amherst, and obtained his Ph.D. degree from the University of Mainz in 2001. After postdoctoral studies at Seoul National University supported by a Feodor Lynen Postdoctoral Research Fellowship and at Stanford University, he completed his Habilitation at the University of Mainz in 2007. From 2009 to 2012 he held a joint appointment at Seoul National University within the WCU program. In 2011, he accepted a prize senior lectureship at the University of Sheffield. Shortly after he moved to the University of Hamburg, where he is currently an associate (W2) professor for polymer chemistry. He has published over 220 publications. His research interests are post-polymerization modifications, stimuli-responsive polymers, surface modifications, energy storage, functional nano-objects, and nanofibers.

Multifunctional, Responsive Hydrogels and Nanoreactors as Selective Compartments and Actuators in Microfluidics

Brigitte Voit*, **Dietmar Appelhans, David Gräfe, Banu Iyisan, Andreas Krause, David Simon, Stefan Zschoche**

IPF-Institute Macromolecular Chemistry, Leibniz-Institute of Polymer Research (IPF) Dresden, Hohe Strasse 6, 01069 Dresden, Germany.

E-mail: voit@ipfdd.de

Responsive polymeric materials are becoming a highly studied field not only in basic synthetic approaches and sophisticated characterization, but especially with regard to their use in technical applications, e.g. as sensors and actuators in microsystems and microfluidic applications. In addition, compartimentalization becomes an issue for being able to localize specific actions and reactions e.g. by preparing nano- and microcapsules equipped with stimuli-responsive shells and membranes.[1] We will report on robust, multi-responsive and multifunctional nanovesicles which are interesting for studies in synthetic biology, but also for application in microsystem devices and in nanotechnology. Here, pH responsive nanocapsules based on photocrosslinked polymersomes have been prepared. In those stable nanoreactors, cascade enzyme reactions could be carried out under pH control using polymersome-encapsulated enzymes in a one-pot arrangement.[2] Furthermore, multifunctional, pH-responsive and photo-crosslinked polymersomes decorated with adamantane and azide groups have been realized for introduction of specific photoresponsive moieties, but also for non-covalent surface immobilization.[3] High membrane stability in our surface-immobilized polymersomes combined with a tunable membrane traffic allows to use our nanocontainers in microfluidic application. Presently, enzyme cascade reactions are adapted to be carried out in a microfluidic set-up preparing suitable hydrogel reaction compartments with and without further compartimentalization using polymersomes.

Multiresponsive hydrogels are further developed making use of an interpenetrating network approach as well as the concept of graft copolymer networks for retaining responsivity for up to four different stimuli in one hydrogel.[4] The hydrogels have been adapted for integration into microfluidic devices as volume-changing gels, adjusting degree of swelling as well as mechanical stability and allowing their use as effective sensors and actuators as needed for a chemical valve function.[5] First examples of their use as chemical transistors and the development of logical modules for achieving basic logic gate functions through a microfluidic setup will be also presented.[5]

References

[1] Gaitzsch, J.; Huang, X.; Voit, B. *Chem. Rev.,* **2016**, 116, 1053.

[2] Gräfe, D.; Gaitzsch, J.; Appelhans, D.; Voit, B.; *Nanoscale,* **2014**, 6, 10752.

[3] Iyisan, B.; Janke, A.; Reichenbach, P.; Eng, L.M.; Appelhans, A.; Voit, B. *ACS Appl. Mater. Interfaces,* **2016**, 8, 15788.

[4] Krause, A.T.; Zschoche, S.; Rohn, M.; Hempel, C.; Richter, A.; Appelhans, D.; Voit, B. *Soft Matter* **2016**, 12, 5529. Gräfe, D.; Appelhans, D.; Voit, B. *RSC Advances*, **2016**, 6, 34809-34817.

[5] Gräfe, D.; Erdmann, T.; Richter, A.; Appelhans D.; Voit, B. *ACS Appl. Mater. Interfaces,* **2017**, 9, 7565.

[6] Frank, P., Gräfe, D.; Probst, C.; Häfner, S.; Elstner, M.; Appelhans, D.; Kohlheyer, D.; Voit, B.; Richter, A., *Adv. Funct. Mater.,* **2017**, 1700430, early view.

BIOGRAPHY

Brigitte Voit received her PhD in Macromolecular Chemistry in 1990 from University Bayreuth, Germany, in the field of photoactive polymers. After postdoctoral work in 1991/1992 at Eastman Kodak in Rochester, USA, in the field of hyperbranched polymers, she joined Technische Universität München continuing the work on dendritic polymers and she received her habilitation degree in Macromolecular Chemistry in 1996. In 1997 Brigitte Voit was appointed head of the Institute of Macromolecular Chemistry at the Leibniz Institute of Polymer Research (IPF) Dresden, as well as full professor for "Organic Chemistry of Polymers" at Technische Universität Dresden (TUD). In addition, since 2002, she is heading the IPF Dresden, a large public research institute hosting about 480 people, as Managing Director and Chief Scientific Officer (Scientific Director).

Her major research interest is in the synthesis of new functional polymer architectures by various synthetic approaches covering topics like dendritic polymers, hydrogels, functional block and graft copolymers, as well as biofunctional polymers and functional polymers for optoelectronic applications. Brigitte Voit is very active in the community and in review work e.g. in the DFG panel Chemistry and in the European Commission (ERC grant evaluation); she is a board member of several polymer journals (e.g. Kuratorium Angewandte Chemie). For the period 2014/2015 Brigitte Voit had been elected as president of the European Polymer Federation and in 2015 she was awarded with membership in acatech (Akademie der Technikwissenschaften). In 2017 she received the Sächsischen Verdienstorden (Saxon Order of Merit) and the Covestro Distinguished Lectureship Award by Texas A&M.

Functional PVA-*co*-PE Nanofiber Membrane Obtained by Surface Chemical Structural Modification

Dong Wang*, **Wenwen Wang, Qinghua Zhao, Mengying Luo and Mufang Li**

School of Materials Science and Engineering, Wuhan Textile University, Wuhan, 430200, China

E-mail: wangdon08@126.com

Nanofiber membrane has been extremely concerned due to its superiorities such as ultrahigh specific surface area and porosity. It has been widely applied on filtrators, lithium ion battery separators and effective electrode substrates.[1, 2] Herein, a Large-Scale producible thermoplastic PVA-*co*-PE nanofiber membrane is prepared through phase separation method. Due to the plentiful reaction sites provided by bare hydroxyls, functional compounds or intermediate groups can be easily immobilized on to the nanofiber through nucleophilic substitution reaction. Along with this theory, nanofiber membranes with heavy metal ion adsorption capacity were designed and prepared by using cyanuric chloride to provide active chlorine and iminodiacetic acid (IDA) acted as the adsorbent. Figure 1a presents the schematic of preparation method. The optimum technology including concentration of IDA and pH of adsorption circumstance has been detail investigated. The results predict that the adsorption capacity to copper ion can reach the maximum 102 mg/g under 2 mol/L IDA and pH of 5.5.

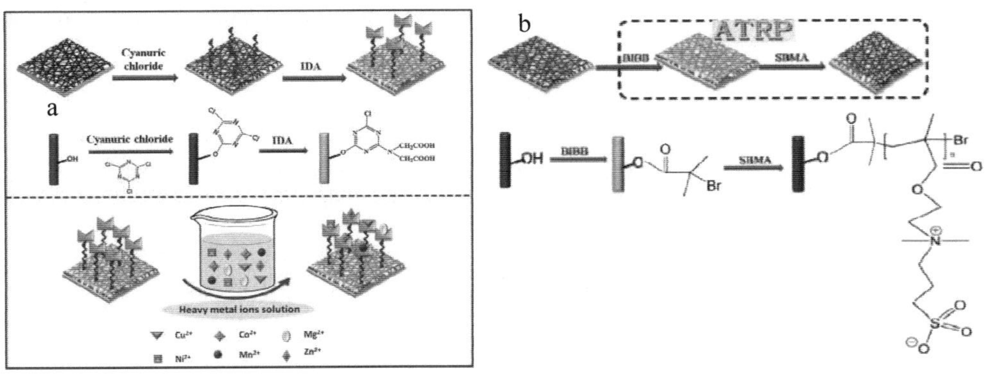

Figure 1. Schematic diagram of preparation of (a) modified PVA-co-PE nanofiber membrane with heavy meatal ion absorption capacity and (b) antibacterial PVA-co-PE nanofiber membrane by ATRP.

Additionally, a modified PVA-*co*-PE nanofiber with antibacterial ability was also introduced. Bromoisobutyryl bromide (BIBB) was bonding on PVA-*co*-PE by immersing the nanofiber membrane into its toluene solution. After that, [2-(Methacryloyloxy)ethyl]dimethyl-(3-sulfopropyl)ammonium hydroxide (SBMA) was successfully grafting on the nanofiber membrane via atom transfer radical polymerization (ATRP) while the immobilized BIBB reacted as initiator. Due to the unique chemical property of SBMA, the functionalized PVA-co-PE nanofiber membrane shows excellent antibacterial performance which was turned out to reach the value of 99.46%. Meanwhile, the membrane displayed specific adsorption of protein which indicates the further applications in biotechnology and medical science.

References

[1] Q. Liu, M. Xia, J. Chen, et al., *Electrochimica Acta*, **2015**, 176, 949-955.

[2] M. Li, X. Xue, Y. Lu, et al., *Desalination*, **2013**, 329, 50-56.

Thermoplastic Nanofibrous Membranes as Controllable Porous Scaffolds for High

Efficient Energy Conversion Applications

Qiongzhen Liu[1], Bo Wang[1], Xue Liu[1], Weibing Zhong[2], Dong Wang*[1,2]

[1]*College of Materials Science and Engineering, Wuhan Textile University, Wuhan, 430200, China;*

[2]*College of Chemistry, Chemical Engineering and Biotechnology, Donghua University Shanghai, 201620, China*

*E-mail: wangdon08@126.com, E-mail: windlqz_2000@163.com

Polymeric nanofibers have generated tremendous interest in the remarkably advancing field of nanoscience and nanotechnology, due to their smaller size and higher surface area, as well as ease of surface functionalization. Herein, thermoplastic nanofibers with hydroxyl or amide groups have been mass-produced via a melt extrusion phase separation method [1]. The nanofibers in continuous bundle-like yarn form can be easily dispersed as a suspension and facilely coated on various substrates, thus can prepare a variety of nanofibrous membranes with controllable porous scaffolds. This work addresses tailoring porous scaffold of nanofibers membranes for energy conversion filed, which can be used as separators for Lithium-ion batteries, anodes for Microbial fuel cells (MFCs), and flexible electrodes for super-capacitor. The hierarchical nanofibrous PVA-*co*-PE and PA6 separators with tortuous pores have shown high electrolyte-affinity and ionic conductivity, as well as improved cycling cycle stability of lithium-ion cells. Besides, a novel and facile chemical polymerization approach is developed to control the fibrillur morphologies of the Polypyrrole (PPy) and Polyaniline (PANI) on the nanofibrous membrane to form flexible textile electrodes for supercapacitors. The PPy@PVA-*co*-PE or PANI@PVA-*co*-PE nanofibers act as inter-bridges among the conductive networks leading to more continuous conductive pathways for ion/electron transfer as well as more electrochemical active sites for faradic reaction. Benefiting from this unique structure, the resultant two-electrode cell achieves outstanding specific capacitances over 300 F g^{-1} at 0.25 A g^{-1} using 2M NaCl as electrolyte [2]. For MFC application, the anodes possess three-dimensional interconnecting porous conductive scaffold and larger surface roughness for microbial metabolism. This MFC achieves an ultra-high power density more than 3200 mW m^{-2} with *E. coli.* as the microbial catalyst. Considering the low cost, low weight, facile fabrication, and good winding, our thermoplastic nanofibers promises great potentials for high-efficient energy conversion filed.

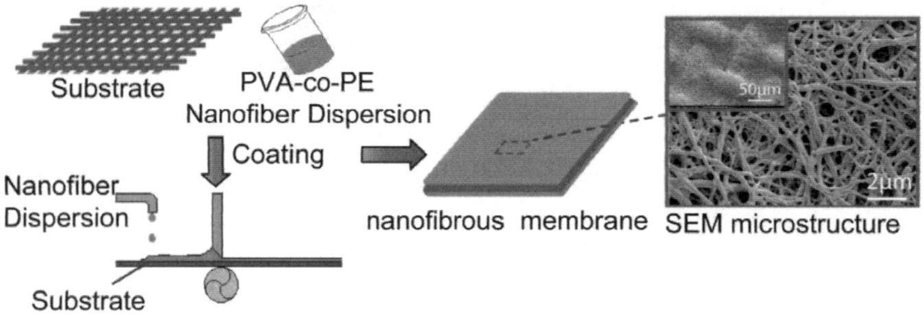

Figure 1. Schematic diagram of preparation of PVA-*co*-PE nanofiber membrane.

References

[1] Q. Liu, M. Xia, J. Chen, et al., *Electrochimica Acta*, **2015**, 176, 949-955.

[2] Q. Liu, B. Wang, J. Chen, et al., *Composites Part A: Applied Science and Manufacturing*, **2017,** 101, 30-40.

BIOGRAPHY

Prof. Dong Wang is the executive dean of the Materials Science and Engineering School, director of Hubei Key Laboratory of Advanced Textile Materials & Application, director of Intelligent Textiles and Interdisciplinary Research International Technology Cooperation Base of Hubei Province. He is also the New Century Talent of China Education Ministry, Chutian Scholar of Hubei Province. He has presided more than twenty ministerial and provincial-level research projects such as National Natural Science Foundation of China, National Science and Technology support program, Plan for Scientific and Technological Innovation Team of Excellent Young Investigator from Education Department of Hubei Province of China and Innovation Team from Science and Technology Department of Hubei Province of China, et al. Till now, he has published more than 60 high level academic papers, applied about 60 patents. He has won the Fund for Distinguished Young Scholars of Hubei Province, 14th Young Science & Technology Award of China, 1st Young Textile Science & Technology Award of China, the Third Prize of Scientific and technological Progress of China, et al.

BIOGRAPHY

Qiongzhen Liu was born in 1980, received her PhD degree from Shanghai Jiaotong University in 2010. She now is an associate professor at Wuhan Textile University. Her research focuses on energy-related nanofibers and textiles, magnetic materials, and new aerogels. She has received several research foundations from National Post Doctoral Committee, NSCF and Hubei province. She has also been active member in many projects, including Science and Technology Innovation Major Projects of Hubei Province, National Science-technology Support Program of China, National Key Research and Development Program, Excellent Innovative Team of Young Researchers of Hubei provincial department of education, Creative Research Group of Science and Technology Department of Hubei province. Till now, she has published more than 25 peer-reviewed journal articles and 6 issued Chinese patents.

Design and Preparation of Textile-based Devices for Medical Applications

Lu Wang, Guoping Guan, Fujun Wang, Jing Gao, Jing Lin, Yan Li and Ting Zou

Key Laboratory of Textile Science and Technology of Ministry of Education, College of Textiles, Donghua University, Shanghai 201620, China

E-mail: wanglu@dhu.edu.cn

This presentation makes a brief introduction on the design and preparation of textile-based medical devices and their applications. Firstly, an overview of textiles materials used in medicine in general is presented. Fibres with different polymers, cross sectional shapes and sizes, can be both bioabsorbable and nonabsorbable and of different hydrolytic stabilities. The basic textile structures and the associated technologies used in medical products, include braided, knitted, woven and nonwoven ones. The textile-based structure fabrics present a wide range of controllable physical and mechanical performance, which adapt to the clinical demands. Secondly, the practices referring to R & D of several novel textile-based medical devices are introduced as the biodegradable ureteral stent, in-situ fenestrated stent-graft, sutureless stent-graft, light weight hernia mesh, antibacterial silk suture, textile pile debridement, biological fibre composite dressing, ankle brace for protecting sprain varus, medical compression stockings, portable colorimetric sensor strips, etc.. The clinical requirements on textile-based medical devices obtained from surgeons are benefit to extract their biological and engineering parameters to develop the functional and biomimetic design. The micro-fabricating textile technology is applied to prepare various 1D, 2D and 3D structures. After the treatment with the heat, plasma, LBL or other processes, the prototype of medical device is formed. Furthermore, several examples of functional innovation method of characterization are introduced as well. The innovations of the textile-based medical devices are closely related to the materials' innovations, fabricatings' innovations as well as the deep understanding of the clinical requirement. This talk provides our excises, and discusses further opportunities.

BIOGRAPHY

Dr. WANG Lu graduated with BSc and MSc in textile engineering from Donghua University, Shanghai, China, and was awarded a Ph.D. in biomedical textile at ENSITM of University of Haute-Alsace, Mulhouse, France. She has been serving as a professor in Textiles Engineering and Biomedical Textiles in Donghua University since 2003. She is current a director of Key Lab. of Textile Science & Technology of Ministry of Education, China. Her main research interests include functional design, forming and evaluation of biomedical textiles as well as green wool textiles. Dr. WANG has been responsible of a number of research projects, including "Biomedical Textile Science and Technology" 111 project of Ministry of Education (B07024), National "863" project, Science & Technology Commission project of Shanghai Municipality, Science & Technology key project of Jiangsu province, National Nature Science Foundation of China, Doctoral Fund of Ministry of Education of China, as well as dozens of industrial co-projects. She also has been involved in several international cooperation projects, including EU Asia-link project, Australia CAWIN project, American NSF project. She has published 91 Chinese patents (65 awarded), more than 210 scientific papers as well as 6 books as co-editor and/or contributing author. She has gotten 6 Science and Technology Progress Prizes and 4 Education Prizes awarded by provincial government or Chinese Textile Industry Federation. WANG has extensive experience in biomedical textile production and close relationship with industries.

Advanced Functional Semiconductor Fibers

Lei Wei

School of Electrical and Electronic Engineering, Nanyang Technological University, 50 Nanyang Avenue, Singapore 639798, Singapore

E-mail: wei.lei@ntu.edu.sg

Advanced functional semiconductor fibers offer a novel degree of freedom to engage applications spanning smart and flexible electronics, nanorobotics, next generation of sensors and transducers, biomedical diagnosis/therapy, and prosthetics. In this talk, we firstly report on the demonstration of an optoelectronic device positioned along a silica fiber by integrating crystalline semiconducting material and metal electrodes in a precisely defined metal-semiconductor-metal architecture. The formation of such an optoelectronic structure inside a fiber relies on a physical breakup mechanism, which significantly broadens the repertoire of accessible materials and geometries towards advanced multifunctional fibers. Secondly, we report on the demonstration of crystalline thermoelectric (TE) fibers by co-drawing hermetically sealed high-quality inorganic TE materials in a flexible fiber-like substrate. The resulting TE fibers are intrinsically crystalline, highly flexible, ultralong, and mechanically stable, while maintaining high TE properties as their inorganic bulk counterparts. Furthermore, a two-dimensional TE fabric is assembled by weaving thin, lightweight, and flexible TE fibers into a textile to provide mW/cm^2-level output power density, while a maximum cooling/regulation temperature of 5 °C is experimentally achieved.

BIOGRAPHY

Dr. Lei Wei is currently a Nanyang Assistant Professor at Nanyang Technological University in Singapore. He received the B.E. degree in Electrical Engineering from Wuhan University of Technology (China) in 2005; and the Ph.D. degree in Photonics Engineering from Technical University of Denmark (Denmark) in 2010. Then he joined the Research Laboratory of Electronics at the Massachusetts Institute of Technology in the U.S. as a postdoctoral associate. In September 2014, he joined the School of Electrical and Electronic Engineering at Nanyang Technological University in Singapore. His main research interests are fiber-based optoelectronic devices, multifunctional fiber structures, bio-fiber interfaces, and in-fiber energy generation and storage.

Fabrication and Properties of Novel Porous Polymer and Carbon Materials

Dingcai Wu

Materials Science Institute, PCFM Lab, School of Chemistry, Sun Yat-sen University, Guangzhou, 510275, P. R. China.

E-mail: wudc@mail.sysu.edu.cn

Currently, most porous polymers and their derived porous carbons present ill-defined pore structures, have low porosities and lack multi-functional characteristics, thus leading to undistinguished properties and limited applications.[1] In order to address these issues, my group has designed and constructed well-defined pore structures by utilizing polymer precursor-template hybrid building blocks to achieve uniform polymer/template interfaces,[2,3] has designed and prepared well-developed pore structures by using novel nanostructured polymers as building blocks or by simply tuning the carbonization process of conjugated polymers,[4-6] and has also developed multi-functional microporous polymers with water-dispersible and responsive characteristics.[7] These advanced pore structures were found to greatly enhance their properties in many applications, including selective adsorption, electrochemical energy storage and catalysis.

References

[1] D. Wu, F. Xu, B. Sun, R. Fu, H. He, K. Matyjaszewski, *Chem. Rev.*, **2012**, 112, 3959.

[2] Z. Li, D. Wu, Y. Liang, R. Fu, K. Matyjaszewski. *J. Am. Chem. Soc.*, **2014**, 136, 4805.

[3] D. Wu, Z. Li, M. Zhong, T. Kowalewski, K. Matyjaszewski. *Angew. Chem. Int. Ed.*, **2014**, 53, 3957.

[4] Y. Liang, L. Chen, D. Zhuang, H. Liu, R. Fu, M. Zhang, D. Wu, K. Matyjaszewski. *Chem. Sci.*, **2017**, 8, 2101.

[5] F. Xu, Z. Tang, S. Huang, L. Chen, Y. Liang, W. Mai, H. Zhong, R. Fu, D. Wu. *Nat. Commun.*, **2015**, 6, 7221.

[6] H. Liu, S. Li, H. Yang, S. Liu, L. Chen, Z. Tang, R. Fu, D. Wu. *Adv. Mater.*, **2017**, 29, 1700723.

[7] W. Mai, B. Sun, L. Chen, F. Xu, H. Liu, Y. Liang, R. Fu, D. Wu, K. Matyjaszewski. *J. Am. Chem. Soc.*, **2015**, 137, 13256.

BIOGRAPHY

Dingcai Wu obtained his Ph.D. in Polymer Chemistry and Physics at Sun Yat-sen University in 2006. He joined the faculty of Sun Yat-sen University in 2006 and became a Full Professor in 2014. His current research interests focus on the design, preparation, and applications of novel functional materials, with emphasis on porous polymer and carbon materials. He was the winners of National Science Foundation for Excellent Young Scholars in 2014, and National Young Top-Notch Talent Support Program in 2014. He currently serves as the deputy director of Materials science Institute of Sun Yat-sen University.

Acknowledgements: The author acknowledge financial support from the projects of NSFC (51422307, U1601206, 51372280).

Fabrication and Application of Forward Osmosis Composite Membrane based on Nanofiber Membrane Support Layer

Zhenliang Xu*, **Xiangyu Chi, Yichen Zuo, Mingxiao Zhang, Baogen Xia**
State Key Laboratory of Chemical Engineering, Membrane Science and Engineering R&D Lab, Chemical Engineering Research Center, East China University of Science and Technology (ECUST), 130 Meilong Road, Shanghai 200237, China
E-mail: chemxuzl@ecust.edu.cn

The preparation of forward osmosis (FO) composite membrane with high performance is one of the key issues in the study and application of FO technology. The nanofiber membrane support with a scaffold-like structure, high porosity and low tortuosity minimizes the internal concentration polarization and improve the FO performance effectively, which is widely used for preparation of the FO membranes. The review introduces the principle of the internal concentration polarization in FO process and structural parameter estimating the performance, summarizing the development process of the nanofiber membrane as the support layers (single material, mixed materials, nanomaterial additive and chemical modification). Moreover, the development trend of support layer is prospected. Besides, FO membranes were used for seawater and the spent electroless nickel plating baths.

Keywords: Forward osmosis; Membrane Support; Internal concentration polarization; Nanofibers

References

[1] Elimelech M,Phillip W A. *Science*, **2011**, 333: 712-717.
[2] Pan S-F, Dong Y, Zheng Y-M, *et al. J Membr Sci,* **2017**, 523: 205-215.

BIOGRAPHY

Dr. Zhen-Liang Xu is the Professor of State Key Laboratory of Chemical Engineering, National Engineering Research Center of Ultrafine Powder and Shanghai Key Laboratory of Multiphase Materials Chemical Engineering, and the Director of Membrane Science and Engineering R&D Lab, and Chemical Engineering Research Center of ECUST. Dr. Xu worked for thirty years in ECUST and two years in Singapore in membrane laboratory. Dr. Xu focused on the research areas of hollow fiber/flat composite membranes, functional membrane materials, nanofibers, interfacial polymerization, separation engineering and desalination engineering and worked in collaboration with University of Hongkong, National University of Singapore and Imperial College London.

Ionic Liquid-Based Polyelectrolyte Membranes: Synthesis & Applications

Feng Yan

Department of Polymer Science and Engineering, College of Chemistry, Chemical Engineering and Materials Science, Soochow University, Suzhou, 215123, China.

E-mail: fyan@suda.edu.cn

Ionic liquids (ILs) are attracting significant attention in many fields of chemistry and industry because of their unique physicochemical properties, such as negligible vapor pressure, low volatility, high thermal stability and ionic conductivity. Recently, the number of polymerizable ILs is steadily increasing, and ionic polymers of polymerizable ionic liquid monomers have been produced as exotic polyelectrolytes.[1-3] This talk will report our recent results on the imidazolium-based ionomer polyelectrolytes for anion-exchange membrane, stimulus-responsive functional material and antibacterial applications.

References

[1] W. Qian, J. Texter and F. Yan*, *Chem. Soc. Rev.*, **2017**, 46, 1124.

[2] J. Guo, C. Yuan, M. Guo, L. Wang, F. Yan*, *Chem. Sci.*, **2014**, 5, 3261.

[3] F. Gu, H. Dong, Y. Li, Z. Sun, F. Yan*, *Macromolecules*, **2014**, 47, 6740.

BIOGRAPHY

Prof. Feng Yan received his Ph.D. degree in polymer science at Nanjing University in 2000, and was appointed as a professor at Soochow University in 2006. His research interests mainly focused on the synthesis and characterization of functionalized ionic liquids, poly(ionic liquids), ionic polymers, polyelectrolytes and carbon materials for electrochemical energy harvesting/generation, and for antimicrobial applications. He has authored more than 120 publications. He received the National Natural Science Foundation of China for Outstanding Young Scholars in 2014.

Development of Novel Nano-in-Nano Drug Delivery Systems

Fang Yang, Jiankang Song, Jinlong Shao, Eva Kolwijck, John A. Jansen

Department of Biomaterials, Radboud University Medical Center,

Philips van Leydenlaan 25, 6525EX Nijmegen, The Netherlands

E-mail: fang.yang@radboudumc.nl

An ideal delivery system of drugs or biomolecules should release the agents at the required local site in a spatiotemporally controlled manner while retaining their bioactivity, thereby optimizing their efficacy and reducing their side effects. For this purpose, numerous nanostructured carriers have been developed, including electrospun nanofibers due to their high drug loading and encapsulation efficiency, cost-effectiveness, and ease of processing. In most cases, the release of biomolecules from electrospun nanofibers is based on diffusion. Consequently, control over the rate of drug delivery is poor due to a lack of interaction between biomolecules and the fiber materials. Therefore, we hypothesized that enrichment of electrospun membranes with additional drug carriers, such as nanoparticles would improve the suitability of nanofibers for drug delivery. Two examples were shown below to demonstrate the efficacy of this nano-in-nano concept.

In the first study[1], we successfully developed a silk fibroin based nano-in-nano drug delivery system. We incorporated the oppositely charged gelatin nanospheres into the silk fibroin nanofibers using a colloid electrospinning technique without using any organic solvent. Through simple adjustment of the weight ratio between the nanospheres and the relative feeding rate of core and shell solutions containing gelatin nanospheres by using single and co-axial nozzle electrospinning, respectively, we arranged the spatial distribution of positively charged gelatin A nanospheres and negatively charged gelatin B nanospheres inside the nanofibers. To prove the concept of sustained drug delivery, we compared the release kinetics of a model drug vancomycin from gelatin nanospheres containing and gelatin nanospheres free membranes. We proved that the incorporation of gelatin nanospheres into the silk fibroin nanofibers strongly prolonged the release period and antibacterial effects of the prepared membranes.

In the other study[2], we prepared chitosan/poly(ethylene oxide) (PEO) nanofibres containing silver nanoparticles (AgNPs) as carrier matrix for the release of silver ions. The nanofibres were prepared by one-step electrospinning of chitosan/PEO in an acetic acid solution containing different amount of $AgNO_3$. Formation of AgNPs was confirmed by transmission electronic microscopy combined energy dispersive X-ray spectroscopy technique. The AgNPs were distributed homogeneously throughout of the fibers. The release study and antibacterial test indicated a sustained release of silver ions and effective antibacterial effect of the membranes containing 5 wt% $AgNO_3$ up to 28 days.

With the positive outcome, we believe that the novel nano-in-nano system will have prosperous applications in the field of antimicrobial therapy and regenerative medicine.

References

[1] J. Song, A. Klymov, J. Shao, Y. Zhang, W. Ji, E. Kolwijck, J. A. Jansen, S. C. G. Leeuwenburgh, F. Yang, *Adv. Healthcare Mater.* **2017**, 1700014.

[2] J. Song, S.J.A. Remmers, J. Shao, E. Kolwijck, X.F. Walboomers, J.A. Jansen, S.C.G. Leeuwenburgh, F. Yang, *Nanomed.-Nanotechnol.*, **2016,** 12, 1357-1364.

BIOGRAPHY

Dr. Fang Yang is a biomaterial scientist and interested in the development and evaluation of novel biomaterials-based approaches for tissue regeneration and the controlled delivery of drugs or growth factors. She obtained her Ph.D. degree from National University of Singapore in 2005 under the supervision of Prof. Seeram Ramakrishna. After that, she joined Prof. John Jansen's group at Department of Biomaterials in Radboud University Medical Centre in the Netherlands until now.

Her current research focuses on the development of biomimetic nanofibrous matrix as an extracellular matrix with the incorporation of therapeutic agents for bone and periodontal tissue regeneration. She is also interested in local drug delivery to prevent implant-associated infections. Her expertise lies in biopolymers and biocomposites, electrospinning technique and biomaterials characterization (both physicochemical and biological evaluation). She has published more than 70 journal articles and served as a peer reviewer for about 10 journals in the related field.

Evaluation of Drugs Permeation through Modified PVA Blended Membranes

Jen-Ming Yang[1,2,3]

[1]*Department of Chemical and Materials Engineering, Chang Gung University, Tao-Yuan 333, Taiwan, ROC.*

[2]*Department of General Dentistry, Chang Gung Memorial Hospital, Tao-Yuan 333, Taiwan.*

[3]*Department of Pediatrics, Chang Gung Memorial Hospital at Linkou*

E-mail: jmyang@mail.cgu.edu.tw

Due to the favorable characteristics of inherent non-toxicity, exceptional biocompatibility, and noncarcinogenicity, poly(vinyl alcohol) (PVA) is one of the most extensively studied hydrogels in biomedical applications[1-3]. In this study, the non-toxicity chitosan, sodium alginate and poly(γ-glutamic acid) were blended with PVA directly and the homogeneous membranes were prepared for the evaluation of drug permeation.

Ascorbic acid is a potent reducing and antioxidant agent for fighting bacterial infections and detoxifying reactions. Acetaminophen is a p-aminophenol derivative with analgesic and antipyretic activities. Fluorouracil (5-FU) can be used for treatment of carcinoma. The permeability of ascorbic acid, acetaminophen, and 5-FU through the modified PVA membranes were presented in this speech.

References

[1] F. Cavalieri, F. Miano, P. D'Antona, G. Paradossi, *Biomacromolecules*, **2004**, 5, 2439-2446.

[2] J. Liu, C. Detrembleur, M. Hurtgen, A. Debuigne, M.C. De Pauw-Gillet, S. Mornet, E. Duguet, C. Jerome, *Polym. Chem.*, **2014**, 5, 5289-5299.

[3] B. Gupta, R. Agarwal, M. Sarwar Alam, J Mater Sci Mater Med., **2014**, 25 1613-1622.

BIOGRAPHY

Jen Ming Yang is currently a Professor in department of Chemical and Materials Engineering at Chang Gung University, Research Fellow (joint appointment) of Chang Gung Memorial Hospital, Linkou, President of Taiwan Society of Blood Biomaterials, and Council of Tissue Engineering and Regenerative Medicine International Society-Asia-Pacific Chapter (TERMIS-AP). He is one of the editors of two international journals, MOJ Polymer Science and Material Science & Engineering International Journal. Prof. Yang received his Bachelor's Degree in Chemistry at National Taiwan University in 1980. Two years later, he obtained a Master Degree in Polymer Science at National Tsing Hua University, and in 1990 he received his PhD in the same field. From 1990 to 1991 he continued his studies at the University of California at Berkeley, and achieved his Postdoctorate degree in Polymer Science. From Aug, 2002 to Jul. 2003 he had been a visiting scholar of Prof. Langer's group at Massachusetts Institute of Technology. Prof. Langer is one of the prestigious and outstanding pioneers in the fields of Tissue Engineering, Biomaterials and Drug Controlled Release. Currently, Prof. Yang's field of research encompasses Polymer Science, Membrane, and Biomaterials. He has been board member of some societies and conferences. He has also presented keynote and invited speeches at many conferences and hosted international conferences. His research interests include Hydrogels, Polyelectrolytes, Polymeric Nanomaterials and Nanostructured Polymers, Polymers at Surfaces and Interfaces, Biomedical Applications of Polymer, and Polymers for Energy Applications.

Polymer Complexation for Multicomponent Fibers

Shuguang Yang
State Key Laboratory for Modification of Chemical Fibers and Polymer Materials & Center for Advanced Low-dimension Materials, Donghua University, Shanghai 201620
E-mail: shgyang@dhu.edu.cn

Polymer complexation is a type of polymer aggregation process. Two initially soluble polymers are bound together through cooperative interactions, such as electrostatic force and hydrogen bonding, along the polymer chain to form polymer complex which generally precipitates out form the solution. Recently, we introduce an approach to prepare hydrogen-bonded complex fibers. First, a spinnable fluid is obtained by restricting hydrogen bonds, and then is extruded through a spinneret into a coagulation bath where hydrogen bonds are built to induce fiber formation. Many hydrogen-bonded complexes have been prepared into fibers, such as PVPON/PAA, PEO/PMAA, and PEO/PAA fibers. Among them, PEO/PAA fiber shows excellent elastic behavior and can be drawn to more than 12 times of its original length without breaking, which is much higher than Spandex fiber or natural rubber fiber. Graphene oxide (GO) as hydrogen bonding donor component and synthetic polymer PEO as hydrogen bonding acceptor component, the fiber also been fabricated through hydrogen bonding complexation approach. After reduction, GO/PEO fiber exhibited good electronic and mechanical properties. Polymer complexation provide a pathway to prepare multicomponent fibers, which have properties such as elastic and conductive and have potential for flexible electronics and smart textile.

BIOGRAPHY

Prof. Shuguang Yang received his B. Sc. in 2002 on Chemistry from Wuhan University and Ph. D. in 2007 on Polymer Science from Institute of Chemistry, Chinese Academy of Sciences (CAS). He was research assistant (2007-2008) in Department of Polymer Science and Engineering, Peking University, and Research Associate Postdoc Fellow (2008-2010) in Department of Polymer Science, University of Akron. He has been professor of polymer physics at Donghua University since 2010 and Associate director of Center for Advanced Low-dimension Material since 2016. He got 2007 President Award of CAS and 2012 Pujiang Talent title of Shanghai Municipal Science and Technology Commission.

Dr. Yang's academic interest is on structure and dynamics of low-dimension macromolecular materials. His research has been in complexes, coacervates, micelle, gels, thin films and fibers. He has published more than 70 peer-reviewed journal papers and is the inventor of 10 patents.

Melt-Spun PLLA-PDLA Fibers with Completely Stereo-Complexed Crystallites

Yiqi Yang[1], **Gangwei Pan**[1,2], **Bingnan Mu**[1], **Helan Xu**[1,2], **Bomou Ma**[2], **Zhiliang Qian**[3], **and Hanzhang Lao**[3]

[1]*Department of Textiles, Merchandising & Fashion Design, and Department of Biological Systems Engineering, University of Nebraska-Lincoln, Lincoln, NE, 68583-0802, USA;*

[2]*Key Laboratory of Science and Technology of Eco-Textiles, Ministry of Education, Jiangnan University, Wuxi 214122, China;*

[3]*Zhejiang Xinning Biomaterials Co. Ltd, Shaoxing 312073, China*

E-mail: yyang2@unlnotes.unl.edu

We report preparation and characterization of PLLA-PDLA fibers with completely stereo-complexed crystallites via melt spinning without solvents or additives. PLLA fibers, after its development in 1990s, have limited applications in textiles due to their poor resistance to hydrolysis, undesirable dyeability and mediocre colorfastness. Dyeing at a low temperature of 110 °C decrease 35% or more of tensile strength of PLLA textiles. A common disperse dyeing temperature of 130 °C is impossible for PLLA, since the textiles dyed at that temperature lose all their tensile properties. Most disperse dyes have less than 70% exhaustion onto PLLA, especially for heavy shades, with low colorfastness to laundering and wet crocking. PLLA textiles even lose strength during repeated wear and home laundering substantially. Another disadvantage of PLLA is its low softening temperature (about 65 °C), which is problematic for heat setting and pressing. Many researchers throughout the world have been working over the last two decades in order to overcome the problems associated with PLLA. Major approaches are PLLA-PDLA stereo-complexation, copolymerization, polyblending, and addition of various chemicals and particles to the fibers. The most encouraging approach is PLLA-PDLA stereo-complexation. However, researches have to use solvents and/or nucleation agents for complete stereo-complexation, and are only successful with low molecular weights (MW). For large scale production of PLA fibers, melt spinning is ideal due to its economics, and high MW polymers, e.g. MW of 3×10^5, should be used due to the demanding mechanical properties. Limited availability of high MW PDLA, and the lack of a scalable method for the production of stereo-complexed PLLA/PDLA fibers probably are the reasons for the minimal progress in fiber and textile applications of PLA. We have developed methods to make PLLA/PDLA fibers of MW up to 6×10^5 via continuous melt spinning with completely stereo-complexed crystallites. The fibers thus formed have remarkably increased softening temperature (from 72 °C to 133 °C) and resistance to hydrolysis, and much improved dyeing behavior. Tensile strength retentions after one hour in water at 110 °C and 130 °C were 94% and 90%, respectively, while that of PLLA fibers' were 63% and zero, respectively. Because of excellent resistance to hydrolysis, the PLLA/PDLA interlocked fibers could be dyed at 130 °C. Using Disperse Blue 79 as an example, % dye exhaustion of PLLA/PDLA interlocked fibers was 91%, compared to only 67% of PLLA fibers at 2% initial dye concentration. Colorfastness of dyed fibers also was much improved.

BIOGRAPHY

Professor Yiqi Yang received his PhD from Purdue University and his undergraduate and master's educations from Donghua University. As a textile chemical engineer, he is a Distinguished Professor at the University of Nebraska-Lincoln, and works for the Department of Textiles, Merchandising and Fashion Design, and the Department of Biological Systems Engineering.

Dr. Yang's research focuses on development of materials from agricultural byproducts and co-products. Examples include development of new lignocellulosic, protein and synthetic fibers from agricultural byproducts and coproducts, development and application of environmentally responsible chemicals for textiles, composites, cosmetics and medical industries.

Bio-medical Application of Rylenediimide-cored Dendritic Polymers

Meizhen Yin*, **Shaobo Zhang, Wenyu Cheng**

State Key Laboratory of Chemical Resource Engineering, Beijing Laboratory of Biomedical Materials, Beijing University of Chemical Technology, Beijing, 100029, China

E-mail: yinmz@mail.buct.edu.cn

In recent years, the tremendous development of dendritic polymers has provided rylenediimides (RDIs) with high biocompatibility that is a key factor in bio-medical application.[1-2] Aggregation of the encapsulated RDI chromophores can be noticeably suppressed by outer branched topological structures, thus leading to the enhancement of optical performances and solubility in aqueous solutions. Our group has reported several water-soluble perylenediimide (PDI)-cored dendrimers[3] and dendritic polymers, which have been successfully demonstrated as fluorescent gene nanocarriers or protein-carriers with high delivery efficacy. The fluorescent cationic dendrimers containing multiple amino groups can delivery gene to implement on the genetic control of insect pests by knocking down key development genes.[4] The fluorescent anionic polymers can efficiently deliver bacillus thuringiensis (Bt) toxin to kill Bt-resistant pests.[5] Thus, we conclude that the PDI-cored dendritic polymers have been verified as therapy-aided vector for the management of biological challenge.

Very recently, terrylenediimides (TDIs), with more longitudinal extension π-conjugation and bathochromic shift in optical spectra than PDI, has been innovatively modified by dendritic polyacrylic acid polymers.[6] This TDI-cored dendritic polymer self-assembled a nanomedicine as an intrinsic theranostic agent and has been used for curing tumor *via* photothermal therapy technology. The TDI nanomedicine develops photothermal agents with eligible optical property and reliable biosecurity in aqueous solution. This is the first time to report the RDI utilized as photothermal agent to carry out the cancer issues. Due to the excellent performance, RDI-cored dendritic polymers will provide more opportunities to apply in bio-medical field.

References

[1] M. Sun, K. Müllen, M. Yin, *Chem. Soc. Rev.*, **2016**, 45, 1513-1528.
[2] K. Liu, Z. Xu, M. Yin, *Prog. Polym. Sci.*, **2015**, 46, 25-54.
[3] Z. Xu, B. He, J. Shen, W. Yang, M. Yin, *Chem. Comm.*, **2013**, 49, 3646-3648.
[4] B. He, Y. Chu, M. Yin, K. Müllen, C. An, J. Shen, *Adv. Mater.*, **2013**, 25, 4580-4584.
[5] Y. Zheng, S. You, C. Ji, M. Yin, W. Yang, J. Shen, *Adv. Mater.*, **2016**, 28, 1375-1380.
[6] S. Zhang, W. Guo, J. Wei, C. Li, X. Liang, M. Yin, *ACS Nano*, **2017**, 11, 3797-3805.

BIOGRAPHY

Prof. Meizhen Yin, from Beijing University of Chemical Technology (BUCT). She got her PhD in Organic Chemistry under the supervision of Prof. Brigitte Voit and Prof. Wolf D. Habicher from Dresden University of Technology (TUD), Germany in 2004. Afterwards, she worked as a post-doctor in Prof. Klaus Müllen group in Max Planck Institute for Polymer Research, Mainz, Germany. Since 2009, she became a full Professor in Beijing University of Chemical Technology (BUCT), China.

Professor Meizhen Yin built a research group of "Fluorescent Functional Materials" in BUCT. She devoted herself to the construction and biological applications of fluorescent topological macromolecules. She designed new functional fluorescent topological macromolecular materials and explores their biological applications including: bio-imaging, cancer therapy, ultra-sensitive detection and fluorescent nano-carrier for agricultural pest control.

Prof. Meizhen Yin was awarded "New Century Excellent Talents" of the Education Ministry in 2010. She appointed as expert for "2011 Cardiovascular Disease Collaborative Innovation Center". In 2016, She was achieved the 13th "Chinese Young Women Scientist Award" and the "President Award" of Beijing University of Chemical Technology. In 2016, she was awarded "Scientific and Technological Innovation Leader" of the Ministry of science and technology.

Prof. Meizhen Yin has published nearly 80 papers as first/corresponding author in high ranking journals such as Chem. Soc. Rev., Prog. Polym. Sci., J. Am. Chem. Soc., Angew. Chem. Int. Ed., Adv. Mater., ACS Nano, Adv. Funct. Mater. and so on. Her research work was highlighted by "Nature Materials", "The Latest Science", "Chemistry World" and other academic journals. She has 2 authorized international patents and 16 authorized Chinese patents.

Functional Biodegradable Elastomers for Biomedical Applications

Zhengwei You, Shuo Chen, Dong Lei, Lijie Sun, Jiahui Liang
State Key Laboratory for Modification of Chemical Fibers and Polymer Materials, College of Materials Science and Engineering, Donghua University, Shanghai, China
E-mail: zyou@dhu.edu.cn

Biodegradable and biocompatible elastomers (bioelastomers) can mimic the mechanical properties of soft tissues, and sustain and recover from various deformations without mechanical irritations to the surrounding tissues when implanted in a mechanically dynamic environment in the body. Thus, bioelastomers have attracted great attentions in regenerative medicine [1]. However, it is still a challenging to produce ideal bioelastomers with good biocompatibility, biodegradability, processability, bioactivity, and mechanical properties. Herein, we will present our recent works on design and development of new bioelastomers with aforementioned advantages. Poly(glycerol sebacate) (PGS) is a representative of bioelastomers and has been widely used [2]. However, the limitations of its synthetic method lead to unsatisfied mechanical strength and inconvenient processing, which retard its further applications. We established acid-induced epoxide ring-opening polymerization to produce poly(sebacoyl diglyceride) (PSeD), a refined version of PGS, with significantly improved structural regularity and mechanical properties [3-4]. Then, we construct supramolecular PSeD elastomers with good thermoplastic processabilities and self-healing properties by introducing dynamic hydrogen bond-based crosslinkings [5]. Furthermore, we yield more tough PSeD bioelastomers via hybrid crosslinking mechanism. In addition, we demonstrate the efficiency to tailor the bioactivity using small functional units in side groups, which represents a powerful way of practically producing therapeutic materials [6-7]. Then, 3D printing technology with special processes has been developed to fabricate elastic scaffolds with well-organized hierarchical porous structures, which exhibit a high efficiency on the treatment of myocardial infarction in a rat model. The design strategies, the synthetic and processing methods developed here are versatile and provide new tools to construct bioelastomers with expected properties for a wide range of biomedical applications.

Reference

[1] You Z, Wang Y. Bioelastomers in tissue engineering. In: Burdick J, Mauck RL, editors. *Biomaterials for tissue engineering applications: a review of the past and future trends*: Springer-Verlag/Wien; **2011**. p. 75-118.
[2] Rai R, Tallawi M, Grigore A, Boccaccini AR. *Prog. Polym. Sci.* **2012**; 37: 1051-78.
[3] You Z, Cao H, Gao J, Shin PH, Day BW, Wang Y. *Biomaterials* **2010**; 31: 3129-38.
[4] You Z, Wang Y. *Adv. Funct. Mater.* **2012**; 22: 2812-20.
[5] Chen S, Bi X, Sun L, Gao J, Huang P, Fan X, You Z, Wang Y. *ACS Appl. Mater. Interfaces* **2016**; 8: 20591-9.
[6] Wang S, Jeffries E, Gao J, Sun L, You Z, Wang Y. *ACS Appl. Mater. Interfaces* **2016**; 8: 9590-9.
[7] Huang P, Bi X, Gao J, Sun L, Wang S, Chen S, Fan X, You Z, Wang Y. *J. Mater. Chem. B* **2016**; 4: 2090-101.

BIOGRAPHY

Zhengwei You is a professor in the State Key Laboratory of Chemical Fibers & Polymer Materials and the Chair of the Department of Composite Materials, College of Material Science & Engineering at Donghua University. He received his degrees of B.S. (2000) from Shanghai Jiao Tong University and Ph.D. (2007) from Shanghai Institute of Organic Chemistry, Chinese Academy of Sciences. From 2007 to 2012, he conducted his postdoctoral research on biomaterials at Georgia Institute of Technology and University of Pittsburgh. Later he joined Bayer MaterialScience as an innovation manager in charge of new technology scouting in Asia-pacific region, and moved to current position at Donghua University in 2013. His current research involves using molecular engineering approaches to precisely control the structures, functionalities, and consequent tailored biological and mechanical properties of synthetic biodegradable polymeric biomaterials. The major projects focus on the development of new bioelastomers and 3D printing technologies to create biomimetic niches for directing the stem cell differentiation and *in situ* tissue regeneration.

A Global Approach to Enhance the Mechanism of Thin Film Deposition from Liquid Solution; Toward the Efficient, Highly Reproducible, Long Lasting, Scalable and Low-cost Photo-Active Thin Films for Thin Film Solar Devices

[a]Fatemeh Zabihi, [a]Hao Xiong, [b,c]Morteza Eslamian, [a]Meifang Zhu*, [a]Hongzhi Wang*

[a]*State Key Laboratory for Modification of Chemical Fibers and Polymer Materials, College of Materials Science and Engineering, Donghua University, Shanghai, 201620, China;*

[b]*University of Michigan-Shanghai Jiao Tong University Joint Institute, Shanghai 200240, China;*

[c]*State Key Laboratory for Composite Materials, School of Materials Science and Engineering, Shanghai Jiao Tong University, Shanghai 200240, China*

E-mail: fzabihi@dhu.edu.cm

Thin film devices, such as thin film solar cells, transistors, capacitators, Organic Light Emission Diodes and organic sensors are emerging devices as the low-dimensions, efficient alternatives to the traditional complicated opto-electronic devices. So far, a multitude of approaches and strategies have been developed for preparation of proficient thin film devices. Deposition of the precursors from a gas phase or in a vacuum chamber are proficient and accurate, but require huge costs and energy consumption (1-2). Recently, solution deposition techniques have outpaced their traditional analogous, due to being facile and cost-effective. However, the practical use of solution deposition is limited by tough challenges such as poor reproducibility and tricky operation. More importantly, a solution casting process is essentially followed by a thermal annealing step, which makes it inappropriate for the thermally degradable materials (1-3). This communication proposes a modified annealing-free solution deposition method on ultrasonic post-treatment. Perovskite materials and corresponding devices subjected to this technology, have demonstrated superior performance with a negligible hysteresis behavior and a prolonged life-time compared with the reference spun and annealed samples. Moreover, an identical photo-to-current efficiency obtained for 8 cells assembled on a single electrode. This approach paves the pathway to the feasible fabrication of large area, low-cost and long-lasting solar devices for the long-time storage and commercial uses.

References

[1] F. Zabihi, Y. Xie, S. Gao, M. Eslamian, Morphology, conductivity, and wetting characteristics of PEDOT:PSS thin films deposited by spin and spray coating, Applied Surface Science 338 (2015) 163-177.

[2] F. Zabihi, M.R. Ahmadian-Yazdi, M. Eslamian, Fundamental study on the fabrication of inverted planar perovskite solar cells using two-step sequential substrate vibration-assisted spray coating (2S-SVASC), Nanoscale Res. Lett. 11 (2016) 71.

[3] F. Zabihi, M. Eslamian, Substrate vibration-assisted spray coating (SVASC): significant improvement in nano-structure, uniformity, and conductivity of PEDOT:PSS thin films for organic solar cells, J Coat Technol. Res. 12, 4 (2015) 711-719.

BIOGRAPHY

Fatemeh Zabihi is a Lecturer in College of Material Science and Engineering in DHU. She has been working as postdoc. fellow and senior researcher in University of Michigan-Shanghai Jiao Tong University Joint Institute from 2012-2014. She received her PhD in Chemical Engineering from Tehran Sci. Res. Branch, IAU in 2008 and served as Faculty in IAU for ten years. Her commitment to the research and academic work has been started since 1997, when she joined the research center of Iran National Oil Co, as researcher. According to her strong chemistry and engineering background her multidisciplinary research is focused on Thin Film Technology and Thin Film Photo-voltaic, Solution Processing and Solution Engineering, Surface Sciences and Surface Phenomena, Coating and Spray Pyrolysis, Thermodynamic and Phase Equilibrium studies.

Electrospun Polymer Fiber Network for Acoustic Metamaterials

Jianfeng Zang

School of Optical and Electronic Information, Huazhong University of Science and Technology, Wuhan 430074, China

E-mail: jfzang@hust.edu.cn

Up to now, the acoustic metamaterials are usually constrained by one of the following limitations: cumbersome configuration, narrow operation band, large acoustic loss, or rigid structure, which limits their further application in practice. Thus, it is of fundamental interest to search for a new material paradigm to solve those problems. Here we demonstrate an approach for one-step fabrication of large area meta-membrane in lightweight, flexible, and ultra-deep subwavelength thickness. The proposed meta-membrane is made of electrospun polymer fiber network. The ultrathin polymer fiber network exhibits high acoustic transmittance and efficient phase delay. Furthermore, we also demonstrate various acoustic applications by tailoring the as-prepared network membrane to various on-demand patterns. Our polymer fiber network shows potential opportunities to implement various ultrathin acoustic devices for manipulating sound waves at will, providing a vast platform for applications in medical imaging, ultrasonic therapy, noise control, as well as acoustic communication.

BIOGRAPHY

Dr. Jianfeng Zang is a Professor of School of Optical and Electronic Information, and Innovation Institute, Huazhong University of Science and Technology (HUST). His scientific research interest is in the area of soft active materials, nanomaterials, flexible electronic/photonic devices, soft metamaterials for acoustics and plasmonics.

Dr. Zang studied materials science (B.E.) and condensed-matter physics (M.S.) in Jilin University, and received his PhD degree in Electrical and Electronic Engineering from Nanyang Technological University. He worked as a Postdoctoral Fellow at University of South Carolina and Duke University. He has received serval prestigious awards including: "HUST Bird Nest Program (2013), National 1000 Young Talent Program (2014), Duke University Kewaunee Postdoctoral Achievement Award (2014), MRS Meeting Session Chair (2014 spring, San Francisco), Program Chair of The 3rd International Symposium of Flexible and Stretchable Electronics 2017 (Wuhan, China), He has published over 30 papers in prestigious journals including Nature Materials, Nano Letters, Advanced Materials, JMPS, EML, etc. His papers have received more than 2000 citations with an H-index 21. He has 4 ISI Essential Science Indicators Top 1% highly cited papers. He has reviewed manuscripts for many scientific journals like Science Advances, Advanced Materials, Advanced Functional Materials, Small, Nanoscale, Scientific Reports, Journal of Materials Chemistry, etc.

Poly(dopamine)-Inspired Surface Functionalization of Polypropylene Tissue Mesh for Prevention of Intra-Peritoneal Adhesion Formation

Tianzhu Zhang[1]*, Wanjun Hu[1], Zhigang Zhang[2], Zhenling Ji[2]

[1] *State Key Laboratory of Bioelectronics, Jiangsu Key Laboratory for Biomaterials and Devices, School of Biological Science and Medical Engineering, Southeast University, Sipailou 2, Nanjing 210096, China;*

[2] *Department of General Surgery, Zhong-Da Hospital, School of Medicine, Southeast University, Dingjiaqiao 87, Nanjing 210009, China*

E-mail: zhangtianzhu@seu.edu.cn

In the present study, we developed several strategies to functionalize PP surface with polyvinyl alcohol (PVA), poly(L-lactic acid) (PLLA), or chitosan (CS) for hernia repair, respectively[1-4]. Antiadhesion efficacy and complications of PLLA, PVA and CS modified PP meshes (PP-g-PLLA, PP-g-PVA and O-PP/PDA/CS) were evaluated through repairing mouse or rabbit abdominal wall defect at different implantation time points. PP-g-PLLA, PP-g-PVA and O-PP/PDA/CS mesh showed an excellent biocompatibility and can effectively reduce adhesion formation in vivo tests, although a slight inflammation reaction was found. Compared with PP-g-PLLA and PP-g-PVA, the PDA-inspired surface functionalization technique is the easiest.

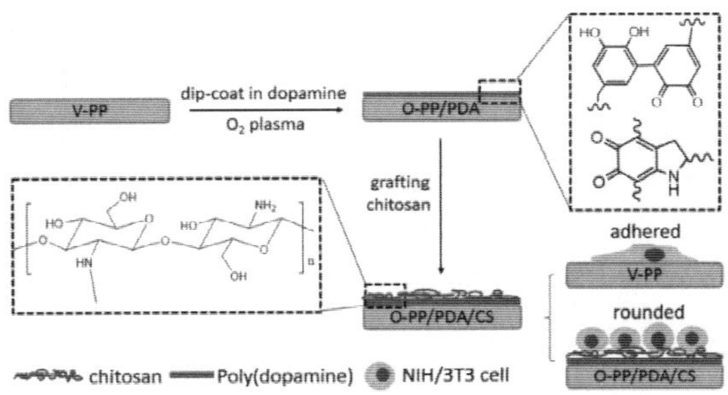

Figure 1. The schematic illustration of poly(dopamine) coating and subsequent chitosan immobilization on PP meshes and comparison of anticell adhesion efficacy. V-PP: virgin PP; O-PP: oxygen plasma treated PP.

References

[1] Xu, Z.; Wang, J.; Shen, L.; et al. *J. Memb. Sci.* **2002**, 196, 221–229.

[2] Zhang, T.; Zhang, Z.; Hu, W.; et al. *Colloids Surfaces A Physicochem. Eng. Asp.* **2016**, 500, 10–16.

[3] Zhang, Z.; Zhang, T.; Li, J.; et al. *J. Biomed. Mater. Res. Part B Appl. Biomater.* **2014**, 102, 12–21.

[4] Hu, W.; Lu, S.; Ma, Y.; et al. *J. Mater. Chem. B* **2017**, 5, 575–585.

BIOGRAPHY

Prof. Dr. Tianzhu Zhang obtained B.S. degree from Xuzhou Normal University in 1993, M.S. degree from Soochow University in 1997 and his Ph.D. degree from the Institute of Chemistry, the Chinese Academy of Sciences in 2003. From 2004 until 2009, he conducted his post-doctoral research at Ghent University (with Prof. Dr. Filip Du Prez), Catholic University of Leuven (with Prof. Dr. Erik Nies) in Belgium, at Technische Universität München and University of Ulm (with Prof. Dr. Bernhard Rieger) in Germany. In 2009, he joined the School of Biological Science and Medical Engineering at Southeast University in China as a full professor. As a head of research group, Dr. Zhang's research interests mainly focus on the basic and applied research of functional polymers in biomedical engineering. In 2009, he was the Winner of Education Ministry's New Century Excellent Talents Supporting Plan for his excellent work. In 2011 he was awarded the first prize of China Petroleum and Chemical Industry Federation of Science and Technology Progress, in 2012 he was awarded the third prize of Jiangsu Medical Science and Technology.

Architecture, Self-assembly and Applications of Hybrid Polymers Based on Polyhedral Oligmeric Silsequioxane (POSS)

Weian Zhang

Shanghai Key Laboratory of Functional Materials Chemistry, East China University of Science and Technology, 130 Meilong Road, Shanghai 200237, China,

E-mail: wazhang@ecust.edu.cn

A variety of well-defined hybrid polymers based on polyhedral oligmeric silsequioxane (POSS) have been constructed *via* living polymerization and "click chemistry" using POSSs as initiators, monomers, RAFT transfer agents or "click chemistry" agents. The self-assembly behavior of these POSS-containing hybrid polymers with different architectures was further studied, which could self-assemble into nano-scaled aggregates in selective solutions, or form nanostructures in bulk. Some of the interesting self-assembly morphologies are remarkably different with those formed from the conventional pure organic amphiphilic polymers. More recently, the application of well-defined POSS-containing hybrid polymers has been explored in biological fields, especially in photodynamic therapy (PDT). For example, we constructed a porphyrin/POSS alternating copolymer, which displayed a high photochemical yield and phototoxicity against cancer cells and therefore a high PDT efficacy.

References

[1] Zhang, Z. H.; Zhang, P. C.; Wang, Y.; Zhang, W. A. *Polym. Chem.*, **2016**, 7, 3950 -3976.

[2] Rui, L. L.; Xue, Y. D.; Wang, Y.; Zhang, W. A. *Chem. Commun.*, **2017**, 53, 3126-3129.

[3] Zhang, Z. H.; Xue, Y. D.; Zhang, P. C.; Müller, A. H. Zhang, W. A., *Macromolecules*, **2016**, 49, 8440-8448.

[4] Liu F.; Zhang, Y.; Xu, L.; Zhang, W. A. *Chem. Eur. J.*, **2015**, 21, 5540-5547.

[5] Zhang, Z. H.; Hong, L. Z.; Gao, Y.; Zhang, W. A. *Polym. Chem.*, **2014**, 5, 4534-4541.

[6] Zhang, W. A.; Müller, A. *Prog. Polym. Sci.* **2013**, 38, 1121-1162.

BIOGRAPHY

Weian Zhang is a professor at the East China University of Science and Technology. He received his PhD from the University of Science and Technology of China in 2003. Prior to his independent work, he worked at the University of Bayreuth in Germany as a Humboldt Fellow and the University of Leeds in UK as a Marie Curie Fellow. His research interests mainly focus on construction, self-assembly and application of well-defined functional polymers.

Functional Nanofibers for Advanced Energy Storage by Centrifugal Spinning

Xiangwu Zhang

College of Textiles, North Carolina State University, Raleigh, NC 27695-8301, USA

E-mail: Xzhang13@ncsu.edu

Nanofibers are an important class of material that is useful in a variety of applications, including filtration, tissue engineering, protective clothing, composites, battery separators, energy storage, etc. So far, electrospinning is the most studied method for producing nanofibers. A literature search using the Web of Science^{TM} database shows that Year 2015 alone had publication of over 3000 articles in the electrospinning of nanofibers. Among these publications, over 50% focused on the investigation of the electrospinning process and the characterization of the resultant nanofibers, and the others mainly address the innovative use of electrospun nanofibers for various applications. However, the wide-spread commercial use of electrospinning is limited due to its low production rate, poor safety, and high cost. Most other nanofiber production methods, such as phase separation, template synthesis, and self-assembly, are complex and can only be used to make nanofibers from limited types of materials. This presentation introduces a simple, yet versatile technique for producing nanofibers of various materials including polymers, carbons, ceramics, metals, and composites. Centrifugal spinning eliminates the limitations encountered by current nanofiber production methods and can produce nanofibers at high speed and low cost. Centrifugally-spun nanofibers can be used in various applications, including filtration, tissue engineering, protective clothing, composites, battery separators, energy storage, etc. The application of centrifugally-spun nanofibers in energy storage applications will be discussed in this presentation.

BIOGRAPHY

Xiangwu Zhang is a Professor and the Associate Dean for Research in the College of Textiles at North Carolina State University. He is also an Alumni Distinguished Graduate Professor and an Inaugural University Faculty Scholar at North Carolina State University. Zhang earned a B.S. in Polymer Materials and Engineering in 1997 and a Ph.D. in Materials Science and Engineering in 2001, both from Zhejiang University, China. Zhang's research interests focus on nanostructured and multifunctional polymer, composite, fiber, and textile materials with an emphasis on practical applications: *i*) energy storage and conversion, *ii*) chemical and biological protection, and *iii*) composites. His research encompasses both fundamental materials studies such as synthesis and physical characterization, as well as system design and fabrication. Zhang has received over seven million dollars of research funding from the US government, foundation, and industry. Zhang has published two books, seven book chapters, and more than 180 peer-reviewed journal articles. Zhang has also delivered over 200 presentations at international and national conferences. Recently, Zhang received Distinguished Achievement Award in Fiber Science (The Fiber Society, 2015), BASF Open Innovation Award on Energy Storage (BASF, 2015), Outstanding Service Award (Sigma XI Society - Scientific Research Society, 2015), Inaugural University Faculty Scholar Award (NC State, 2013), Alumni Association Outstanding Research Award (NC State, 2013), and Outstanding Teacher Award (NC State, 2013). Zhang also hold the titles of Yangguang Chair Professor at Wuhan Textile University and Guest Professor at Qingdao University and Zhejiang Sci-Tec University, respectively.

Electrospun Fibrous Scaffolds with Shape Memory Effect for *in situ* Bone Regeneration

Yanzhong Zhang

Department of Biomedical Engineering, College of Chemistry, Chemical Engineering & Biotechnology, Donghua University, Shanghai 201620, China

E-mail: yzzhang@dhu.edu.cn

In bone tissue engineering, a significant challenge remains in the design and fabrication of biomaterial scaffolds capable of recapitulating the cellular microenvironments, both biologically and biomechanically. While electrospinning has been well documented for enabling to biomimetically construct the cell-residing extracellular matrix, how to exert *in situ* mechanical stimuli upon scaffolds post-implantation is a concern to be addressed. In this regard, biodegradable shape memory polymers-based scaffolds may offer a solution to this problem. This talk will present a brief overview of our recent work in developing electrospun fibrous scaffolds with shape memory effect for potential use in *in situ* bone regeneration. Our proposed strategy may afford to achieve enhanced efficacy in engineering bone tissues.

BIOGRAPHY

Yanzhong Zhang is currently a Professor of Biomaterials at the Donghua University (DHU), located in Shanghai, China. He received his B.Eng. degree in Polymeric Materials from the Northwestern Polytechnical University (Xi'an, China) and his Ph.D. in Materials Science (Biomaterials) from the National University of Singapore (NUS). He did his postdoctoral training at the University of Bristol (U.K.) and the NUS. In Nov. 2009, he joined DHU as a Full Professor / Principal Investigator to lead the *Biomimetic Materials for Regenerative Medicine* research theme at the Institute of Biological Sciences and Biotechnology, DHU. Dr. Zhang has more than 20 years of research experiences in polymer-based composites and biomaterials. His team at DHU mainly focuses on multidisciplinary research relating to electrospun biomimetic fibrous biomaterials for tissue engineering and regenerative medicine applications. He has authored or co-authored more than 100 peer-reviewed journal papers with 7 patents granted, among which 12 papers were listed as ESI Highly Cited Papers (top 1% papers). His research papers have so far been well cited over 8,000 times with an H-index of 31.

Construction of Silk Materials

Yaopeng Zhang

State Key Laboratory for Modification of Chemical Fibers and Polymer Materials, College of Materials Science and Engineering, Donghua University, Shanghai 201620, China

E-mail: zyp@dhu.edu.cn

Animal silks produced by insects and spiders exhibit high strength and toughness and are of great interests for various medical applications. Inspired by the spinning process, spinning apparatus of silkworms and structure of silkworm silk, artificial silks tougher than natural silkworm silk were spun from regenerated silk fibroin (RSF) aqueous solutions in air by using a dry spinning process. Recombinant spider silk protein aqueous solution was also adopted to spin tough artificial spider silk via a bio-inspired microfluidic chip. Moreover, nanofillers such as carbon nanotubes, TiO_2 nanoparticles, graphene oxide nanosheets were applied to further reinforce the artificial silks and understand the structural evolution of the hybrid fibers. A green, sustainable ,and promising route was also developed to produce reinforced and ultraviolet resistant silkworm silk by feeding silkworms food containing titanium dioxide nanoparticles. Synchrotron radiation wide-angle X-ray diffraction (WAXD) and small-angle X-ray scattering (SAXS) were adopted to investigate the aggregation mechanism of natural silk fibroin in silkworm gland and RSF in spinning process. Various RSF scaffolds with different structures, improved mechanical properties and bioactive functions were prepared and applied for urethral reconstruction, nerve regeneration and liver function recovery.

BIOGRAPHY

Prof. Yaopeng Zhang received his BS degree in Polymer Material and Engineering from Hubei University of Technology, China, in 1998, and his PhD degree in Material Science from Donghua University, China, in 2002. He had been a postdoctoral fellow of Kawamura Institute of Chemical Research, Japan from 2004 to 2007. In 2010 and 2016, he was appointed a Visiting Professor in Akita University and in Stony Brook University, respectively. In 2012, he became a professor in material science in Donghua University. Since 2014, He has served as a Deputy Director of State Key Laboratory for Modification of Chemical Fibers and Polymer Materials. Prof Zhang's research has been awarded the Second prize of China National Textile Apparel Council Scientific Technology Award (2007). He also received a Second Class Award in Research Achievements (Science and Technology), Ministry of Education, China, 2013, Second Prize of Shanghai Sci & Tech Progress Awards in 2014. His research deals with biomimetic materials (silk worm silk and spider silk), silk fibroin based biomaterials and microfluidics. Dr. Zhang has published more than 50 SCI papers and two book chapters. He has applied 43 patents, in which 21 patents have been authorized.

Advanced Carbon and Silk Fibers/Fabric for Wearable Electronics

Yingying Zhang

Department of Chemistry, Tsinghua University, Beijing 100084, China

E-mail: yingyingzhang@tsinghua.edu.cn

The development of flexible electronics and equipment attracts significant interests in recent years. It is of great importance to explore low cost and scalable materials for high performance flexible smart electronics. Low-dimensional carbon materials are one kind of ideal materials for flexible electronics. Besides, silkworm silk, with five thousand years' usage history, is a popular natural material for clothes or wearing accessories. In this talk, I will present our recent work on exploring the application of nanocarbon materials and silk-derived carbon materials in flexible electronics. Transparent and flexible pressure sensors based on carbon nanotubes, graphene and silk-derived carbon nanofibers will be presented. And highly sensitive strain sensors made from traditional fabrics will be discussed and their applications in wearable electronics and intelligent robots will be demonstrated. These strategy paves new ways for the low cost and large scale fabrication of high performance wearable devices.

References

[1] Chunya Wang, Xiao Li, Enlai Gao, Muqiang Jian, Kailun Xia, Qi Wang, Zhiping Xu, Tianling Ren, Yingying Zhang*. *Advanced Materials,* **2016**, 28, 6640.

[2] Qi Wang, Muqiang Jian, Chunya Wang, Yingying Zhang*. *Advanced Functional Materials*, **2017**, 27, 1605657.

[3] Muqiang Jian, Kailun Xia, Qi Wang, Zhe Yin, Huimin Wang, Chunya Wang, Huanhuan Xie, Mingchao Zhang, and Yingying Zhang*. *Advanced Functional Materials,* **2017**, 27, 1606066.

BIOGRAPHY

Dr. Zhang is a joint faculty member in Department of Chemistry and in the Center for Nano and Micro Mechanics (CNMM) of Tsinghua University since 2011. Her research interest is the developing of advanced carbon and silk materials toward the next generation of flexible smart electronics. She has authored more than 70 journal papers, including *Nat Nanotechnol*, *Nat Commun*, *Adv Mater*, *JACS*, *Nano Lett*, *ACS Nano*, *Adv Funct Mater* etc, with more than 1300 citations (Web of Science). Besides, she has authored 2 book chapters, and 20 patents (7 awarded). She also serves as a reviewer for more than 20 international journals. She has been awarded as "National Science Fund for Excellent Young Scholar" (2014) of National Natural Science Foundation of China (NSFC) and "Young Cutting-Edge Nanochemistry Researcher Award" (2014).

Assembly of Polymer-Tethered Inorganic Nanoparticles in Confined Spaces

Ke Wang, Ruijing Liang, Weikun Li, and Jintao Zhu

School of Chemistry and Chemical Engineering, Huazhong University of Science and Technology, Wuhan 430074, China.

E-mail: jtzhu@mail.hust.edu.cn

Functional nanoparticles (NPs) have great potential applications in many fields, such as microelectronic devices, bio-sensors, and catalytic reactor. The properties and applications of the NPs are not only dictated by the size, shape, and chemical composition of the NPs but also dependent on their order, spatial arrangement and orientation. External field manipulation and confined assembly provided the effective route for fabricating well-ordered NPs assemblies.

We will present the formation of inorganic NP assemblies with well-ordered structures in confined geometries.[1-3] We will show that well-ordered inorganic NPs superlattice with tunable lattice constant and a series of assemblies with novel structures (e.g., helix, double helix) can be generated by tuning confined effect and/or applying external stimuli. In addition, packing style and interparticles distance could be adjusted by varying molecular weight of the polymer ligand, diameter of NPs and others, which triggered the corresponding variations in surface plasmon resonance spectrum. Furthermore, we have taken as-prepared polymer-tethered assemblies to fabricate the pentacene-based organic nano-floating-gate memory (NFGM) device, which demonstrated significantly improved performances.

References

[1] K. Wang, S. M., Jin, J. Xu, R. Liang, Z. Xue, X. Xie, E. Lee, J. Zhu, *ACS Nano* **2016**, 10, 4954.

[2] W. Li, K. Wang, P. Zhang, J. He, S. Xu, J. Zhu, X. Xie, Z. Nie. *Small* **2016**, 12, 499.

[3] R. Liang, J. Xu, R. Deng, K. Wang, S. Liu, J. Li, J. Zhu, *ACS Macro Lett.* **2014**, 3, 486.

BIOGRAPHY

Jintao Zhu received his B.S. and M.S. degrees from Hunan University in 2000 and 2002, respectively. He received his Ph.D. degree from Changchun Institute of Applied Chemistry (CIAC), Chinese Academy of Sciences in 2005. Afterwards, he carried out postdoctoral research at the University of Alberta, Canada and University of Massachusetts Amherst, USA. In 2009, he joined as a Full Professor in the School of Chemistry and Chemical Engineering at the Huazhong University of Science and Technology. His current research interests include design and synthesis of hybrid polymer materials with well-defined sizes, structures and built-in functionalities for optics, drug delivery and imaging.

The Design of Wearable Smart Mouthguard in Detecting Saliva and Study on the Properties of its Electrochemical Biosensors

Yu Jiang, Hongting Ma, Junlin Ma, and Nan Zhu*

Zhang Dayu School of Chemistry, Dalian University of Technology, Dalian, 116024, China

E-mail add: nanzhu@dlut.edu.cn

One non-invasion wearable electrochemical smart mouthguard biosensor has been proposed by this work, which could simply continuous detect ions (Na^+, K^+, et al) and pH, metabolites (glucose, lactate, uric, et al) and special metabolites (cortisol, et al) in the saliva. In addition, combining Wifi or blue tooth technology with this biosensor in the mouthguard, the detected electrochemical signal could be transformed into concentration signal by smart phone or laptop. The parameters in body's saliva will be in-site monitored by this unlimited remotely controller. Furthermore, this biosensor plays a role in sporting, reflecting or alarming the body situation when the body lacking of energy. Exploring a practical wearable smart mouthguard device, with the function of detecting the in-site parameters of body in bad health or emergent, is an important issue in the modern micro home remotely healthcare or personal healthcare society.

References

[1] Nan Zhu, Shuang Han, Jens Ulstrup and Qijin Chi*, "Graphene Paper Doped with Chemically Compatible Prussian Blue Nanoparticles as Nanohybrid Electrocatalyst", *Adv. Funct. Mater.* **2013**, 23, 5297-5306.

[2] Nan Zhu, Xian Hao, Jens Ulstrup and Qijin Chi*, "Single-Nanoparticle Resolved Biomimetic Long-Range Electron Transfer and Electrocatalysis of Mixed-Valence Nanoparticles", *ACS Catalysis* **2016**, 6 (4), 2728–2738.

BIOGRAPHY

Dr. Nan Zhu, Professor, Zhang Dayu School of Chemistry, Dalian University of Technology (DUT). 2017 joined in DUT as an independent PI, and initiates the research group of "Flexible wearable electrochemical biosensors and Nano energy device". Dr. Zhu works as a Postdoc Researcher at University of California, San Diego, USA during 2016-2017. And 2013-2015 as Postdoc Fellow at University of Copenhagen, Denmark. 2013 obtained Ph.D. Degree at Technical University of Denmark. Dr. Zhu received some world-level results on the research fields on "Flexible wearable electrochemical biosensors and Nano energy device". Experience on Publishing top journals in the research fields, including Adv. Funct. Mater., ACS Catalysis, Nat. Comm., J.A.C.S., et. al, as well as some European Patents and Chine Patents. Participating in application and completing scientific research projects and programs in European Union, USA, Sweden and Denmark. Cooperation with world famous companies (including LEGO, Philips, Mekoprint, et al), transforming research results and achievements into products.

Precision Synthesis and Industrialization Efforts for Hyperbranched Polymers

Xinyuan Zhu

School of Chemistry and Chemical Engineering, State Key Laboratory of Metal Matrix Composites, Shanghai Jiao Tong University, Shanghai 200240, China

E-mail: xyzhu@sjtu.edu.cn

Due to their unique topological structure and interesting physical/chemical properties, hyperbranched polymers have attracted more and more attention.[1,2] In this presentation, we summarize our recent progress on the precision synthesis of various hyperbranched polymers and their industrial applications.[3] New strategies have been developed for controlling the architecture and properties of hyperbranched polymers. More importantly, in recent years, we have devoted our efforts on the industrialization of hyperbranched polymers and would like to find their practical applications.

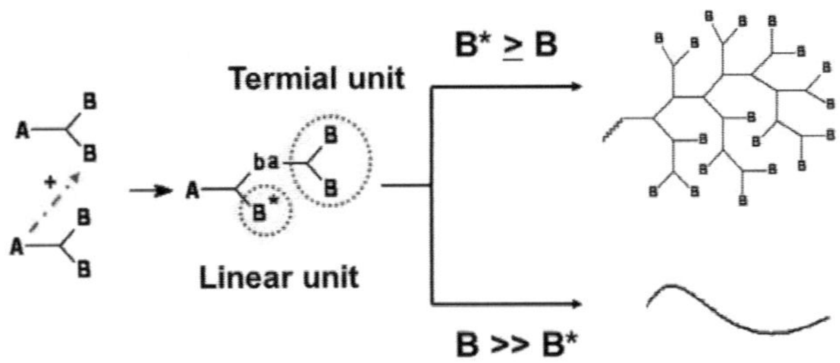

Figure 1. Branching structure control of polymerized products.

References

[1] Kim, Y. H.; Webster, O. W. *J. Am. Chem. Soc.* **1990**, 112, 4592.

[2] Wang, D. L.; Jin, Y.; Zhu, X. Y.; Yan, D. Y. *Prog. Polym. Sci.* **2017**, *64*, 114; Wang, D. L.; Zhao, T. Y.; Zhu, X. Y.; Yan, D. Y.; Wang, W. X. *Chem. Soc. Rev.* **2015**, *44*, 4023; Dong, R. J.; Zhou, Y. F.; Zhu, X. Y. *Acc. Chem. Res.* **2014**, *47*, 2006; Jin, H. B.; Huang, W.; Zhu, X. Y.; Zhou, Y. F.; Yan, D. Y. *Chem. Soc. Rev.* **2012**, 41, 5986.

[3] Dong, R. J.; Zhou, Y. F.; Huang, X. H.; Zhu, X. Y.; Lu, Y. F.; Shen, J. *Adv. Mater.* **2015**, 27, 498;

[4] Zhou, Y. F.; Huang, W.; Liu, J. Y.; Zhu, X. Y.; Yan, D. Y. *Adv. Mater.* **2010**, 22, 4567.

BIOGRAPHY

Xinyuan Zhu is the dean of School of Chemistry and Chemical Engineering at Shanghai Jiao Tong University. He serves currently as the editor of Colloids and Surfaces B: Biointerfaces, the editorial advisory board member of Biomacromolecules, the board member of the Journal of Macromolecular Science, Pure and Applied Chemistry and Chinese Journal of Polymer Science.

3.Abstracts of Oral Presentation

Cellulose Nanofibers and their Polymer Composites from Indonesian Agricultural Wastes

Farah Fahma

Department of Agroindustrial Technology, Bogor Agricultural University, Indonesia

E-mail: farah_fahma@apps.ipb.ac.id

Much attention on the development of eco-friendly products has encouraged by many researchers to exploit the resources of cellulose, such as woods and agricultural wastes. Cellulose is the most abundant renewable and biodegradable polymer on earth. It is distributed in plants, animals, and some bacteria. Because of their great abundance, agricultural wastes have been very interesting raw material for the isolation of nanofibers. Many researchers have studied on the isolation of nano-sized fibers from various agricultural wastes, such as oil palm empty-fruit-bunches (OPEFB), sugar palm fiber, soybean stock, rice straw, potato tuber, banana fibers, sugarcane baggase, wheat straw and soy hulls, coconut husk and a bio-residue from wood bioethanol production. The use of a variety of agricultural wastes as raw material will lead to increase the supply of cellulose material that can be used as nanofillers for nanocomposite materials.

In higher plants, semicrystalline cellulose microfibril reinforced amorphous matrices made of hemicelluloses, lignin, waxes, extractive and trace elements which can be regarded as a bio-nanocomposite. Meanwhile, in nature, cellulose does not occur as an isolated individual molecule, but it is found as assemblies of individual cellulose chain-forming fibers. Cellulose nanofibers can be extracted from this natural polymer using a top-down mechanically or chemically induced deconstructing treatment.

The cellulose nanofibers are potential nanofillers for polymer matrix, if good dispersion into polymer matrix can be achieved. Generally the cellulose nanofibers favor water-based systems, because of hydrophilic nature of cellulose due to its OH groups. The use of nanofibers is mostly restricted to water-soluble-polymers. Therefore, in order to expand the utilization of them for nanocomposites, it is important to reduce the entanglement of fibrils and increase their dispersion into non-polar polymer matrixes by modification of their hydrophilic surface. The modification treatment (such as acetylation) and the gradual exchange to a non-polar solvent aims to let the hydrophobic groups move out to the surface of the nanofibers, and hide the hydrophilic groups inside the nanofibers. Self-aggregation can be prevented and simultaneously a good dispersion of nanofibers in non-polar solvent can be reached. Ultimately, good interfacial adhesion between the cellulose nanofiber and polymer matrix can be obtained.Because of the nano-size effect, the nanocomposites have some unique outstanding properties with respect to their conventional microcomposite counterparts. The addition of small quantities of cellulose nanofibers from OPEFB into poly(vinyl alcohol) (PVA), poly(methyl methacrylate) (PMMA) and polychloroprene elastomer (PCR) matrices improved their properties.

References

[1] Fahma F, Hori N, Iwata T, Takemura A, *Emirates Journal of Food and Agriculture*, **2017**, doi: 10.9755/ejfa.2016-02-215.

[2] Kendri W, Evi Savitri Irian, Fahma F, *Indonesian Journal Chemistry*, **2016**, 16 (2), 181-189.

[3] Fahma F, Hori N, Iwata T, Takemura A, *Emirates Journal of Food and Agriculture*, **2016**, 28 (8), 566-571.

[4] Fahma F, Takemura A, Saito Y, *Cellulose*, **2014**, 21, 2519-2527.

[5] Fahma F, Hori N, Iwata T, Takemura A, *Journal Applied Polymer Science*, **2014**, 131, 40159-40159.

[6] Fahma F, Hori N, Iwata T, Takemura A, *Journal Applied Polymer Science*, **2013**, 128, 1563-1568.

[7] Fahma F, Iwamoto S, Hori N, Iwata T, Takemura A, *Cellulose*, **2011**, 18, 443–450.

[8] Fahma F, Iwamoto S, Hori N, Iwata T, Takemura A, *Cellulose*, **2010**, 17, 977-985.

Advanced Oxide Ceramic Composites Based on Graphene

Yuchi Fan[1], Wan Jiang[1], Akira Kawasaki[2]

[1]*Institute of Functional Materials, Donghua University, Shanghai, 201620, China;*

[2]*Department of Materials Processing, Graduate School of Engineering, Tohoku University, Sendai, 980-8579, Japan*

E-mail:yuchifan@dhu.edu.cn

Graphene is a 2D carbonaceous material with extraordinary mechanical and electrical properties which is very suitable to be exploited as reinforcement in ceramic matrix composites (CMC) for achieving superior performance. Different from previous intensively studied carbon based reinforcement such as carbon fiber and carbon nanotube, the extremely thin 2D feature of graphene can profoundly change the properties of CMC especially in the case of oxides. In alumina composite, a percolation threshold as low as 0.38 vol.% is achieved and the electrical conductivity surpasses 10^3 Sm $^{-1}$ when graphene content is only 2.35 vol.%. More interestingly, it is found that the charge carrier type changes from p- to n-type as graphene content becomes higher, as a result of synergistic doping effect between graphene and oxide matrix.[1] Further study reveals that the doping level can be manipulated by tuning the concentration of oxygen vacancy of matrix, which is realized in yttria stabilized zirconia composite as a proof of concept.[2] In addition, the graphene based oxide CMC also shows outstanding mechanical properties, as exemplified by alumina composite in which an unprecedented increase of strain tolerance by ~40%, completely retained fracture strength (417MPa) and enhanced fracture toughness (5.3MPam$^{1/2}$) are simultaneously realized at merely 2.18vol.% of filler loading.[3] Therefore, it is believed that the graphene based oxide ceramic composites are very promising material for the application requiring both advanced functional and mechanical properties.

References

[1] Fan Y C, Jiang W, Kawasaki A, *Advanced Functional Materials,* **2012**, 22 (18), 3882-3889.

[2] Fan Y C, Kang L J, Zhou W W, et al., *Carbon,* **2015**, 81, 83-90.

[3] Fan Y C, Igarashi G, Jiang W, et al., *Carbon*, **2015**, 90, 274-283.

Advanced Double-network Hydrogel Fibers from Hydrophilic Polyurethane

Bin Fei*, **Feng Wu, Yangling Li**
Institute of Textiles & Clothing, Hong Kong Polytechnic University, Hong Kong, China.
E-mail: tcfeib@polyu.edu.hk

Recoverable hydrogels with high stretch and toughness have been synthesized by a one-step radical polymerization. They consist of covalently crosslinked polyacrylamide (PAm) and hydrophilic thermoplastic polyurethane (HTPU).[1] Such double-network (DN) hydrogels can be stretched beyond 30 times their initial length, and their energy of fraction reached a high value of 19.14 MJ/m^3 (Fig. 1). Their modulus and strength were in the range of 0.12 - 1.42 MPa and 0.014 – 0.143 MPa, adjusted by the mass ratio of the PAm and HTPU. By comparing their recover tensile properties at different tensile strains, the contribution of physical network to the mechanical properties was analyzed.[2] Unlike the early reported DN hydrogels,[3] these hydrogels recovered completely at 100% extension strain, and healed quickly after higher strain at room temperature. They were also fabricated into filaments in lab.

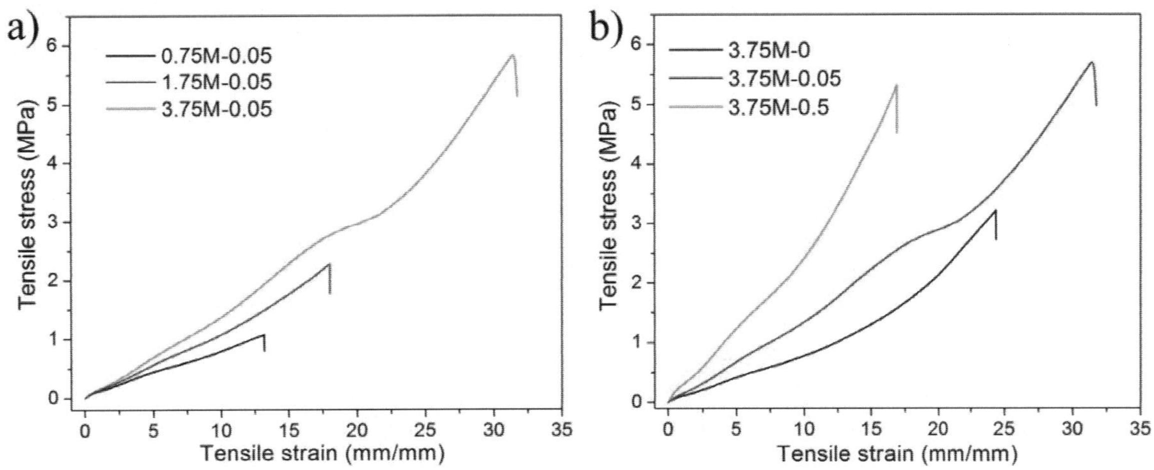

Figure 1. The tensile property of as-prepared hydrogels with (a) different AAm concentrations (MBA 0.05wt%) and (b) different MBA concentrations (AAm 3.75M).

References

[1] F Wu, L Chen, Y Li, KI Lee and B Fei. Super-tough hydrogels from shape-memory polyurethane with wide-adjustable mechanical properties. *J. Mater. Sci.* **2017**, doi:10.1007/s10853-016-0689-7.

[2] X Lu, CY Chan, KI Lee, B Fei and JH Xin. Super-tough and thermo-healable hydrogel – promising for shape-memory absorbent fiber. *J. Mater. Chem. B* **2014**, 2, 7631.

[3] JP Gong, Y Katsuyama, T Kurokawa and Y Osada. Double-Network Hydrogels with Extremely High Mechanical Strength. *Advanced Materials* **2003**, 15, 1155.

Acknowledgment: we thank the supports by HK GRF 15204614 and PolyU fund 4-BCBR.

Multifunctional Photonic Fibers: Sensing through Structure and Material

Xian Feng

Laser Institute of Engineering, Beijing University of Technology, Beijing 100124, China

E-mail: xfeng2015@bjut.edu.cn

The microstructured photonic crystal fiber [1] and the emerging multi-material multifunctional fiber [2,3] represent the trend of the next generation of photonic fibers. To construct the fiber with structures and materials, novel functionalities can be realized, compared to the conventional optical fiber. Using the structure approach, the performance of fiber waveguide can be widely tailored through building microstructure or nanostructure in the fiber. By introducing various non-glass, non-dielectric materials, e.g., semiconductors, metals, and so on, into dielectric glass or polymer matrix, multiple functionalities beyond the traditional optics capabilities can be achieved in a single integrated fiber.

I present here with our progress of multi-functional photonic fibers through structure and material approaches. First, an air-suspended-core (ASC) photonic crystal fiber (see Fig.1a) is employed as high-sensitivity sensors for monitoring bio-chemical substances in the aqueous environment around the fiber core. Due to the strong evanescent field around the small core, ppm-level of chemical impurities can be detected. The fluorescence signal or Raman signal has been utilized for characterizing the species and the concentration of various impurities. Nano-liter of aqueous solution is sufficient for the sampling in such a holey fiber microfluidic device. Second, a glass-clad semiconductor-core fiber (see Fig.1b) has been demonstrated as a photo-detector with fast response of 1MHz. Semiconductor germanium is chosen as the core material. In the future, a combination of structure and material would provide even complex and desirable sensing functions in photonic fibers.

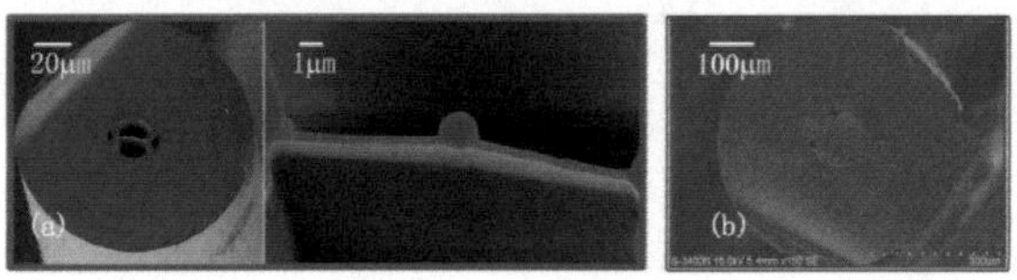

Figure 1. SEM photographs of (a) ASC fiber with 1μm core. (b) Ge fiber with 125 μm core.

References

[1] P. St. J. Russell, *Science*, **2003**, 299, 358-362.

[2] A. F. Abouraddy, et al., *Nature Materials,* **2007**, 6, 336-347.

[3] G. Tao, et al., *I. J. Appl. Glass Sci.,* **2012**, 3, 349-368.

Highly Conductive One-Dimensional Nano fibers: Silvered Electrospun Silica Nanofibers via Poly(dopamine) Functionalization

Ye Fu, **Liqun Zhang, and Wencai Wang***

State Key Laboratory of Organic-Inorganic Composites, Beijing University of Chemical Technology, Beijing, 100029, China.

E-mail: wangw@mail.buct.edu.cn

Using tetraethyl orthosilicate as a main raw material, silica nano fi bers (SiNFs) were prepared through the combination of a sol-gel process and an electrospinning technique followed by pyrolysis. Surface modi fi ed electrospun SiNFs developed by self-polymerization of polydopamine on the surface (SiNFs-PDA) served as templates for the electroless plating of silver nanoparticles (Ag NPs), using glucose as a reducing agent. The electrical resistivity of silver coated SiNPs-PDA (SiNFs-PDA/Ag) was measured by the four-point probe method and was found to be as low as 0.02 mΩ•cm at room temperature. The morphology of SiNFs-PDA/Ag before and after the blending with silicon rubber indicated a strong interaction between the silver layer and the SiNFs-PDA. The electrical and mechanical properties of the silicon rubber fi lled with SiNFs-PDA/Ag were studied to demonstrate the conductive performance application of SiNFs-PDA/Ag.

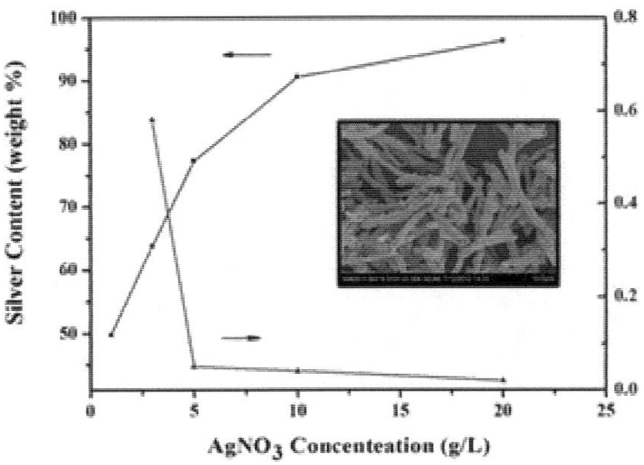

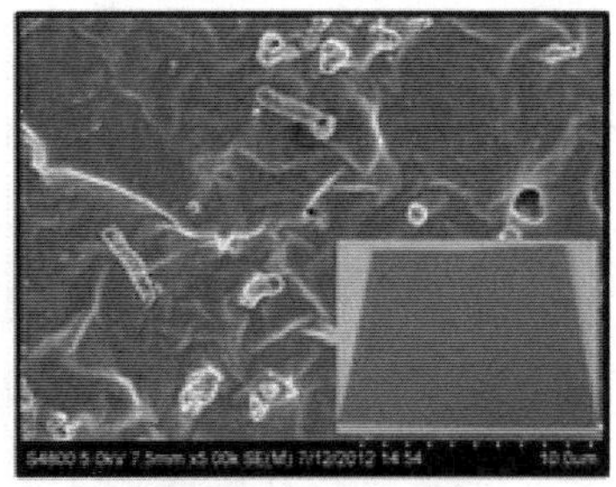

▲**Conductive elastomer composited with silvered electrospun silica nanofibers**

References

[1] Fu, Y.; Liu, L.; Zhang, L.; and Wang, W., *ACS Applied Materials & Interfaces* **2014**, 6, 5105-5112.

[2] Fu, Y.; Li, G.; Tian, M.; Wang, X.; Zhang, L.; and Wang, W., *Journal of Applied Polymer Science* **2014**, 131.

[3] Xu, C.; Tian, M.; Liu, L.; Zou, H.; Zhang, L.; and Wang, W. *Journal of The Electrochemical Society* **2012**, 159, D217-D224.

Nanofiber of Nanocellulose/ Sulfophenylatedpoly(ether ether ketone ketone) for Proton Exchange Membranes

Xiaowei Hou, Qi Zhao, Chuangjiang Ni, Wei Hu*

College of Chemical Engineering, Changchun University of Technology, 2055 Yan'an Street, Changchun 130012, P.R. China

E-mail: hw7527@sina.com

Electrospinning has been recognized as an efficient technique for the fabrication of polymer superfine fibers. Due to the nanosize of electrospun fibers, they exhibit outstanding properties which make them to be optimal candidates for many important fields, including proton exchange membranes (PEMs) applied in fuel cells [1, 2]. In this work, phenylatedpoly(ether ether ketone ketone) (Ph-PEEKK) was initially synthesized by polycondensation from 2-phenylhydroquinone and 1,4-bis(4-fluorobenzoyl)benzene. Then, post-treatment with sulfuric acid was carried out to prepare Ph-SPEEKK. On the other hand, nanocellulose (NCC), consisted of highly crystalline rod-shaped particles with a high aspect ratio and specific area, was obtained by the acid hydrolysis of microcrystalline cellulose (MCC). Different content (0, 1, 1.5, 2, 4 and 6%) of NCC was inserted into the polymer solution to prepare Ph-SPEEKK/NCC blend solution. A series of nanofibrous composite membranes were prepared by electrospun method. FESEM images revealed well-distributed nanofibers with uniform size, as shown in Figure 1(a) and (b) for both Ph-SPEEKK and Ph-SPEEKK/NCC 2 (NCC ratio was 2% in the nanocomposite PEMs). The diameter size of Ph-SPEEKK/NCC 2 nanofiber was about 50 nm thicker than that of Ph-SPEEKK nanofiber. The diameter of nanofibers was about 350 nm. Furthermore, NCC was fully embedded in the Ph-SPEEKK matrix according to the TEM images shown in Figure 1 (c) and (d). NCC could not be clearly observed from the nanofibers in TEM images.

Figure 1. SEM images of Ph-SPEEKK (a) and Ph-SPEEKK/NCC 2 (b), TEM images of Ph-SPEEKK (c) and Ph-SPEEKK/NCC 2 (d).

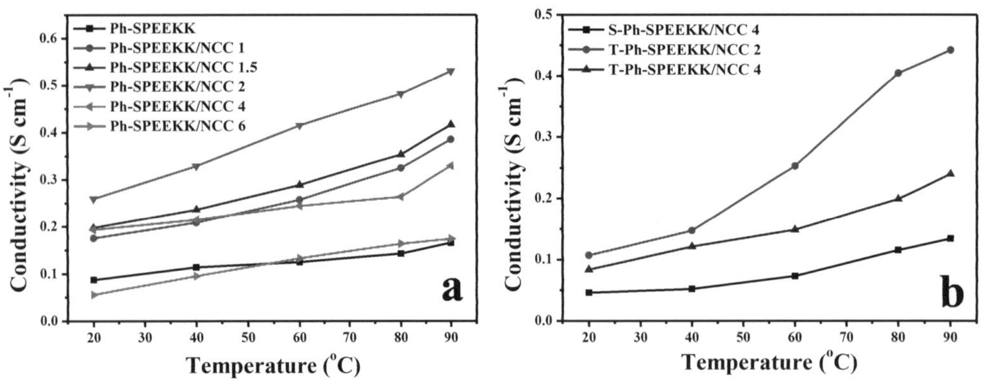

Figure 2. Proton conductivity of the membranes.

NCC was incorporated into Ph-SPEEKK nanofibers to enhance the proton conductivity and mechanical properties of the electrospun Ph-SPEEKK nanofibers because of the hydrogen bond interaction between the $-SO_3H$ and hydroxyl groups (-OH) on the surface of NCC in the nanofiber and the water retention of NCC. Figure 2 (a) showed the proton conductivity of the membrane based on the method of electrospun. The proton conduction was increased from 0.165 S cm^{-1} of the pure Ph-SPEEKK to 0.53 S cm^{-1} of Ph-SPEEKK/NCC 2 at 90 °C under fully hydrated condition. This high vale should be due to the porous structure of the nanofiber mats.

To prevent the fuel permeation and increase the mechanical properties, the composite nanofiber mat was further hot compressed with pure Ph-SPEEKK membranes with the thickness of 30 m on both sides of the mat to obtain a thick PEM (T-Ph-SPEEKK/NCC) with the thickness of about 100 m. The effect of different concentration of NCC and treatment method on the performance of composite nanofiber mat PEMs was studied. As shown in Figure 2 (b), the proton conductivity of T-Ph-SPEEKK/NCC 2 at 90 °C was as high as 0.442 S cm^{-1}. Though it was lower than that of Ph-SPEEKK/NCC 2, the membrane became much stronger than the membranes obtained by electrospinning. And the conductivity of T-Ph-SPEEKK/NCC 4 at 90 °C was higher than the membrane prepared by solution casting method (S-Ph-SPEEKK/NCC 4), which indicated the aligned nanofibers played an important role. The attractive advantages of the nanofiber rendered more sulfonic groups ($-SO_3H$) expose on its surface, and the inserted NCC constructed an efficient proton transfer channel along with nanofiber in three-dimensional scheme [3, 4].

References

[1] B. K. Park, I. C. Um, *International Journal of Biological Macromolecules*, **2017**, 95, 8-13.

[2] J. Chio, K. M. Lee, R. Wycisk, P. N. Pintauro, and P. T. Mather, *Macromolecules*, **2008**, 41, 4569-4572.

[3] T. Tamura, R. Takemori, H. Kawakami, *Journal of Power Source*, **2012**, 217, 135-141.

[4] L. D. Santos, S. Rose, O. Sel, M. Maréchal, H. Perrot, C. Laberty-Robert, *Journal of Membrane Science*, **2016**, 513, 12-19.

Controllable Synthesis of 1D Hierarchical Nanostructures for Efficient Electrochemical Catalysis

Yunpeng Huang, **Tianxi Liu,* Huaming Li***

Institute for Energy Research, Jiangsu University, Zhenjiang, 212013, PR China;

State Key Laboratory for Modification of Chemical Fibers and Polymer Materials, College of Materials Science and Engineering, Donghua University, Shanghai 201620, China.

E-mail: lhm@ujs.edu.cn, txliu@dhu.edu.cn

Exploring non-precious metal-based ORR electrocatalysts featuring high efficiency, low cost and environmental friendliness is of great importance for the broad applications of fuel cells and metal-air batteries. In this work, one-dimensional (1D) SnO_2 nanotubes (SNT) deposited with ultra-thin $NiCo_2O_4$ nanosheets was successfully fabricated through a productive electrospinning technique followed by sintering and low-temperature co-precipitation strategy. This hierarchically engineered architecture has ultrathin $NiCo_2O_4$ nanosheets uniformly and fully erecting on both walls of tubular SNT, which manifests improved electrochemical activity as oxygen reduction reaction (ORR) catalyst in terms of positive onset potential and high current density, as well as superior tolerance to crossover effects and long-term durability with respect to the commercial Pt/C catalyst. The excellent performance of SNT@$NiCo_2O_4$ composites may originate from their rationally designed hierarchical tubular nanostructure with completely exposed active sites and interconnected 1D networks for efficient electron and electrolyte transfer, making this composite nanotubes promising candidate to replace Pt-based catalysts for practical fuel cell and metal-air battery applications.

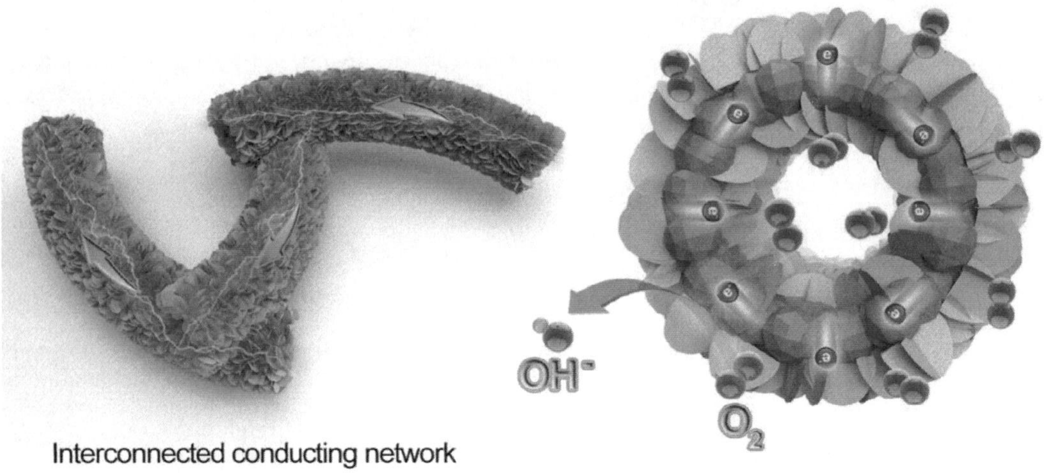

Scheme 1. Schematic representation of the fast charge transfer and efficient electrolyte diffusion during the ORR process of SNT@$NiCo_2O_4$ composite nanotubes.

References

[1] C. Z. Zhu, H. Li, S. F. Fu, D. Du, Y. H. Lin, *Chem. Soc. Rev.* **2016**, 45, 517-531.

[2] F. X. Ma, L. Yu, C. Y. Xu, X. W. D. Lou, *Energy Environ. Sci.* **2016**, 9, 862-866.

Electrospun Silver Ions-loaded Antibacterial Composite Fibrous Membranes for Guided Bone Regeneration

Shue Jin[1], **Fang Yang**[2], **Fuhua Sun**[1], **Yi Zuo**[2], **Yubao Li**[1], **Jidong Li**[1] *

[1]*Research Center for Nano-Biomaterials, Analytical and Testing Center, Sichuan University, Chengdu 610064, China;*

[2]*Department of Biomaterials, Radboud University Medical Center, Nijmegen 6500 HB, the Netherlands*

E-mail: nic1979@scu.edu.cn

Introduction: An ideal bone repair material is supposed to imitate the native extracellular matrix (ECM) from both the structure and the components so as to promote cell adhesion and proliferation [1]. Guided bone regeneration (GBR) membrane with antibacterial ability can reduce the risk of a biomaterial-centered infection, especially if the membrane is exposed to infected tissue or an area with high bacteria concentrations such as the oral cavity [2]. Our strategy is to design and fabricate a biomimetic and bioactive silver ions-loaded calcium phosphate/chitosan (Ag-CaP/CS) membranes with antibacterial ability using electrospinning, which resemble the ECM so as to create conducive living milieu to induce cells to function naturally.

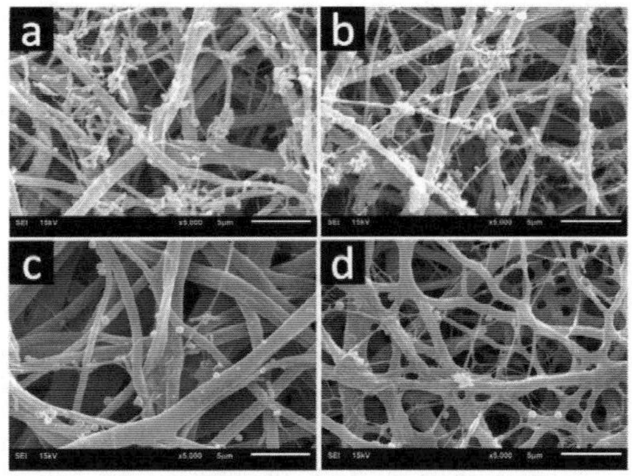

Materials and methods: The composite fibrous membranes were prepared via electrospinning. Subsequently, the morphology, composition, antibacterial properties and cytocompatibility of the fabricated membrane were evaluated respectively.

Results and discussion: The results suggest that the antibacterial composite fibrous GBR membranes were successfully developed by incorporation of silver ions and calcium phosphate into electrospinning CS fiber and the prepared membranes showed a high spatial correlation. The diameter of fabricated microfibers were in two-level, the larger fibers was about from 200 nm to 500 nm and the thinner fibers was about from 50 nm to 100 nm. Importantly, the Ag-CaP/CS composite fibrous membranes exhibited sustainable release of Ag^+ so as to inhibit the adhesion and growth of *Staphylococcus mutans* (Figure 1). Cell adhesion and MTT assay of bone morrow stromal cells (BMSCs) cultured on the membranes revealed that the Ag-CaP/CS composite fibrous membranes were cytocompatible. These results suggested that the fabricated fibrous membranes mimicked the ECM and possessed commendable biocompatibility and sufficient antibacterial properties.

Figure 1. SEM images of bacteria adhesion on different membranes after 24 hours of incubation with *S. mutans*. (a) pure CS fibrous membrane, (b) CaP/CS composite fibrous membrane containing 20% calcium phosphate, and Ag-CaP/CS composite fibrous membrane containing (c) 0.075wt% and (d) 0.144wt% silver. A large number of bacteria adhered to the surface of the membranes without silver ions(a, b), while only a small amount of bacteria were present on the surface with silver ions(c, d).

Conclusion: The fabricated electrospinning Ag-CaP/CS nanofiber membranes with excellent biocompatibility and sufficient antimicrobial property have great potential to be used for guided bone regeneration.

References

[1] Song, J. K., Remmers, S. J. A., Shao, J. L., Kolwijck, E., Walboomers, X. F., Jansen, J. A., Leeuwenburgh, S. C. G., & Yang, F., *Nanomedicine*, **2016**, 12, 1357-1364.

[2] Zheng, F. Y., Wang, S. G., Wen, S. H., Shen, M. W., Zhu, M. F., & Shi, X., *Biomaterials*, **2013,** 34, 1402-1412.

Sponsoring fund: The authors acknowledge support of the National Key Research and Development Program of China (2016YFA0201703/2016YFA0201700).

Multifunctional Electrospun Nanofiber for Air Filter

Yanyun Ju, Zhiwei Lu, Songying Lan, Lijie Dong*

School of Materials Science and Engineering, Wuhan University of Technology, Wuhan 430070, China

E-mail:juyanyun@whut.edu.cn

It's well-known that air pollution is a major threat to the human health[1-2]. The fine particles, the main component for air pollution, are also carriers of bacteria, which is vary harmful to human. Air filter materials are considered to be the effective barrier to resist air pollution, but the traditional filter material can not effectively filter the submicron particles, which makes it incapable to meet people's expectations. So, a kind of air filtration material is of great theoretical significance and application value with high filter efficiency, low pressure drop, good antibacterial performance and haze resistance. Electrospun nanofibrious membrane(ENM) with fine diameter, small pore structure and high surface-to-volume ration can significantly intercept the fine particulate[3-4].

In this paper, PA6@Ag composite nanofibers were prepared by electrospinning method, which aimed to fabricate antibacterial and efficient air filtration material. Under the operating condition, the PA6 ENM exhibits uniform nanofiber structure and the mean diameter is ~100 nm. The microstructure and chemical structure of PA6 ENM were confirmed by FTIR and XRD, which indicates that part of α crystal form transforms to γ crystal form. Moreover, the PA6 ENM displays excellent $PM_{2.5}$ filtration efficiency, low pressure drop, which can meet the requirements of low resistance and high efficience as an air filter.

Besides fine particles and gase ous pollutants, the bacteria carried on particle matter have more notable influence on human health. It is reported that Ag nanoparticles exhibit excellent antibacterial performance[5]. The Ag nanoparticles were prepared by aqueous phase oxidation-reduction method with the presence of the reducing agent of sodium borohydride and the stabilizer of citric acid. After binding Ag nanoparticles on PA6 nanofibers, the ENM shows high filtration efficiency and excellent antibacterial activity and good biosecurity.

Further more, the ENMs were applied to the actual air filter device. PA6@Ag ENM was replaced with nonwoven fabric of the disposable mask, and the filtration performance was improved remarkably. $PM_{0.3}$ filtration efficiency increases notably. The PA6 ENM was installed in air conditioning. After the cigarette burned, $PM_{2.5}$ concentrations of room can be reduced from severe harmful air to standard air. Based on the electrospinning technique, a kind of air filtration material, with high filter efficiency, low pressure drop, good antibacterial performance and haze resistance, is promising in air filter area.

References

[1] M. A. Kioumourtzoglou, J. D. Schwartz, M. G.Weisskopf, S. J. Melly, Y. Wang, F. Dominici, A. Zanobetti. *Environ. Heal. Persp.*, **2016**, 124, 23-29.

[2] Y. Shang, Z. Sun, J. Cao, X. Wang, L. Zhong, X. Bi, H. Li , W. Liu , T. Zhu , W. Huang. *Environ. Int.*, **2013**, 54, 100-111.

[3] A. Podgórski , A. Bałazy , L. Gradoń. *Chem. Eng. Sci.*, **2006**, 61, 6804-6815.

[4] C. Liu , P. C. Hsu, H. W. Lee, M. Ye, G. Zheng, N. Liu, W. Li, Y. Cui. *Nature commun.*, **2015**, 6: 6205.

[5] S. Prabhu, E. K. Poulose. *Int. Nano Letters*, **2012**, 2(1): 32.

Electrospun Fibers for Electrocatalytic Oxygen Reduction Reaction

Qi Shi[2], Nan Wu[b], Yongpeng Lei[1,3*], Yingde Wang[2*]

[1]*College of Basic Education, National University of Defense Technology, Changsha, 410073, China;*

[2]*Science and Technology on Advanced Ceramic Fiber and Composites Laboratory National University of Defense Technology, Changsha, 410073, China;*

[3]*Science and Technology on High Strength Structural Materials Laboratory & School of Aeronautics and Astronautics, Central South University, Changsha 410083, China*

E-mail: lypkd@163.com & wyd502@163.com

Oxygen reduction reaction (ORR) has aroused considerable attention due to its significance in fuel cell applications. The strategy of combining highly conductive frameworks with abundant active sites is desirable to prepare alternative catalysts to commercial Pt/C. Three different strategies have been applied using electrospinning. Firstly, B, N-codoped carbon nanofibers (CNF) have been massively prepared by heat treatment of electrospun CNF with boric acid/urea in N_2 (BNC_f-N) and subsequently activated in NH_3 (BNC_f-NA). The directly electrospun self-standing 3D non-woven structure with void spaces between each fiber facilitates the mass transport of reactant and resulted molecules. Further NH_3 activation gives BNC_f-NA a high surface area of 306.3 $m^2\ g^{-1}$ with micro/mesoporous structure, providing abundant passageway for proton transfer. Simultaneously, NH_3 activation also realizes the optimization of surface functionalities, such as more B-N-C and pyridinic-N. These intriguing features render BNC_f-NA excellent catalytic behaviour with nearly four-electron ORR process in alkaline media; Secondly, N-doped graphene (NG) and CNTs were in situ grown on cobalt-containing CNF to form 3D interconnected networks. The NG and CNT bound the interlaced CNF together, facilitating electron transfer and providing more active sites. The 3D interconnected fiber networks show excellent ORR catalytic behaviour with onset potential of 0.924 V (vs. RHE) and higher current density than Pt/C beyond 0.720 V. Thirdly, the interconnected Fe-N/C nanofiber networks (Fe-N/C NNs) with low electrical resistance have been simply constructed by simply maturing the electrospun nanofibers and subsequent pyrolysis. The Fe-N/C NNs show excellent ORR activity with onset and half-wave potential of 55 and 108 mV less than those of Pt/C catalyst in 0.5 M H_2SO_4. The resulting Fe-N/C NNs exhibit 34% higher peak current density and superior durability than generic Fe-N/C ones with similar microstructure and chemical compositions. The higher electro-activity is mainly due to the more effective electron transport and transfer between the interconnected nanofibers. In summary, the above work using electrospinning technology show potential in advanced clean energy application.

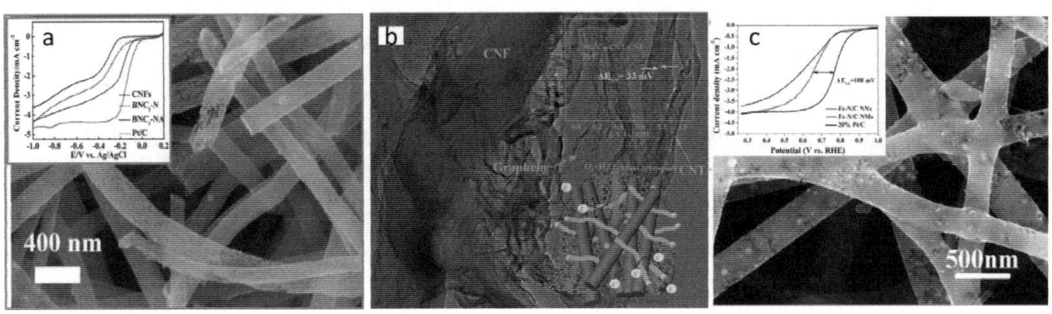

Figure 1. Three routes to prepare electrospun fibers for electrocatalytic ORR.

References

[1] Qi Shi, Yongpeng Lei, Yingde Wang, et al. *Current Applied Physics*, **2015**, 15, 1606-1614.

[2] Qi Shi, Yingde Wang, Zhongmin Wang, et al, *Nano Research*, **2016**, 9, 317-328.

[3] Nan Wu, Yingde Wang, Yongpeng Lei, et al. *Scientific Reports*, **2015**, 173.

Functionalization of Natural Wood Fiber Toward Highly Transparent Nanofibrous Paper

Detao Liu*, Pengbo Lu, Yanghao Ou, Meiyan Lin, Lingfeng Su
State Key Laboratory of Pulp and Paper Engineering, South China University of Technology, Guangzhou 510640, China
*E-mail: dtliu@scut.edu.cn

As a novel degradable biological material, natural wood fibers have been widely applied in the field of advanced materials or nanocomposites due to its excellent optical and mechanical properties which provides perfect basis for fabricating flexible electronics and devices. Among known strategies for generating highly transparent nanocomposites, the modification of wood fibers, generated by separation of cellulose nanofibrils, is arguably the most versatile and easily scalable method. However, high energy consumption and/or long producing cycle during present production procedure of cellulose nanofibrils from wood resources still remains unsettled. In past years, our research team and the coworkers reported the implementations of low costs, clean production, highly flexible and transparent nanocomposites or films from wood fibers obtained by mainly three approaches as followed; 1) Rapid ILs-polishing technology; 2) Rapid Dissolving-debonding Strategy; 3) Partial dissolution technology; In the paper, we summarized and introduced a rapid, clean and economical pathways to fabricate highly flexible and transparent nanocomposites with their applications in flexible electronics and devices. The optical, mechanical and electrical properties were investigated their relationships between its nanostructure and morphology. This work is capable of fabricating next-generation flexible and highly transparent nanocomposite substrates by a high-speed roll-to-roll manufacturing process with a much lower cost.

One-step Solvent Media Process Toward the Production of Cellulose Nanofibers

Pengbo Lu, Yanghao Ou, Meiyan Lin, Lingfeng Su, Detao Liu*
State Key Laboratory of Pulp and Paper Engineering, South China University of Technology, Guangzhou 510640, China
E-mail: dtliu@scut.edu.cn

In this paper, one-step solvent media process was firstly used to enable the carboxymethylation of eucalyptus wood dissolving pulp and disintegrate them into the nanometer-sized fibrils. The surface morphology, mechanical properties, optical properties and electrical properties of nanopaper and nona conductive paper were characterized by AFM, SEM, UV-Vis, tensile strength tester, four-probe resistance tester. When the content of silver nanowires is zero, the transmittance of nanopaper reaches 92%, and when the transmittance is 80%, the square resistance is only 15 Ω/cm^2; The maximum tensile stress can reach 93.0 MPa, and the surface roughness is about 10 nm. This kind of transparent flexible conductive paper with smooth surface, high transmittance, high strength and good conductivity has great potential in the field of optoelectronics.

Design, Preparation and Application of One-dimensional Carbonaceous Electrocatalytic Nanocomposites

Yuee Miao, Tianxi Liu

Key Laboratory for Modification of Chemical Fibers and Polymer Materials, College of Materials Science and Engineering, Donghua University, Shanghai 201620, P. R. China.

E-mail: yuee_miao@dhu.edu.cn

High-performance electrocatalysts with excellent catalytic activity and long durability competitive to Pt are urgently necessary for the fast development of energy storage and conversion systems. Electrospinning has been proved to be an efficient technique to produce advanced fibrous materials with high specific surface area, high porosity and three-dimensional network structure for diverse applications. Herein, nitrogen-doped carbon nanofiber (NCNF) membrane has been developed via electrospinning, acting as a three-dimensionally networked conductive template for immobilization of electrochemically active non-precious metal, such as MoS_2-based nanoparticles (Fig. 1). Thus, non-precious metal/NCNF composite fibers with hierarchical structures are obtained, which subtly combine the synergistic effects between the electroactive nanoparticles, efficient surface nitrogen doping and highly conductive NCNF network. Therefore, the non-precious metal/NCNF composite exhibits excellent catalytic activity towards oxygen reduction reactions or hydrogen evolution reactions with high catalytic activity, large current density and superior durability, being promising noble metal-free catalysts for practical electrocatalytic applications.

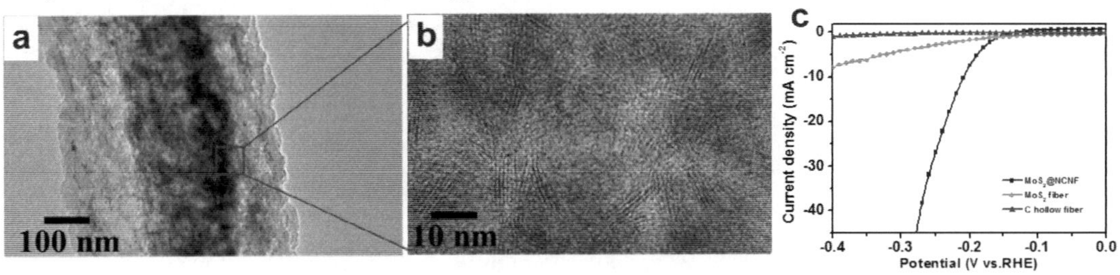

Figure 1. (a, b) TEM images and (c) the corresponding electrocatalytic performance of MoS_2@NCNF composite.

Keywords: Nanofiber composite, hierarchical structure, non-precious metal, electrocatalytic reaction

References

[1] Yan JJ, Zhang YF, Huang YP, Miao YE, Liu TX. *Adv. Mater. Interfaces* **2017**, 4, 1600825.
[2] Huang YP, Zhang LS, Lu HY, Lai FL, Miao YE, Liu TX. *Inorg. Chem. Front.* **2016**, 3(7), 969.
[3] Lai FL, Miao YE, Huang YP, Zhang YF, Liu TX. *ACS Appl. Mater. Interfaces* **2016**, 8, 3558.
[4] Yan JJ, Lu HY, Huang YP, Fu J, Mo SY, Wei C, Miao YE, Liu TX. *J. Mater. Chem. A* **2015**, 3(46), 23299.

Heterogeneous Catalysis Supports Based on Synthetic Fibers

Xianlei Shi, Yongju Chen, Qianqian Hu
Henan Polytechnic University, Jiaozuo, Henan 454003, China
E-mail: shixl@tju.edu.cn

The heterogenization of homogeneous catalysts onto solid supports has been well investigated and proven to be more crucial to chemical technology, and it will continue to influence and motivate the development of catalysts for environmental remediation. Given many advantages such as low cost, high strength, light density and great flexibility, fibers provide an appealing option to other polymers as the support for heterogeneous catalysis. In our work, the synthetic fibers such as polyacrylonitrile (PAN), polypropylene fiber (PP), polybenzimidazoles (PBI) and polyphenylene sulfide (PPS) *etc.* were used as supports to the preparation of fiber-supported base or acid catalysts, fiber-supported ionic liquids and fiber-supported metal active sites. As well as, the merits and details of the above fiber-supported catalysts were investigated on different types of reactions in different systems. Moreover, this newly developed fiber-based catalytic method provides a new strategy to design and develop greener, safer and more economical processes in the chemical industry.

Figure 1. The diagrammatic sketchs of fiber supported catalyst and continuous-flow processing.

References

[1] X.-L. Shi, Q. Hu, F. Wang, W. Zhang, P. Duan, *J. Catal.*, **2016**, 337, 233-239.
[2] X.-L. Shi, M. Zhang, Y. Li, W. Zhang, *Green Chem.*, **2013**, 15, 3438-3445.
[3] X.-L. Shi, X. Xing, H. Lin, W. Zhang, *Adv. Synth. Catal.*, **2014**, 356, 2349-2354.
[4] X.-L. Shi, H. Lin, P. Li, W. Zhang, *ChemCatChem*, **2014**, 6, 2947-2953.
[5] X.-L. Shi, M. Zhang, H. Lin, M. Tao, Y. Li, W. Zhang, *Chem-Asian J.*, **2015**, 10, 752-758.
[6] X.-L. Shi, Y. Chen, Q. Hu, et al. *J. Ind. Eng. Chem.*, **2017**, 53, 134-142.
[7] Q. Hu, X.-L. Shi, Y. Chen, et al. *J. Ind. Eng. Chem.*, **2017**, 54, 75-81.

Rational Design for Copper Chalcogenide to Enhance the Photothermal Conversion Efficiency and their Application

Qiwei Tian*, Chenglin Yan, Shiping Yang

Department of chemistry, College of Life and Environmental Sciences, Shanghai Normal University, Shanghai, 201234, China

E-mail: qiweitian@shnu.edu.cn

Copper chalcogenide nanoparticles have received increased attention as photothermal agents due to their low cost and easy preparation. Nevertheless, the underdeveloped photothermal conversion efficiency of the copper chalcogenide has limited its further application for cancer therapy. Thus, enhancing the photothermal conversion efficiency of copper chalcogenide through rational design is desired. To now, three kinds of rational design and mechanism for copper chalcogenide to enhance the heat conversion efficiency have been proposed. First is the design and synthesis of the superstructure nanoparticle agents to enhance the photon refraction. Then, tuning the degree of copper deficiency enhances the localized surface plasmon resonances absorption caused by the copper chalcogenide. The rest is constructing the hybrid composite to further enhance the localized surface plasmon resonances absorption through coupling effect of different functional units. Here, we will discuss the mechanism for the three kinds of rational design, also introduce some application for the copper chalcogenide with enhanced photothermal conversion efficiency.

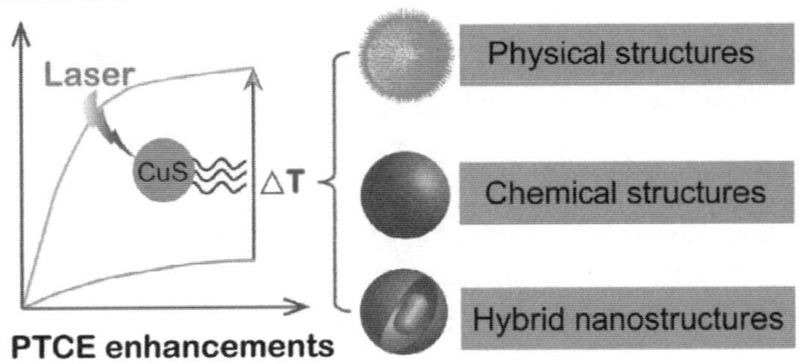

Figure 1. Three rational designs and the mechanism for copper chalcogenide to enhance the heat conversion efficiency

References

[1] C. Yan, Q. Tian, S. Yang, *RSC Adv.*, **2017,** DOI: 10.1039/c7ra05468h.

[2] Q. Tian, F. Jiang, R. Zou, Q. Liu, Z. Chen, M. Zhu, S. Yang, J. Wang, J. Wang and J. Hu, *ACS Nano*, **2011**, 9761-9771.

[3] Q. Tian, M. Tang, Y. Sun, R. Zou, Z. Chen, M. Zhu, S. Yang, J. Wang, J. Wang and J. Hu, *Adv. Mater.*, **2011**, 3542-3547.

[4] Q. Tian, J. Hu, Y. Zhu, R. Zou, Z. Chen, S. Yang, R. Li, Q. Su, Y. Han and X. Liu, *J. Am. Chem. Soc.*, **2013**, 8571-8577.

Nanofibrous Scaffold for Highly Efficient Adsorption of Lanthanide Ions and its Photoluminescence

Min Wang, Xuefen Wang*

State Key Laboratory for Modification of Chemical Fibers and Polymer materials, College of Material Science and Engineering, Donghua University, Shanghai, 201620, China

*E-mail: wangxf@dhu.edu.cn

Currently, rare earth elements (REEs) have been widely applied in mundane and high-tech fields relying on their specific optical, electrical, magnetic, catalytic and outstanding photoluminescent properties[1-2]. However, with the rapid depletion of REEs global reserves and the severe environment pollution of the REEs effluent, developing a highly efficient, environmentally friendly and material to separate REEs has become an urgent task[3]. Significantly, electrospun nanofibers with remarkable structural features like large surface-to-volume ratio, facile functionalization, convenient recyclability, and especially safety to be used, exhibited great potential in the field of REEs recovery[4]. Polyacrylonitrile-co-acrylic acid (PAN-AA) with a viscosity averaged molecular weight (Mn) of 2.4×10^5 g/mol was synthesized, and then porous PAN-AA nanofibers were fabricated by wet-electrospinning PAN-AA and poly(vinyl pyrrolidone) (PVP) blended solution using hot water as extractor. The obtained porous PAN-AA nanofibrous membrane showed highly adsorption capability for La^{3+} of 57.37 mg/g at pH 5, and it could also be regenerated easily. Nevertheless, this work only focused excessively on the increase the adsorption sites on the nanofiber surface to promote the recover efficiency, ignoring the potentially abundant available adsorption sites in the interior of nanofibers. Thus, a novel choreographed poly(acrylic acid)-silica hydrogel nanofibers (PAA-S HNFs) scaffold was fabricated by a facile route consisting of colloid-electrospinning of PAA/SiO_2 precursor solution, moderate thermal cross-linking of PAA-S nanofiber matrix and fully swelling under the water circumstance. The resultant PAA-S HNFs with loose and spongy porous network structure exhibited a remarkable adsorption capacity of Ln^{3+} (268.8 mg/g) triggered by the penetration of Ln^{3+} from the nanofiber surface to interior through the abundant water channels. Moreover, the Ln-PAA-S HNFs simultaneously exhibited versatile advantages including good photoluminescent performance, tunable emission color and excellent flexibility and processability. Significantly, a three-dimensional (3D) nanofiber-assembled spongy hydrogel exhibits several attractive properties, such as superelasticity, ultrahigh-water-content, high porosity, low density and multi-functionality, which will provide a new platform for adsorption of target molecules. Therefore, we described a robust strategy for constructing 3D ultrahigh-water-content and superelastic sodium alginate nanofiber spongy hydrogel (SA NSH) by reconstructing the SA nanofibers into 3D bulk scaffold combined with the electrospinning, homogenizing and freeze-drying techniques, and then cross-linked with phthalic anhydride (PMDA). The resultant SA NSH possessed large adsorption capacity for Ln^{3+} of 487.25 mg/g and excellent regeneration property. Additionally, we also demonstrated the good photoluminescent performance and tunable emission color of the Ln^{3+}-3D SA NSH.

References

[1] Zheng, X.; Wu, D.; Su, T.; Bao, S.; Liao, C.; Wang, Q, *ACS Appl. Mater. Interfaces,* **2014**, 6, 19840-19849.

[2] Han, W.; Ding, B.; Park, M.; Cui, F.; Ghouri, Z. K.; Saud, P. S.; Kim, H. Y., *Nanoscale*, **2015**, 7, 14248-14253.

[3] Roosen, J.; Binnemans, K., *J. Mater. Chem. A*, 2014, 2, 153021540.

[4] Hong, G.; Shen, L.; Wang, M.; Yang, Y.; Wang, X.; Zhu, M.; Hsiao, B. S., *Chem. Eng. J.,* **2014**, 244, 307-316.

[5] Okajima, M. K.; Masatoshi, N.; Tetsu, M.; Tatsuo, K., *Biomacromolecules,* **2010**, 11, 1773-1778.

Effect of Alkali Oxide on the Fluid Flow of Glass Fibers Based on the Poiseuille Equation

Qingwei Wang[1], Hong Li[2], Lida Luo[1], Weizhong Jiang[1], Hongzhi Wang[1]

[1] *State Key Laboratory for Modification of Chemical Fiber and Polymer Materials, Donghua University, Shanghai, 201620, P.R. China;*

[2]*Fiber Glass Science and Technology, PPG Industries, Inc., Cheswick, PA 15024, USA*

E-mail: wqwq888@dhu.edu.cn

Glass fibers are used to reinforce polymer-based composites that require specific mechanical, chemical, and electrical performance[1]. The most common glass fiber is E-glass fibers containing less than 0.8wt.% alkali oxide and C-glass fibers containing about 12wt.% alkali oxide[2].

As we know, the drawing fiber flow of E-glass fiber is about twice as much as that of C-glass fiber in the same condition. In order to increase the output of the C-glass fiber, the bigger Platinum alloy bushings were necessary. The high-temperature strength and chemical consumption of noble metal bushings results in the higher manufacturing cost of C glass fiber.

Poiseuille's law is found to be in reasonable agreement with experiment for glass molten (called Newtonian fluids) in cases where there is no appreciable turbulence.

$$F = \frac{P_1 - P_2}{R} = \frac{\pi \Delta P r^4}{8\eta L}$$

Where ΔP is the pressure difference of the drawing outlet, η is the viscosity of the glass molten, L is the length of the tips on the bushing, r is the radius of the drawing hole in the bushing.

Results show the experimental flow of E glass fiber is in accordance with calculating results, but the C glass fiber present nearly only half of the theory flow. The contacting angle of E-glass and C-glass to platinum was 44 ° and 41 ° at the temperature with viscosity of 10^3 Pa.s, so C-glass fiber is easier to adhere in the bushing. The results demonstrate that alkali oxide would contribute to the thicker boundary layer in the drawing hole, which would significant decrease the drawing flow of C-glass fiber.

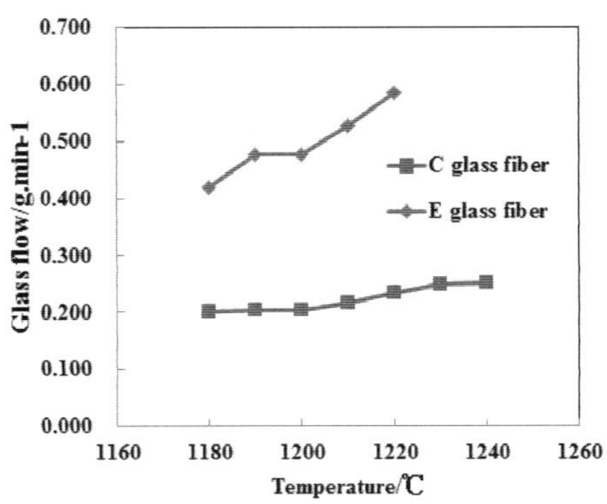

Figure 1. the different fluid flow chart corresponding to different temperature for E- and C-glass fiber.

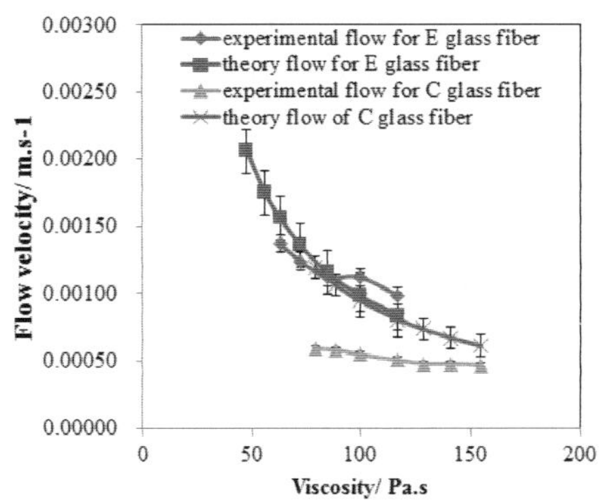

Figure 2. the theory and experimental flow velocity dependent on the viscosity for E- and C-glass fiber.

References

[1] Ma H, Jia Z,Lau K, et al. *Composites B,* **2016**,92(2),210-217.

[2] Wang Q., Brow R., Li H. et al. *Eu J Glass Sci Tech A,* **2016**, 57 (2), 58-61.

Smart Fibers Based on Low Dimensional Conductive Materials

Ranran Wang, Yin Cheng, Xiao Wang, Tao Wang, Zherui, Cao and Jing Sun*

Shanghai Institute of Ceramics, Chinese Academy of Sciences 1295 Dingxi Road, Changning District, Shanghai, China

E-mail: wangranran@mail.sic.ac.cn

The fragility of traditional metallic or semi-conductive materials hinders their application in flexible electronics. Low dimensional materials including carbon nanotubes, graphene and metal nanowires own outstanding flexibility and have been wildly used to fabricate flexible devices. Bendable/stretchable substrate is another key component of flexible electronics. Various thin polymer films made of polyethylene terephthalate, polyimide, polydimethylsiloxane et. al. were adopted. However, the air impermeability of these substrates will cause discomfort of humanbeing if applied in wearable electronics. Fiber is an ideal substrate for flexible and wearable electronics due to its excellent flexibility/stretchability, superior breathability, abundant microstructure and low cost. Herein, a series of conductive elastomers and strain sensors were fabricated by combining the low dimensional conductive materials with fiber substrates and regulating the microstructure on the interface. With the help of "twining spring" hierarchical architecture, silver nanowire-double covered yarn (Ag NW-DCY) composite fibers with ultrahigh stretchability were obtained. The conductivity of the composite fibers reached up to 4018 S/cm and remained as high as 688 S/cm at 500% tensile strain. Commercial electronic components (LED arrays) were integrated onto a transparent, foldable and stretchable substrate using the composite fibers as stretchable electric wiring, demonstrating the potential application in large-area stretchable electronics. When AgNWs were replaced with graphene, strain sensing fiber with high sensitivity and large working range (0.2%-100% strain) were fabricated, which enabled the detection of multiple deformation forms, including tensile strain, bending, and torsion. We employ the fibers as wearable sensors, realizing the monitoring of full-range human activities and intricate movement combinations of a robot. These smart fibers are of great meaning to the development of flexible and wearable electronics.

References

[1] Yin Cheng, Ranran Wang,* Haitao Zhai and Jing Sun*, *Nanoscale*, **2017**, 9, 3834–3842.

[2] Yin Cheng; Hange Zhang; Ranran Wang*; Xiao Wang; Haitao Zhai; Tao Wang; Qinghui Jin; Jing Sun*. *ACS Appl. Mater. Interfaces*. **2016**, 8, 32925-32933.

[3] Yin Cheng; Ranran Wang*; Jing Sun, *; Lian Gao, *Adv. Mater.*, **2015**, 27, 7365-7371.

[4] Yin Cheng; Ranran Wang*; Jing Sun, *; Lian Gao, *ACS Nano*, **2015**, 9. 3887-3895.

Esterification of Polymorphs of Cellulose Nanocrystals: Crystallinity and Thermal Decomposition

Jun Xu[1,2], Zhenhua Wu[1], Jie Gong[1], Jun Li[1*], Kefu Chen[1]

[1] *State Key Laboratory of Pulp and Paper Engineering, South China University of Technology, Guangzhou 510640, China;*

[2] *Key Laboratory of Pulp and Paper Science & Technology of Ministry of Education of China, Qilu University of Technology, Jinan 250353, China*

E-mail: ppjunli@scut.edu.cn

Esterification of CNC has been proposed as an available route for transforming the surface of CNC from hydrophilicity to hydrophobicity by replacing hydroxyl groups with hydrophobic ester groups. This can effectively improve their dispersibility/compatibility with polymers when used as a reinforcing agent. But CNC are polymorphic materials and easy to be converted to different polymorphs. The mechanical and thermal properties of CNC can be altered through polymorph interconversion.

In this study, the heterogeneous acetylation of three different polymorphs of CNC I/II/III were investigated under mild reaction conditions. The difference of crystallinity and thermal stability between the three acetylated polymorphs of CNC were studied by Fourier transform infrared spectrometry (FTIR), X-ray diffraction (XRD) and thermogravimetric analysis (TGA). It was shown that all of the obtained acetylated CNC kept the original intact crystal. And, the crystallinities of all the samples were lower than the initial CNC I due to the destruction of the surface morphology or the changes in the internal orientation of CNC I. The thermal stability increased during the transformation from CNC I to CNC II and CNC III as well as the acetylation. However, the thermal decomposition temperature of acetylated CNC II was lower than the initial CNC II. The thermal stability of acetylated CNC III changed relatively smaller. In addition, the pyrolysis process of most samples mainly included two steps due to the existence of double bonds in sulfonic group or acetyl group on the CNC surface. This study about crystallinity and thermal stability of the polymorphs of cellulose nanocrystal may have great significant meaning in the fields like reinforcing fillers for composite materials.

Keywords: cellulose nanocrystal; polymorph; surface modification; crystallinity; thermal decomposition

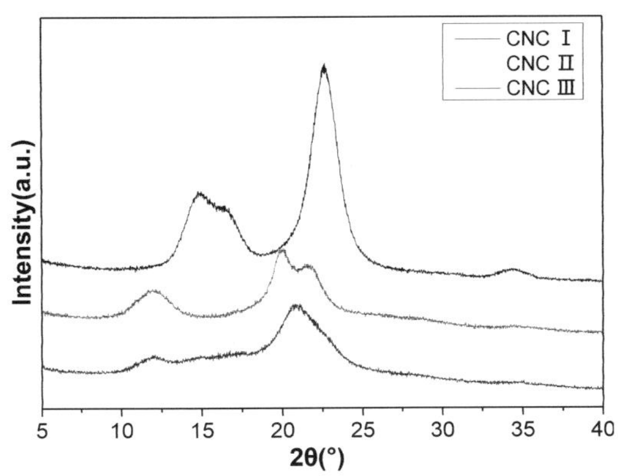

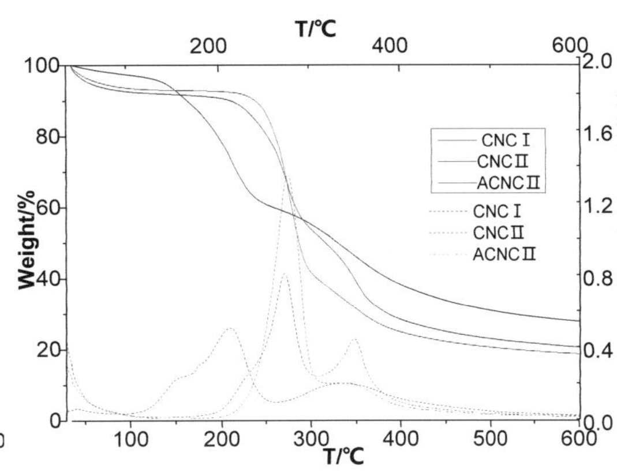

Zwitterion-Functionalized PEI/PVA Nanofibers Embedded in a Microfluidic Chip for Efficient Capture and Intact Release of Circulating Tumor Cells

Yunchao Xiao, Mengyuan Wang, Xiangyang Shi*

State Key Laboratory for Modification of Chemical Fibers and Polymer Materials, College of Chemistry, Chemical Engineering and Biotechnology, Donghua University, Shanghai, 201620, China

E-mail: xshi@dhu.edu.cn

Detection of circulating tumor cells (CTCs) in peripheral blood is of great significance for the early diagnosis, the prediction of cancer development, and individualized cancer therapy. However, due to the extremely low concentration of CTCs in blood, it is very difficult to enrich and separate CTCs via conventional means. In this study, zwitterion-functionalized electrospun polyethyleneimine (PEI)/polyvinyl alcohol (PVA) nanofibers were fabricated and embedded in a microfluidic chip to achieve efficient capture and the non-destructive release of CTCs. First, electrospun PEI/PVA nanofibers with a diameter of 342 nm were formed and then modified with zwitterion 2-methacryloyloxyethyl phosphorylcholine (MPC) *via* atom transfer radical polymerization and targeting ligand arginine-glycine-aspartate (RGD) peptide *via* a pH sensitive linker of benzoic imine. We show that the formed nanofibers exhibit an excellent antifouling property. After integration with a herringbone microfluidic channel, the microfluidic system displays a cancer cell (A549 cells) capture efficiency of 93.2% and the purity of the captured cancer cells is as high as 67.83% at a flow rate of 4 mL/h. Additionally, intact release of cancer cells was achieved with a release efficiency of 92.6% after treated with PBS (pH = 6.8) for 30 min. Furthermore, we prove that CTCs from patients suffering a middle- or late-stage lung cancer are able to be captured and the average number of CTCs per millilitre of blood is 5. Our results suggest that the developed microfluidic system integrated with functionalized nanofibers may be used for efficient capture of CTCs for clinical diagnosis applications.

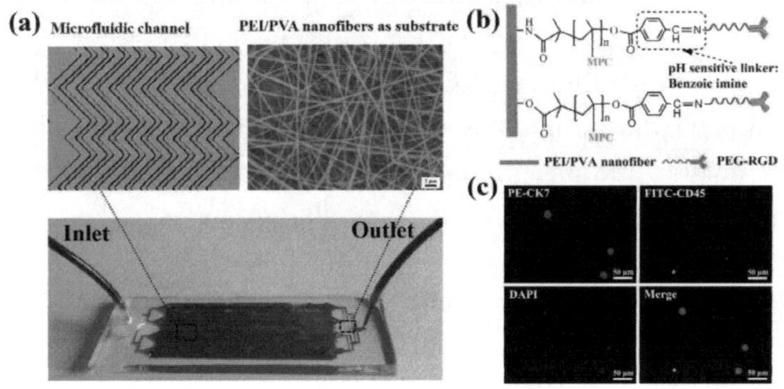

Figure 1. (a) Illustration of the microfluidic device; (b) Representation of MPC- and RGD-modified PEI/PVA nanofibers; (c) Fluorescence images of CTCs from lung cancer patient sample captured using the microfluidic device. The CTCs were identified by predefined criteria (DAPI positive, blue; CD45 negative, green; and cytokeratin positive, red). The CD45 positive cells are white blood cells.

References

[1] Sun N, Liu M, Wang J, Wang Z, Li X, Jiang B, Pei R, *Small*, **2016**, 12, 5090-5097.
[2] Zhao Y, Fan Z, Shen M, Shi X, *Adv. Mater. Interfaces*, **2015**, 2, 1500256.
[3] Chang Y, Shih Y, Lai C, Kung H, Jiang S. *Adv. Funct. Mater.*, **2013**, 23, 1100-1110.
[4] Xu G, Tan Y, Xu T, Yin D, Wang M, Shen M, Chen X, Shi X, Zhu X, *Biomater. Sci.*, **2017**, 5, 752-761.

Transmission Electron Microscopy of Low-Dimensional Nanomaterials

Fangfang Xu

The State Key Lab of High Performance Ceramics and Superfine Microstructure, Shanghai Institute of Ceramics, Chinese Academy of Sciences, Shanghai 200050, China

E-mail: ffxu@mail.sic.ac.cn

Structures and structural characteristics of low-dimensional nanomaterials could be determined by transmission electron microscopy (TEM), some of which could hardly be revealed by other characterization techniques such as X-ray diffractometry or optical spectrometry. Here, two TEM characterization examples are introduced including growth mechanism of one-dimensional (1D) nanomaterials in a vapor-phase (VP) synthesis, and structures of a specific tubular nanomaterials with Archimedes spiral geometry. One-dimensional growth mechanism in a VP process where no catalysts or templates were used has been argued for many years. In our study, we observed polyhedral growth unit by high resolution TEM. Through critical characterizations in TEM including growth direction and surface structure, we finally propose a 1D growth mechanism involving pseudoscrew packing of polyhedral growth units. The proposed mechanism well interprets the growth characteristics of Si_3N_4 and Si nanowires in a VP process. Hexagonal boron nitride (BN) has similar honeycomb structures to graphite. Therefore, the BN filament can bend to form tubular structures like carbon nanotubes. Here, we observed a specific tubular structures of BN, which appears conical helix or Archimedes spiral geometry. By combining TEM observation and crystallographic consideration, we finally built complete structural models of conical helix by taking into account of many structural parameters including disclination angle, apex angle, rotation center, coincidence site lattice and packing patterns. Such specific tubular structure presents special deformation behavior involving both elastic and plastic character, which were observed in in-situ TEM.

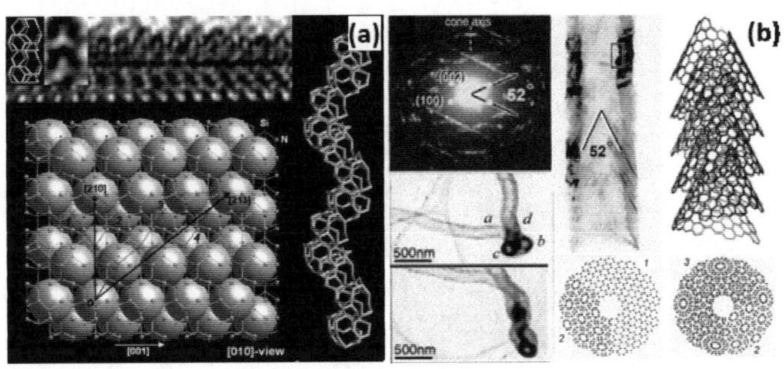

Figure 1. (a) HREM of Si_3N_4 nanowire and structural models for 1D growth; (b) Structures and deformation of BN tubular conical helix.

References

[1] F. F. Xu, L. L. Zhang, X. F. Du and Y. C. Zhu, *Cryst. Growth Des.*, **2008**, 8, 2574-2580.

[2] F. F. Xu, Y. Bando, R. Z. Ma, D. Golberg, Y. B. Li and M. Mitome, *J. Am. Chem. Soc.*, **2003**, 125, 8032-8038.

[3] F. F. Xu and Y. Bando, *Acta Crystallogr. A*, **2003**, 59, 168-171.

Molecular Mobility and Physical Aging of Superglassy Polymer Membranes as Revealed by Dielectric Spectroscopy

Huejie Yin, Nora Konnertz, Martin Böhning, and Andreas Schönhals

Bundesanstalt für Materialforschung und –prüfung, Unter den Eichen 87, 12205 Berlin, Germany

E-mail: huajie.yin@bam.de

Polymeric membranes represent a cost- and energy efficient solution for gas separation. Recently superglassy polymers with high free volume outperform many conventional dense polymers in terms of gas permeability and selectivity. However, such materials with a high fractional free volume (FFV) are prone to pronounced physical aging. The initial microporous structures approach a denser state via local chain rearrangements which results in a dramatic reduction in the gas permeability. For the first time, dielectric spectroscopy with state-of-the-art high-resolution analyzers was employed to investigate the molecular mobility and physical aging of two representative groups of superglassy polymers: PIMs (PIM-1 & PIM-EA-TB)[1-3] and Si-containing polynobornenes (PTCNSi1 & PTCNSi2)[4-6]. The dielectric behavior of the solution-cast polymeric films was measured by isothermal frequency scans during the different heating cycles in a broad temperature range. Structural relaxation of the films was observed during the measurements. Multiple dielectric processes following Arrhenius behavior were observed for all the investigated polymers. Moreover, they all showed conductivity in the glassy state. A 3D plot of the dielectric results of PTCNSi2g is given in Figure 1, showing both β-, β^*-process and conductivity.[6] The significant increase in the conductivity with increasing temperature especially for PIMs is explained in terms of the formation of local intermolecular agglomerated structures due to interaction of π-electrons in aromatic moieties of the polymer backbone.

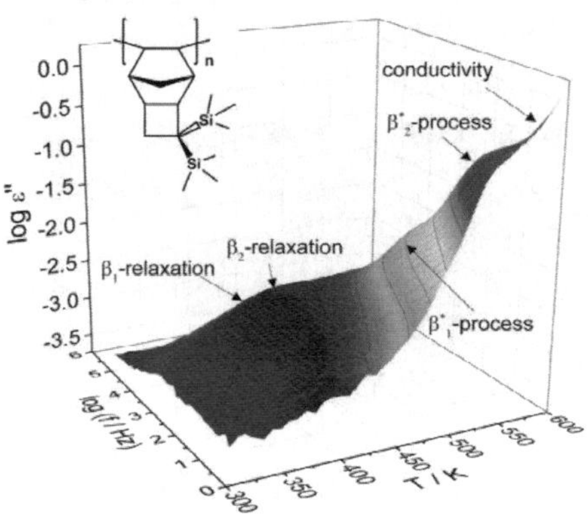

Figure 1. 3D plot of the dielectric results for PTCNSi2g. Inset: Chemical structure of PTCNSi2g.[6]

References

[1] P. M. Budd et al., *Chem. Commun.*, **2004,** 0, 230-231.

[2] N. Konnertz et al., *ACS Macro Lett.,* **2016,** 5, 528-532.

[3] M. Carta et al., *Science,* **2013,** 339, 303-307.

[4] M. Gringolts et al., *Macromolecules,* **2010,** 43, 7165-7172.

[5] P. P. Chapala et al., *Macromolecules,* **2015,** 48, 8055-8061.

[6] H. Yin et al., *ACS Macro Lett.,* **2017,** 6, 813−818.

Improvement of Dyeability of Aromatic co-Polysulfonamide Fibers Using γ-ray Irradiation Induced Graft Polymerization

Jinchao Yu[1], Minglei Wang[2], Jiangtao Hu[2], Shenghui Chen[3], Guozhong Wu[2], Yumei Zhang[1], Huaping Wang[1]

[1]*State Key Laboratory for Modification of Chemical Fibers and Polymer Materials, College of Materials Science and Engineering, Donghua University, Shanghai, 201620, China;*

[2]*Shanghai Institute of Applied Physics, Chinese Academy of Sciences, Shanghai 201800, China;*

[3] *Shanghai Tanlon Fiber Co., Ltd, Shanghai, 201419, China.*

E-mail: yujinchaoqd@163.com

Aromatic co-polyfulfonamide (co-PSA) fibers exhibit excellent flame and heat resistance. However, they are impossible to be dyed with traditional dyeing methods because of their high glass transition temperautres (>350 °C), which limit their use as fire protection clothing and protective uniforms. To improve the dyeability of aromatic co-polyfulfonamide fibers, the graft polymerization of acrylic acid usingγ-ray irradiation was carried out with the aim of improving the dyeability with cationic dyes. It can be found that the grafted co-PSA fibers have better dyeing ability under normal pressure compared to the non-grafted co-PSA fibers. Meanwhile, the tenacity of co-PSA fibers has a little decreased after graft polymerization, while the flame retardancy is not affected. In addition, the hygroscopicity of the fibers is improved obviously. This work will provide some useful information for developing the new dyeing technology of inherent flame retardant heat-resistant fibers.

Key words: Aromatic co-polyfulfonamide, graft polymerization, dyeing properties, microstructure

Fractral Carbon Nanotube Fibers with Mesoporous Crystalline Structure

Hangbo Yue[1,2], V. Reguero[2], Juan J. Vilatela[2]

[1]*School of Chemical Engineering & Light Industry, Guangdong University of Technology, Guangzhou 510006;*

[2]*IMDEA Materials Institute, Madrid 28906, Spain*

E-mail: juanjose.vilatela@imdea.org

Macroscopic fibres of carbon nanotubes are hierarchical structures combining long building blocks preferentially oriented along the fibre axis and a large porosity arising from the imperfect packing of bundles. Synchrotron small-angle X-ray scattering measurements show that such structure is a surface fractal with fractal dimension of 2.5 for MWCNT fibres and 2.8 for SWCNT fibres. N_2 adsorption measurements give similar values of 2.54 and 2.50, respectively. The fractal dimension and deviation from Porod's law are related to density fluctuations associated with the wide distribution of separations between CNTs. These fluctuations are also evident as diffuse wide-angle X-ray scattering from CNTs at distances above turbostratic separation. The structure of CNT fibres produced at different draw ratios is compared in terms of degree of orientation and characteristic lengths parallel and perpendicular to the fibre. Drawing not only increases alignment but also the fraction of graphitic planes forming coherent domains capable of taking part in stress transfer by shear; thus increasing both tensile modulus and strength.

References

[1] Yue H, Reguero V, Senokos E, Monreal-Bernal A, Mas B, Fernández-Blázquez JP, Marcilla R, Vilatela JJ. Fractal carbon nanotube fibers with mesoporous crystalline structure[J]. *Carbon*, **2017**, 122: 47-53

[2] Yue H: Polymer interaction with macroscopic carbon nanotube fibres and fabrication of nanostructured composites[D]. Technical University of Madrid; **2015.**

Biodegradable PLA Non-woven Fabric with Enhanced Functionalization of Controllable Wettability for Efficient Water Purification

Jiangwei Shi, Lei Zhang, Jincui Gu, Jiawei Zhang, Tao Chen

Division of Polymer and Composite Materials, Ningbo Institute of Material Technology and Engineering, Chinese Academy of Science, Ningbo, 315201, China

E-mail: zhangjiawei@nimte.ac.cn

The increasing industrial discharge of organic pollutants has caused tremendous threats to ecosystem, although many efforts have been devoted to construct super-wetting separation materials for waste water treatment, the post-processing of these used materials is still a daunting task because of their non-degradable essence. Moreover, water pollution has presented particularly serious due to the presence of variety of soluble organic pollutants. Therefore, it is highly desirable to develop new functional surfaces possessing with versatility, biodegradability and easy-controllability to realize the effective treatment of sewage sludge and organic waste water. Herein, we have developed a convenient strategy to fabricate cost-effective super-wetting PLA non-woven fabrics for the effective separation of oil/water mixtures. The wettability of such fabrics can be regulated by adjusting the relative proportion of chemical component. The resulting superwetting PLA non-woven fabrics exhibit high absorption capacity and high selectivity when they are employed as absorptive materials and oil/water separation materials. In addition, the superhydrophilic PLA non-woven fabric possesses excellent photo-catalysis degradation for variety of soluble organic pollutants property under UV irradiation. With the versatility and biodegradability, these advanced PLA non-woven fabrics may provide effective solutions to oily water treatment.

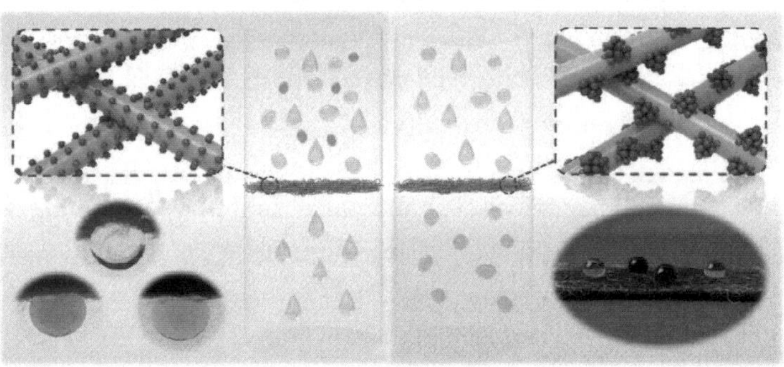

Figure 1. PLA non-woven fabrics with controllable wettability are applied in oil/water separation, respectively.

References

[1] Jincui Gu, Peng Xiao, Peng Chen, Lei Zhang,* Hanlin Wang, Liwei Dai, Liping Song, Youju Huang, Jiawei Zhang,* Tao Chen*, *ACS Appl. Mater. Interfaces,* **2017**, 9, 5968-5973.

[2] Jincui Gu, Peng Xiao, Youju Huang, Jiawei Zhang*, Tao Chen*. *J. Mater. Chem. A*, **2015**, 3, 4124-4128.

[3] Jincui Gu, Peng Xiao, Jing Chen, Jiawei Zhang,* Youju Huang, Tao Chen*. *ACS Appl. Mater. Interfaces*, **2014**, 6, 16204-16209.

[4] Jincui Gu, Peng Xiao, Jing Chen, Fu Liu, Youju Huang, Gaiye Li, Jiawei Zhang*, Tao Chen*. *J. Mater. Chem. A*, **2014**, 2, 15268-15272.

4.Abstracts of Poster Presentation

Preparation and Electromagnetic Wave Absorption Properties of Cobalt/Carbon Composite Nanofiber Membranes

Ibrahim Abdalla[1], Li Zhao-ling[2,3*], Yu Jian-yong[3], Ding Bin[1,2,3*]

[1]State Key Laboratory for Modification of Chemical Fibers and Polymer Materials, College of Materials Science & Engineering, Donghua University, Shanghai 201620, China;

[2]Key Laboratory of Textile Science and Technology, Ministry of Education, College of Textiles, Donghua University, Shanghai 201620, China;

[3]Innovation Center for Textile Science and Technology, Donghua University, Shanghai 200051, China

E-mail: zli@dhu.edu.cn (Prof. Zhaoling Li), binding@dhu.edu.cn (Prof. Bin Ding)

Currently, with the speedy improvement aerospace technology, microwave absorption materials have gained great attention due to concerns about the negative effect of electromagnetic interference on different fields. Electromagnetic wave absorption materials are significantly wanted that can well absorb electromagnetic waves and convert it into thermal energy. This kind of products is commonly categorized to dielectric and magnetic materials. Improvement of these materials is environment friendly microwave absorbing performance releasing from some apparatus especially into the workplace.[1] Composite of carbon nanofiber and cobalt oxide nanoparticles have been successfully synthesized. A reflection loss exceeding -10 dB was obtained in the frequency range of 4.5-18 GHz for absorber material thickness 1-5 mm. The reflection loss of the sample reaches -16.5 dB at 5.3 GHz with thickness 5 mm. Results illustrate that the composite sample is favorable for high-performance due to a good electromagnetic wave absorption properties. Cobalt oxide loaded into the porous carbon nanofibers are important refer to their several suitable applications. It is well-known that electromagnetic wave absorbing properties of absorption materials are highly associated with its permittivity and permeability, where the real parts of permittivity and permeability denote the storage ability of electric and magnetic energy.[2] As a significant ferromagnetic material, cobalt oxides nanoparticle owns high saturation magnetization, making it a hopeful broadband electromagnetic wave absorber. The imaginary parts of permittivity and permeability characterize the loss capability of electric and magnetic energy.[3,4] The electromagnetic wave absorption ability depends on some factors such as nature, shape, and size of an absorber. Consequently, it is highly significant and favorable to improve new approaches for electromagnetic wave absorption composite materials with low cost and enhanced properties. As enhanced electromagnetic wave absorption materials, they should not only have lower reflection loss completed a broad frequency range, but also should be as thin as possible for the purpose of gaining the electromagnetic wave absorber based on composite materials.

In this work, we presented a new absorber composed of flexible carbon nanofibers and magnetic cobalt oxide nanoparticles by electrospinning. The as-prepared composite showed improved electromagnetic wave absorption properties in frequency range with optimal material layer thickness compared with previously reported composites.

Keywords: Electrospinning, electromagnetic wave, permittivity, permeability, reflection loss.

Materials & methods

Materials:

Polyacrylonitrile polymer, cobalt acetylacetonate powder and dimethylformamide solvent were purchased from Aladdin Chemical Reagents Co., Ltd., China.

Methods:

The composite membranes were fabricated by electrospinning of polyacrylonitrile polymer and cobalt acetylacetonate. This polymer was selected for it is good spinnability in electrospinning. Cobalt acetylacetonate also was selected due to it's a common precursor.

Results & Discussion

The temperature for drying precursor fibers was 80°C for 2 h to remove the solvent, and then the material was stabilized. Subsequently, the obtained stabilized samples were carbonized for 2 h. After cooling down to room temperature, the resulting samples were smooth and flexible, and the carbonized final samples were denoted as CA3-800. The sample was prepared by using 1.5 g of acetyl acetone cobalt in the precursor solution. The sample was prepared by composing paraffin with nanofibers. The sample was added into paraffin in 5 wt.%, 10 wt.% and 20 wt.% to obtain electromagnetic wave absorption test. **Figure 1** displays the schematic diagram showing the synthetic steps of composite sample. It is clear that the precursor solution with cobalt salt was first prepared and then the composite membrane was formed by electrospinning technique. **Figure 2** demonstrations (a) Reflection loss curves of the samples and (b) Table of microwave absorption performance of CA3-800-5, CA3-800-10 samples (d=3 mm) and CA3-800-20 sample (d=3 or 5). The results indicate that the improved reflection loss of the sample was gained when the sample was added into 20 wt.% paraffin compared with 5 wt.% and 10 wt.%, as shown in Figure 2a and Table 2b.

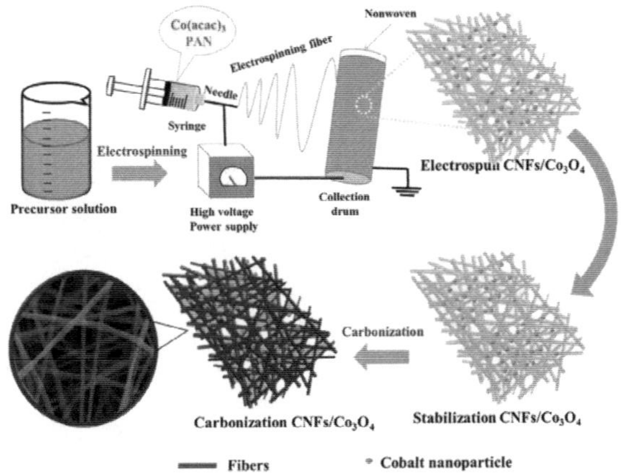

Figure 1. Schematic diagram showing the synthetic steps of composite sample.

(a)

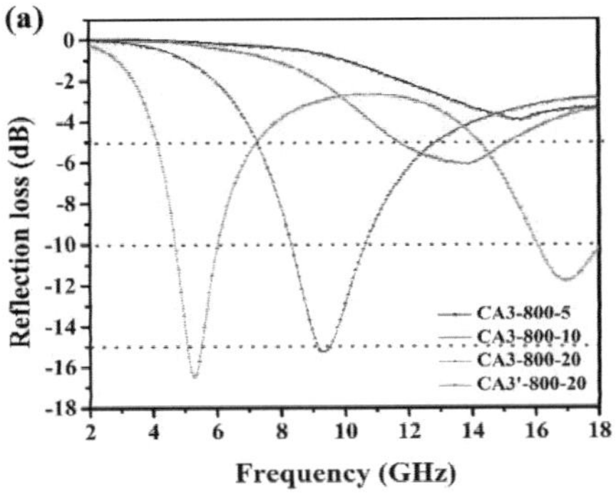

(b)

Sample	Max R_L value (dB)	*d* (mm)
CA3-800-5	3.9	3
CA3-800-10	6.1	3
CA3-800-20	15.3	3
CA3-800-20	16.5	5

Figure 2. (a) Reflection loss curves of the samples and (b) Table of microwave absorption performance of CA3-800-5, CA3-800-10 samples (d=3 mm) and CA3-800-20 sample (d=3 or 5).

Conclusions

In summary, composite of carbon nanofibers and cobalt oxide nanoparticles have been successfully synthesized. A reflection loss exceeding -10 dB was obtained in the frequency range of 4.5-18 GHz for absorber material thickness 1-5 mm. The reflection loss of the sample reaches -16.5 dB at 5.3 GHz with thickness 5 mm. Results illustrate that the composite sample is favorable for high-performance due to a good electromagnetic wave absorption properties. The as-prepared composite showed improved electromagnetic wave absorption properties in frequency range with optimal material layer thickness compared with previously reported composites.

References

[1] X. Liu, Y. Gao, G. Yang, *Nanoscale*, **2016**, 8, 4227.

[2] C. Cui, Y. Du, T. Li, X. Zheng, X. Wang, X. Han, P. Xu, *The Journal of Physical Chemistry B,* **2012**, 116, 9523.

[3] T. Wang, H. Wang, X. Chi, R. Li, J. Wang, *Carbon,* **2014**, 74, 312.

[4] Y. Li, Y. Zhao, X. Lu, Y. Zhu, L. Jiang, *Nano Research,* **2016**, 9, 2034.

Fabrication of Flexible Carbon Nanofiber Composite Electrode Embedded with Porous Silica Nanoparticles for Lithium Ion Batteries

Ahmed A. Aboalhassan[1], Jianhua Yan[2, 3*], Jianyong Yu[3], Bin Ding[1, 2, 3*]

[1]*State Key Laboratory for Modification of Chemical Fibers and Polymer Materials, College of Materials Science & Engineering, Donghua University, Shanghai 201620, China*

[2]*College of Textile, Donghua University, Songjiang District, Shanghai 200051, China*

[3]*Modern Textile Institute, Donghua University, Changning District, Shanghai 200051, China*

E-mail: ch_abdelkawy@hotmail.com

2.Key Laboratory of Textile Science and Technology, Ministry of Education, College of Textiles, Donghua University, Shanghai 201620, China.

3. Innovation Center for Textile Science and Technology, Donghua University, Shanghai 200051, China

1.Introduction

In our modern life, wearable electronics are urgently required for a lot of applications and as a consequence long-lived flexible electrodes of Li-ion batteries (LIBs) become highly necessitated. Lithium-ion batteries (LIBs) have been widely considered as advanced power devices for progressively miscellaneous ranges of applications from automobiles to microchips due to their high energy density, long lifespan [1,2]. Anode materials for next-generation LIBs such as and silica nanoparticles (SiO_2 NPs) have attracted a lot of researchers' interests due to its low cost and easy to be obtained and high theoretical specific capacity of 1965 mA h/g [3]. Nevertheless, SiO_2 is still suffering from a problem of having poor cycling performance, which caused by the huge volumetric change during cycling (like other alloying-type anode materials). Additionally, SiO_2 is suffering from poor electronic conductivity which affects its performance. So that, a lot of researchers have been done to fabricate SiO_2 /Carbon composites in order to accommodate the volumetric changes and to enhance the electronic conductivity but it remains a challenge to fabricate the SiO_2/carbon flexible electrode [4-7]. In this research paper, we intended to fabricate flexible and freestanding porous SiO_2/carbon composite nanofibrous membrane (porous SiO_2/CNFM) by facile encapsulating porous SiO_2 NPs via utilizing the electrospinning technique followed by optimized thermal treatments.

2. Materials and Methodology

2.1 Materials

Polyacrylonitrile (PAN, Mw = 90 000) was supplied by Spectrum Chemicals & Laboratory Products Co., Ltd, USA. Dimethyl formamide (DMF) and Hydrochloric acid (HCl) were obtained from the Shanghai Chemical Reagents Co., Ltd, China. A SiO_2 NPs was commercially purchased from Aladdin Chemistry Co. Ltd, China. Tetraethoxysilane (TEOS) and Cetyltrimethyl ammonium bromide (CTAB) were purchased from Sinopharm Chemical Reagent Co. Ltd. (Shanghai, China). Ultrapure water was obtained from a Heal-Force system for water purification. All chemicals were of analytical grade and were used as received without further purification unless otherwise stated.

2.2 Methodology

The synthesis of porous silica nanoparticles was carried out via the acid-catalyzed hydrolysis and condensation of (TEOS) in water solvent with the presence of (CTAB) as a surfactant. The as-synthesized nanoparticles were washed and dried, then used for calculation at the 550 ℃ for 5 hours. The porous particles after calcination were then loaded on PAN solution in DMF and the carbon nanofibrous membrane was fabricated via electrospinning technique followed by optimized thermal treatments. For comparison commercial SiO_2 NPs were used by the same procedures and parameters. To investigate the electrochemical performance of the fabricated porous SiO_2 NPs / CNFM electrode, coin-type 2025 cells in which our electrode act as a freestanding anode and Li metal counter electrodes and the electrolyte comprising 1M $LiPF_6$ in ethylene carbonate, diethyl carbonate and ethyl methyl carbonate (EC: DEC: EMC = 1: 1: 1 v/v). The cells were assembled within an argon-filled Mikrouna lab glove box. The charge/discharge voltage profiles of the electrode was checked for different cycles at a constant current density of 100 mA/g in the potential range of 0–3 V versus Li/Li^+.

3. Results and Discussion

Morphology and bonding structure considered to be one of the most important factors that affect the mechanical properties of the produced CNFM. Figure 1 (a) shows the SEM images of pure PAN based CNFM (inset the digital photo image of pure CNFM which displays its rigidity); while figure 1 (b) shows the SEM images of embedded commercial SiO_2/CNFM which demonstrates the well interconnected bonded nanofibrous structure incorporated with SiO_2 NPs. Figure 1 (c) shows the SEM of porous SiO_2/CNFM which also displays outstanding flexibility as showed inset of the figure.

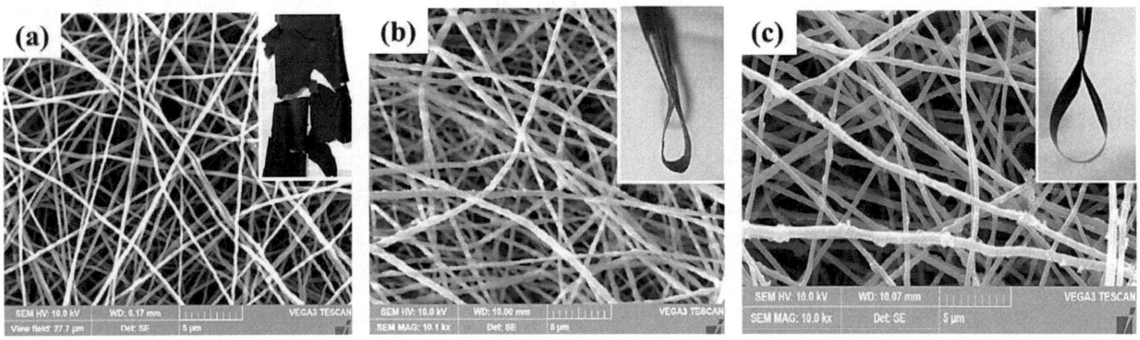

Figure 1 SEM images of (a) pure PAN CNFM, (b) commercial SiO_2/CNFM and (c) porous SiO_2/CNFM; with the photo images of each corresponding sample to display its flexibility.

To determine the exact ratio of incorporated particles within the CNFM thermal gravimetrical analysis (TGA) was employed as shown in figure 2 (a) for pure PAN based CNFM and others loaded with commercial SiO_2 NPs and porous SiO_2 NPs. The crystallographic structure of the obtained porous SiO_2 was analyzed by using an X-ray diffraction technique (XRD), as shown in figure 2 (b) in which there is a weak broadened band at about 23°, revealing the typical amorphous nature of porous SiO_2. Moreover, figure 2 (b) displays XRD of the porous SiO_2 /CNFM and incorporated one.

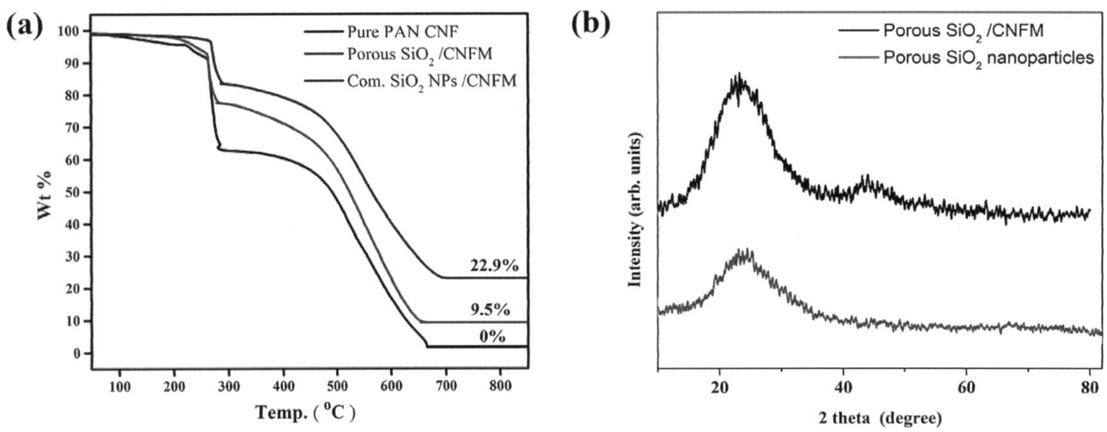

Figure 2 (a) TGA of pure PAN CNF, commercial SiO_2 NPs /CNFM and porous SiO_2 /CNFM; **(b)**XRD of the as-synthesized porous SiO_2 nanoparticles and the embedded porous SiO_2/ CNFM.

Equations (1-3) illustrate the mechanism of SiO_2 as an alloying type anode material for LIBs.

$$SiO_2 + 4Li^+ + 4e^- \longrightarrow Si + 2\,Li_2O \quad (1)$$

$$4\,Li^+ + 2\,SiO_2 + 4e^- \longrightarrow Li_4SiO_4 + Si \quad (2)$$

$$x\,Li^+ + Si + x\,e^- \longleftrightarrow 2\,Li_xSi \quad (3)$$

The importance of porous SiO_2 NPs as the electrode materials of LIBs in shortening the Li^+ diffusion path and accommodating the volume expansion during the lithiation/delithiation process, which was also demonstrated by some reported work with [8,9] . The advantages of carbon nanofibrous structure are that it can effectively buffer the volumetric changes within embedded silica; in addition to it acts as a well-interconnected electronically conductive network which leads to improvement in the cyclic performance.

The combination of porous SiO_2/CNFM possesses excellent mechanical flexibility in addition to the enhanced cyclaibilty as anode for LIBs So that fabricated porous SiO_2/CNFM can be considered as one of the promising flexible, freestanding, long-life and low cost anode for wearable LIBs.

4. Conclusion

In summary, porous SiO_2 NPs were synthesized successfully using a sol–gel approach. The as-obtained porous SiO_2 were then simply loaded on PAN solution and then electrospun CNFM was fabricated. The as-prepared porous SiO_2/CNFM act as anode for LIBs delivers a stable specific capacity at a current density of 100 mA/g after cycles which is better electrochemical performances than commercial SiO_2 NPs loaded on CNFM. The unique porous SiO_2/CNFM composite assists in the improvement of the electronic conductivity of SiO_2 due to carbonous overlapping with it and a consequence enhanced capacity retention.

References

[1] J. B. Goodenough and K.-S. Park, *J. Am. Chem. Soc.,* **2013**, *135*, 1167–1176.

[2] V. Etacheri, R. Marom, R. Elazari, G. Salitra and D. Aurbach, *Energy Environ. Sci.,* **2011**, *4*, 3243–3262.

[3] B. Gao, S. Sinha, L. Fleming and O. Zhou, *Adv. Mater.,* **2001**,13, 816–819.

[4] X. Cao, X. Chuan, S. Li, D. Huang and G. Cao, *Part. Part. Syst. Charact.,* **2016**, 33, 110–117.

[5] B. Guo, J. Shu, Z. Wang, H. Yang, L. Shi, Y. Liu and L. Chen, *Electrochem. Commun.,* **2008**, 10, 1876–1878.

[6] P. Lv, H. Zhao, J. Wang, X. Liu, T. Zhang and Q. Xia, *J. Power Sources,* **2013**, *237*, 291–294.

[7] Y. Yao, J. Zhang, L. Xue, T. Huang and A. Yu, *J. Power Sources,* **2011**, *196*, 10240–10243.

[8] N. Yan, F. Wang, H. Zhong, Y. Li, Y. Wang, L. Hu and Q. Chen, *Sci. Rep.,* **2013**, *3*, 01568.

[9] Z. Favors, W. Wang, H. H. Bay, A. George, M. Ozkan and C. S. Ozkan, *Sci. Rep.,* **2014**, *4*, 04605.

Electrochemical Growth of Polyaniline on Active Carbon / Carbon Fiber Cloth with Enhanced Electrochemical Performance for Flexible Supercapacitors

Yuerong Ba[a,b], Yaomin Zhao[a], Dong Liu[a], Wei Pan[a,b*]

[a]*School of Materials and Chemical Engineering, Zhongyuan University of Technology, Zhengzhou, 451191, China;*

[b]*Collaborative Innovation Center of Textile and garment industry, Henan Province, Zhengzhou, 451191, China*

E-mail: doctorpan0152@163.com

With the fast development of flexible electronics such as roll-up displays, smart mobile devices, and implantable bio-sensors, etc., the exploration of flexible energy storage devices as energy suppliers has drawn much research attention recently[1-3]. Among various energy storage devices, flexible supercapacitors can be charged/discharged quickly and possess ultra-long lifespan, making them an important kind of flexible energy storage devices[4-5]. A freestanding binder-free electrode with favorable mechanical strength and high capacitance is a vital factor for flexible supercapacitors. To achieve such electrodes, one needs to have proper electrode materials and current collectors[6-8].

Herein, the facile fabrication strategy towards flexible supercapacitor electrodes with polyaniline (PANI) loading on active carbon (AC) coated carbon fiber cloth (CFC) is developed via a simple carbonization process and the following electrodeposition of PANI. And the AC is obtained by carbonization of chitosan and polyethylene glycol (PEG). In this process, due to the thermal degradation of PEG, the samples possess ample pore structures. Such a porous structure renders the electrolyte able to penetrate into the bulk particles to form three-dimensional channels for ions' transportation[9]. Directly growing on conductive substrates results in lower contact resistance between the electroactive materials and the current collector which provides a superb highway for fast electron transfer and electrolyte ion diffusion[10]. For comparison, the AC/CFC and PANI/CFC electrodes were prepared with the same method. The supercapacitive behavior of the electrodes was investigated using cyclic voltammetry, galvanostatic charge–discharge and electrochemical impedance spectroscopy techniques. The result indicated that the capacitance and rate performance of the electrodes are improved by the addition of active carbon. Under the current density of 0.5 mA/cm^2, the PANI/AC/CFC electrode exhibited a specific capacitance of 186 mF/cm^2 and retains 89% when current density is increased 10-fold to 5mA/cm^2. Two same PANI/AC/CFC electrodes were used to assemble a symmetric supercapacitor device with PVA/H_3PO_4 hydrogel as separator and electrolyte. The flexible supercapacitor showed a specific capacitance of 35 mF/cm^2 at the current density of 0.2 mA/cm^2 and 24 mF/cm^2 at the current density of 2 mA/cm^2. And the supercapacitor has excellent flexibility with almost no capacity fading under different bending angles. In addition, the assembled symmetric supercapacitor demonstrated a high stability retained almost 100% of the initial capacitance after 4000 cycles of charge-discharge at a current density of 0.25 mA/cm^2. This work demonstrates that the PANI/AC/CFC is promising electrode for high-performance supercapacitors.

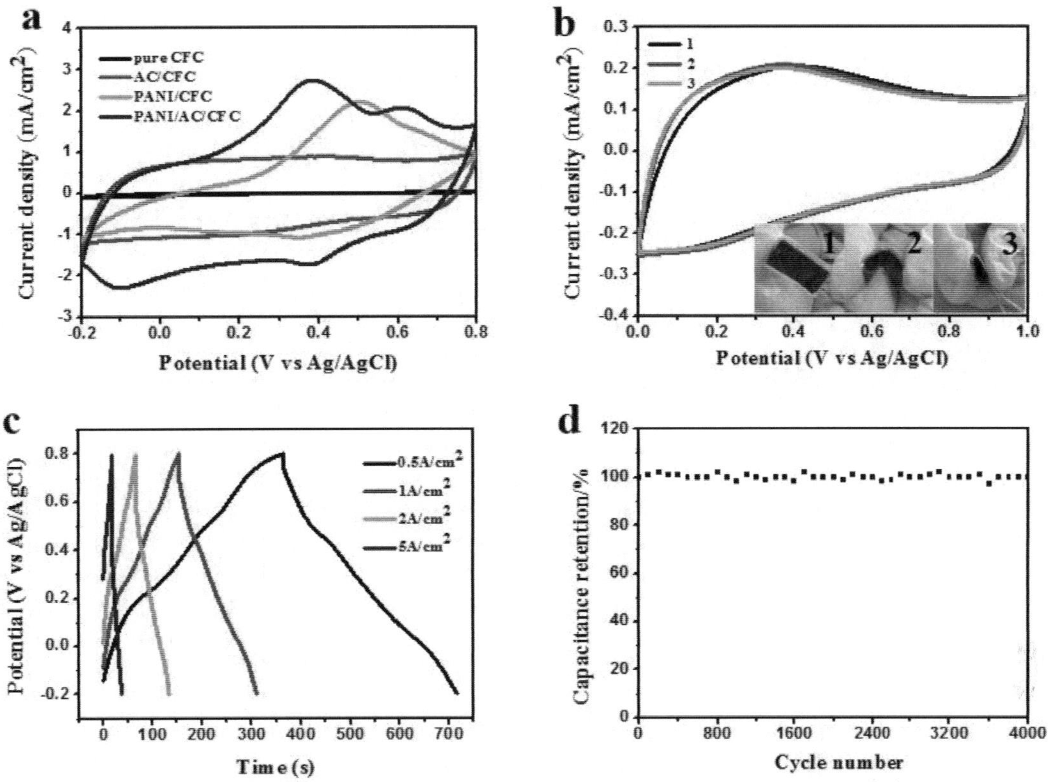

Figure 1. (a) CV curves of pure CFC, AC/CFC, PANI/CFC and PANI/AC/CFC at a constant scan rate of 10mV/s; (b) CV curves of the PANI/AC/CFC symmetric supercapacitor under different bending angles at 10 mV/s; (c) GCD curves of PANI/AC/CFC at different current densities; (d) cycling stability of the PANI/AC/CFC symmetric supercapacitor at current density of 0.25mA/cm^2.

Keywords: polyaniline, active carbon, carbon fiber cloth, flexible supercapacitors

References

[1] Peng Zhang, Zhimeng Liu, Yuping Liu, et al, *Electrochimica Acta*, **2015**, 184, 1-7.

[2] M.A. Bavio, G.G. Acosta, T. Kessler, et al, *Energy*, **2017**, 130, 22-28.

[3] Li Wang, Han Yang, Guangxing Pan, et al, *Electrochimica Acta*, **2017**, 240, 16-23.

[4] Liubing Dong, Gemeng Liang, Chengjun xu, et al, *Nano Energy*, **2017**, 34, 242-248.

[5] Zhichao Zhao, Yingbing Xie, *Journal of Power Sources*, **2017**,337, 54-64.

[6] Jing Cao, Tao Huang, Ruli Liu, et al, *Electrochimica Acta*, **2017**, 230, 265-270.

[7] Xiuling Zhang, Li Ma, Mengyu Gan, et al, *Journal of Power Sources*, **2017**, 340, 22-31.

[8] Xinping He, Bangning Sun, Bo Gao, et al, *Journal of Electroanalytical Chemistry*, **2017**, 792, 88-94.

[9] Leiming Guo, Xiaoting Wang, Yong Wang, *Chemical Engineering Journal*, **2017**, 313, 1295-1301

[10] Jie Yang, Yuliang Yuan, Weicheng Wang, et al, *Journal of Power Sources*, **2017**, 340, 6-13.

Antibacterial Silk Fibroin Scaffolds Modified with 2-Hydroxypropyl Trimethyl Ammonium Chloride Chitosan

Kai Chen, Huili Shao, Xuechao Hu, Yaopeng Zhang*

State Key Laboratory for Modification of Chemical Fibers and Polymer Materials, College of Materials Science and Engineering, Donghua University, Shanghai, 201620, China

E-mail: zyp@dhu.edu.cn

Silk fibroin (SF) has been extensively applied in biological and medical fields due to its excellent biocompatibility, biodegradability and high drug loading capacity. However, the bacterial infection is an urgent issue in the operation process of transplanting SF scaffolds. As a derivative of chitosan, 2-hydroxypropyl trimethyl ammonium chloride chitosan (HACC) possesses significant antibacterial activity and biological compatibility. Compared with pure chitosan, HACC has wider application due to its positive charge and solubility in water over a broad range of pH values. The aim of this present work is to develop antibacterial SF scaffolds, which combine the advantages of both SF and HACC.

The composite nanofibrous SF/HACC scaffolds were prepared by electrospinning method. With the increase of HACC content, the viscosity, the surface tension and the electrical conductivity of SF/HACC composite solution increase obviously. To prepare nanofibrous SF/HACC scaffolds, the maximum mass ratio of HACC/SF in the spinning dope is 4% (w/w). Other than round pure SF nanofibers by electrospining, the shape of the SF/HACC nanofibers is flat. The diameter distribution of the SF/HACC fibers is quite narrow and its average diameter is 967±224 nm when the mass ratio of HACC/SF reaches 1% (w/w). When the range of the HACC content increases from 0% to 3%, the average diameter of the scaffolds tends to decrease. The SF/HACC scaffolds possess high porosity and enhanced breaking energy, breaking strength and elongation at break than SF scaffolds. The cell experiment indicates that the SF/HACC scaffolds have better biocompatibility in vitro than SF scaffolds. Compared with pure SF scaffold, the SF/HACC scaffold shows improved antibacterial effects against S. aureus (Gram-positive) and E. coli (Gram-negative).

The obtained SF/HACC nanofibrous scaffolds with outstanding biological compatibility and antibacterial performance may have promising application potentials in tissue engineering.

Study on the Relationship between Creep Behavior and Structure of Polyester Industrial Yarns

Kang Chen, Jinchao Yu, Yumei Zhang, Huaping Wang

State Key Laboratory for Modification of Chemical Fibers and Polymer Materials, College of Materials Science and Engineering, Donghua University, Shanghai, 201620, China

E-mail: 979131404@qq.com

Polyester (polyethylene terephthalate) industrial yarns, as a cost-effective fiber, can be found their extensive range of applications in the fields of automobile and security protection due to their excellent properties and particular structure [1]. The knowledge of the basic structure and properties of PET yarns had already revealed [2]. But the creep of PET yarns has not been disclosed and the study on the relationship between structure and properties of PET fibers is also insufficient [3], which bring great blindness for practical use. In order to investigate the relationship between the creep behavior and structure of PET industrial yarns, the typical samples were selected to evaluate the creep properties at different temperatures and applied loads. And the creep rate was used to illustrate the deformation rate as a function of time. The structure of fibers was characterized by XRD and birefringence. The results showed that a great initial creep deformation of these fibers occurred when the load applied, however, the deformation rate with load during the creep process showed different tendency. For HT and HMLS yarns, the creep rate increased continually with the increase of applied load, while that of LS and SLS decreased first and then increased. Based on the structural information, it can be found that the different creep behavior is mainly related to the amorphous structure. Compared with LS and SLS yarns, the HT and HMLS yarns have higher amorphous orientation and less molecular entanglement, resulting in the chain slippage at higher load and the creep rate increase. In contrast, the LS and SLS yarns have disorder amorphous structure and serious molecular entanglement. The disordered molecular aggregates was stretched first at a certain rang of load, which can weaken the chain slippage and decease the creep rate. At higher load, the molecule disentanglement occurred and accelerated the chain slippage, leading to the increase of creep rate.

References

[1] Rim P B, Nelson C J. Properties of PET fibers with high modulus and low shrinkage (HMLS). I. Yarn properties and morphology [J]. *Journal of Applied Polymer Science*, **1991**, 42(7):1807-1813.

[2] Samui B K, Prakasan M P, Ramesh C,. Structure-property relationship of different types of polyester industrial yarns [J]. *Journal of the Textile Institute*, **2013**, 104(1): 35-45.

[3] Samui B K, Dasgupta S, Mukhopadhyay. Studies on the static and dynamic properties of different types of polyester industrial yarns [J]. *Journal of the Textile Institute*, **2015**, 1-10.

Physically Chemically Hybrid Crosslinking: An Efficient Way to Construct Strong, Tough, While Soft, Biodegradable Elastomers

Shuo Chen, **Lijie Sun, Zenghe Liu, Ziying Lv, Luzhi Zhang, Yifan Guo, Zhengwei You***

State Key Laboratory for Modification of Chemical Fibers and Polymer Materials, College of Materials Science and Engineering, Donghua University, 2999 North Renmin Road, Shanghai 201620, P. R. China

E-mail: zyou@dhu.edu.cn

Biodegradable and biocompatible elastomers (bioelastomers) could resemble the mechanical properties of extracellular matrix and soft tissues, thus, are very useful for many biomedical applications. Despite the significant advances, tough bioelastomers with fatigue resistance, self-recovery, shape-memory properties, good biocompatible and biodegradability has remained elusive. Here, we report a new design concept of bioelastomers with aforementioned advantages via bioinspired sacrificial bonds. We graft 2-ureido-4[1H]-pyrimidinones (UPy) units with strong self-complementary quadruple hydrogen bonds into covalently crosslinked poly(sebacoyl diglyceride) (PSeD) networks,[1, 2] refined version of a widely used bioelastomer poly(glycerol sebacate) (PGS). The covalent crosslinks maintain the structural integrity of resultant PSeD-U bioelastomers compliant to dynamic mechanical environment. The reversible hydrogen bonds perform several mechanical functions simultaneously: toughening the bioelasomters and enhancing the tearing resistance (tearing energy of 3800 J/m^2) via energy dissipation during bond rupture, and enabling the fatigue resistance and nearly 100% self-recovery during cyclic deformations by bond re-formation. PSeD-U bioelastomers can be tuned over a wider range by changing covalent crosslinking density and the grafting ratio of UPy unit. The resultant bioelastomer could hold 2500 × its own weight. The toughness increased to 18 times higher than the one of PSeD bioelastomer, while the moduli still remained relatively low close to those of soft tissues. The synergetic chemical and physical (intramolecular and intermolecular) interactions might contribute to these unique mechanical properties. In addition, the resultant bioelastomers have good biodegradability and biocompatibility. The free hydroxyl groups also enabled facile functionalization to tailor properties for specific applications.

References

[1] You Z, Cao H, Gao J, Shin PH, Day BW, Wang Y. *Biomaterials*. **2010,** *31*, 3129-38.

[2] You Z, Wang Y. *Advanced Functional Materials*. **2012**, *22*, 2812-20.

Surface Modification of PPTA Fibers by UV Irradiation-Accelerated Deposition of Catechol/Polyamine and Secondary Functionalization for High-Performance Polymer Composites

Sixian Chen, Wencai Wang, Rina Sa, Ming Tian, Liqun Zhang

State Key Laboratory of Organic-Inorganic Composites, College of Materials Science and Engineering, Beijing University of Chemical Technology, Beijing, 100029, PR China

E-mail:13011238088@126.com

A highly efficient mussel-like inspired modification of aramid fibers by UV-accelerated catechol/polyamine deposition and secondary functionalization is an efficient method to improve the interfacial adhesion between aramid fiber and rubber matrix. With UV irradiation, the process of oxidation self-polymerization and the deposition modification rate of catechol and tetraethylenepentamine (TEPA) on para-aramid fiber (PPTA) surface were accelerated. After poly(catechol/polyamine) (PCPA) was deposited on the PPTA fiber surface, the fibers coated with PCPA were further grafted with ethylene glycol diglycidyl ether (EGDE) to introduce epoxy groups onto aramid fibers surface. This remarkably increase in adhesion property was mainly attributed to the introduction of epoxy groups on PPTA fibers surface, which formed efficient chemical interaction between fibers and rubber matrix. The reaction progress and deposition rate of catechol / TEPA were characterized by UV-Vis measurements. It was proved that the pre-deposition time of PCPA on the surface of aramid fiber was shortened to 1 h(3h less than dopamine pre-deposition time). The chemical structural changes of the fibers surface before and after PCPA pre-deposition and the secondary functionalization were investigated by ATR-FTIR measurements. Thermogravimetric analysis (TGA) was carried out to determine the grafting degree (GD) for the PPTA fibers, indicating that the grafting degree of PCPA and EGDE was ~3.7% and ~14.8% respectively. The effects of EDGE concentration and grafting time on the interfacial adhesion strength between aramid fiber and rubber matrix were investigated by H-pull test method. The results showed that the optimal EGDE concentration and grafting time for EGDE grafting was 1.0 vol% and 4 h, respectively and that compared with the unmodified PPTA fiber, the interfacial adhesion force between modified fiber and rubber matrix was increased by 85.6%, and even better than that of dopamine method in our previous study (a 67.5% improvement). The microscopic structure and surface chemical composition of the fiber surface were characterized respectively by scanning electron microscopy (SEM) and X-ray photoelectron spectroscopy (XPS), suggesting that poly(catechol/polyamine) (PCPA) can be deposited evenly on the surface of the PPTA fiber, and the epoxy groups are introduced on the surface of the PCPA-coated fiber by graft modification of EGDE. Compared with the method based on dopamine chemistry, this method has advantages of lower cost (less than 1% of the price of dopamine) and shorter deposition time. This study provides a novelty efficient strategy for fibers surface modification, which has promising application in rubber industry of high performance fiber reinforced composites.

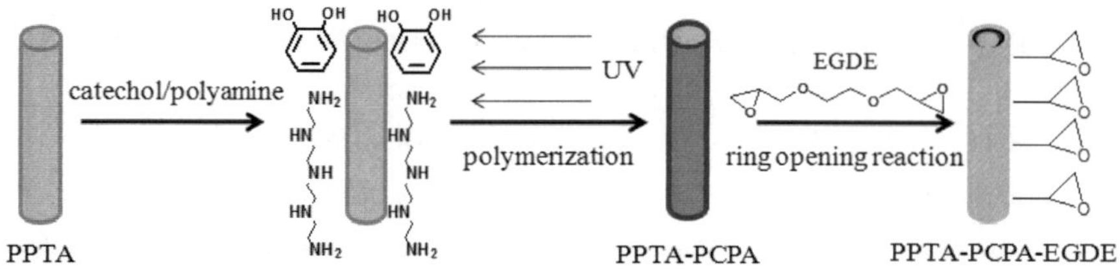

Figure 1. Schematic description of surface modification procedure of PPTA fibers.

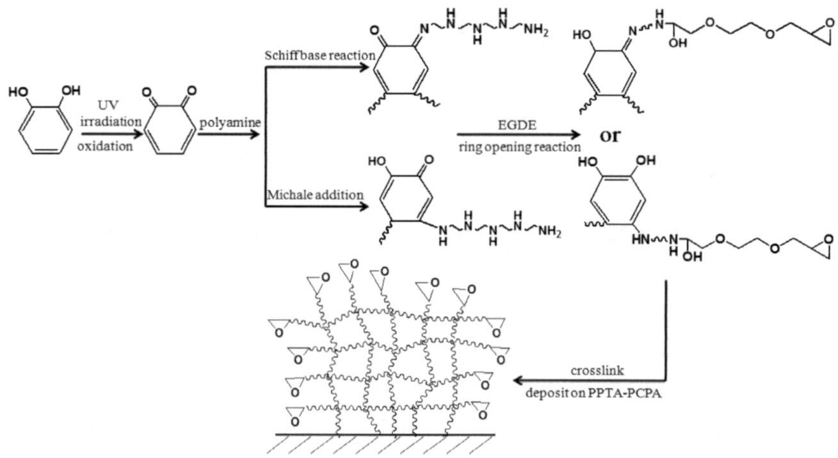

Figure 2. Suggested reaction mechanism for PPTA-PCPA-EGDE.

References

[1] Lei Wang, Yongxiang Shi, Sixian Chen, Wencai Wang, Ming Tian, Nanying Ning, Liqun Zhang, *Chemical Engineering Journal*, **2017**, 314, 583-593.

[2] Lei Wang, Yongxiang Shi, Rina Sa, Nanying Ning, Wencai Wang, Ming Tian, and Liqun Zhang, *Industrial & Engineering Chemistry Research,* **2016**, *55*, 12547-12556.

Monodisperse Silver Nanoparticles loaded Thermo-responsive Nanofibrous membrane as a Free-standing SERS Substrate with Thermo-enhanced Efficiency

Siyuan Chen, **Xiaoyun Liu**, **Liusheng Zha**

State Key Laboratory for Modification of Chemical Fiber and Polymer Materials& College of Materials Science and Engineering, Analysis and Measurement Center, Donghua University, Shanghai 201620, P. R. China

E-mail: lszha@dhu.edu.cn

To develop a smart free-standing SERS substrate with thermo-enhanced efficiency, monodisperse silver nanoparticles (AgNPs) with the size of high surface-enhanced Raman scattering (SERS) efficiency were loaded into thermo-responsive nanofibers by electrospinning the mixed solution containing poly(N-isopropylacrylamide-co-N-hydroxymethylacrylamide- co-stearyl acrylate) and the metal nanoparticles, followed by heat treatment to produce chemical crosslinking points inside the fibers. The ordered chain-like alignment of the loaded AgNPs along the fiber axis was observed by transmission electron microscope (Fig.1(a)). The formed composite nanofibrous membrane has high stability in aqueous mediium, and notable thermo-responsiveness with volume phase temperature at 33 °C. It takes less than 10 seconds to reach its deswelling or swelling equilibrium state with the temperature alternating between 15 °C and 45 °C. The smart composite nanofibrous membrane was used as a free-standing SERS substrate to detect trace amount of methyl benzimidazol-2-ylcarbamate (MBC) in aqueous solution, and the relative standard deviation of fifteen detection results is 1.9%. Its low detection limit at 50 °C is one order of magnitude lower than the one at 20 °C, and the linear detection range at 50 °C is from 10^{-10} M to 10^{-6} M (Fig.1(b)).

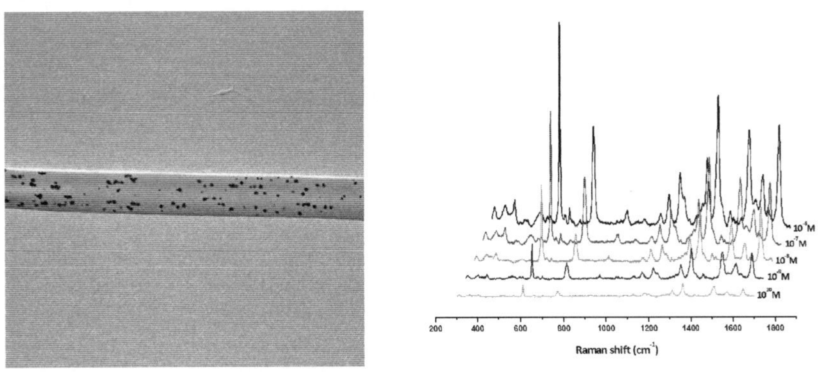

(a) (b)

Figure 1. (a) TEM image of the AgNPs loaded nanofiber; (b) SERS spectra of different concentrations of MBC in aqueous solution at 50 °C.

References

[1] Bich Ha Nguyen, Van Hieu Nguyen, Hong Nhung Tran, *Advances in Natural Sciences-Nanoscience and Nanotechnology*, **2016**, 7, 033001.

[2] Siyuan Chen, Liying Wang, Xu Dong, Xiaoyun Liu, Jianfeng Zhou, Jianmao Yang, Liusheng Zha, *RSC Advances*, **2016**, 6, 48479–48488.

The authors are grateful to the National Natural Science Foundation of China (51373030, 51503033)

Nanomedicine for Tumor Based on pH and Reduction-Responsive Polypeptide

Yue Ding, **Changming Dong***

School of Chemistry and Chemical Engineering, Shanghai Key Laboratory of Electrical Insulation and Thermal Aging, Shanghai Jiao Tong University, Shanghai 200240, P. R. China

E-mail: 296528294@qq.com

The tumor microenvironment-responsive polymeric nanodrugs including physically drug-loaded system and drug-conjugate are at the frontier of nanomedicine and controlled drug delivery. First of all, a tumor microenvironment pHe-labile amphiphilic polypeptide copolymer was developed to achieve the efficient drug release in tumor target. Secondly, the dynamic disulfide bonds can maintain the nanostructure in the extracellular microenvironment and would be cleaved to release the loaded drug at a cellular GSH level, due to the significant redox potential difference between the extracellular (oxidation) and intracellular (reduction) microenvironment. Based on the above two points, we have designed and fabricated dual-responsive pHe-sensitive PEG sheddable and disulfide cross-linked polypeptide micelles, in which amphiphilic diblock polypeptide copolymer containing PEG and poly(*S*-(*o*-nitrobenzyl)-*L*-cysteine) was synthesized via ring-opening polymerization. These micelles can largely minimize drug leakage under physiological conditions, whilst exhibiting accelerated drug release under mildly acidic or reductive microenvironments.

Keywords: Tumor Microenvironment, Dual Responsive, Polypeptide Micelles, Drug Release

References

[1] C. Y. Sun, Y. Liu, J. Z. Du, Z. T. Cao, C. F. Xu and J. Wang, *Angew. Chem. Int. Ed.*, **2016**, 55,1010.

[2] X. L. Hu, H. Li, S. Z. Luo, T. Liu, Y. Y. Jiang and S. Y. Liu, *Polym. Chem.*, **2013**, 4, 695–706.

[3] S. Mura, J. Nicolas and P. Couvreur, *Nat. Mater.*, **2013**, 12, 991.

[4] V. P. Torchilin, *Nat. Rev. Drug Discovery*, **2014,** 13, 813–827.

[5] Liu, W. Liu and C. M. Dong, *Polym. Chem.*, **2013**, 4, 3431-3443.

[6] Wu, L. Zhou, Y. Su and C.-M. Dong, *Polym. Chem.*, **2015**, 6, 6857.

Study on the Stab-resistant Performance of Composite Fabric with Resin Flakes

Jiping Dong[1,2], **Xiaoyan Liu**[1,2]

[1]*Department of Textile materials and textile design, College of Textiles, Donghua University, Shanghai, 201620, China;*

[2]*Key Laboratory of Textile Science & Technology, Ministry of Education, Donghua University, Shanghai, 201620, China*

E-mail:xiaoyanliu@dhu.edu.cn

Stab-resistant body armor is a kind of protective material. When knife and spiked weapon assault on the human body from a variety of angles, the important part of the human body could be protected by stab-resistant body armor. At present, most of stab-resistant body armors on the market are consisted of entire puncture-proof panel or fish-scale hard pieces, which are fabricated by metal alloy, ceramic or other lightweight hard material. But this type of stab-resistant body armor is heavy and impedes human activities. High-density woven, nonwoven, knitted fabrics and spacer fabrics with high performance fibers are used to produce a soft stab-resistant armor. However, conventional soft body armor is relatively bulky and inflexible since the layers of fabrics are required. In order to improve the comfort and flexibility, there are some composite approaches applied on fabric armors, such as polymer coatings, laminated films into fabrics or resin flakes bonded on fabrics. In this paper, UHMW-PE(ultra-high molecular weight polyethylene) plain woven fabric was the substrate. Polycarbonate (PC) hexagonal resin flakes were glued on the base cloth by hot melt adhesive powder to enhance the effective protection. The optimal thickness of the binder was selected through the series of peel strength tests. What's more, the static stab-resistant ability of composite fabrics was tested when the knife was punctured into the different part (the center of resin flake, the edge of resin flake and the gap between the resin flakes). The experimental results revealed when the thickness of the hot melt adhesive powder was 2mm, the peel strength between the fabric and the resin flakes was the largest. There were levels of puncture strength when the knife was punctured into the different parts of composite material. And it had the best stab-resistance when piercing from the center of hexagonal resin flake.

Keywords: Stab-resistantance; Fabric; Resin flake; Composite; Puncture strength

A Biopolymer-Drug Conjugate Nanotheranostics for Multimodal Imaging-Guided Synergistic Cancer Photothermal-Chemotherapy

Chang Du, Chang-Ming Dong*

School of Chemistry and Chemical Engineering, Shanghai Key Laboratory of Electrical Insulation and Thermal Aging, Shanghai Jiao Tong University, Shanghai 200240, P. R. China

E-mail: 546582360@qq.com

Some of biomedical polymer-drug conjugates are being translated into clinical trials; however, they intrinsically lack photothermal and multi-imaging capabilities, hindering them from imaging-guided precision cancer therapy and complete tumor regression. Hence, to make the polymer-drug conjugates with intrinsic photothermal and imaging attributes that combine photothermal therapy with chemotherapy to produce synergistic antitumor effect holds great potential for next generation precision cancer therapy. Herein we introduce a new concept of all-in-one biopolymer-drug conjugate nanotheranostics and prepare a kind of biomimetic and intracellular pH-sensitive polydopamine-doxorubicin conjugate nanoparticles (PDCNs) under mild conditions. Importantly, this strategy integrates near infrared light-mediated photothermal therapy, polymeric prodrug-release induced chemotherapy, and triple modalities including doxorubicin self-fluorescence, photothermal and photoacoustic imaging, into one conjugate nanoparticle. In vitro and in vivo biological assays convincingly demonstrated that intravenous photothermal-chemotherapy of PDCNs produced synergistic antitumor activity, achieving complete tumor ablation during the evaluation period.

Keywords: Polymer-drug conjugate, prodrug, theranostics, photothermal-chemotherapy, synergistic effect.

References

[1] J. Shi, P. W. Kantoff, R. Wooster, O. C. Farokhzad, *Nature Rev. Cancer* **2017**, 17, 20-37.

[2] N. Kamaly, B. Yameen, J. Wu, O. C. Farokhzad, *Chem. Rev.* **2016**, 116, 2602-2663.

[3] T. Sun, Y. S. Zhang, B. Pang, D. C. Hyun, M. Yang, Y. Xia, *Angew. Chem. Int. Ed.* **2014**, 53, 12320-12364.

[4] R. Cheng, F. Meng, C. Deng, Z. Zhong, *Nano Today* **2015**, 10, 656-670.

[5] X. Hu, G. Liu, Y. Li, X. Wang, S. Y. Liu, *J. Am. Chem. Soc.* **2015**, 137, 362-368.

[6] R. Namgung, Y. MiLee, J. Kim, Y. Jang, B.-H. Lee, I.-S. Kim, P. Sokkar, Y. M. Rhee, A. S. Hoffman, W. J. Kim, *Nature Commun.* **2014**, 5, 3702.

[7] P. Huang, D. Wang, Y. Su, W. Huang, Y. Zhou, D. Cui, X. Zhu, D. Yan, *J. Am. Chem. Soc.* **2014**, 136, 11748-11756.

[8] V. Shanmugam, S. Selvakumar, C.-S. Yeh, *Chem. Soc. Rev.* **2014**, 43, 6254-6287.

[9] H. Deng, F. Y. Dai, G. H. Ma, X. Zhang, *Adv. Mater.* **2015**, 27, 3645-3653.

[10] Y. Zhang, C. Teh, M. Li, C. Y. Ang, S. Y. Tan, Q. Qu, V. Korzh, Y. Zhao, C*hem. Mater.* **2016**, 28, 7039-7050.

[11] C. C. Ho, S. J. Ding, J. Biomed. *Nanotech.* **2014**, 10, 3063-3084.

[12] Y. Gao, X. Wu, L. Zhou, Y. Su, C. M. Dong, Macromol. *Rapid Commun.* **2015**, 36, 916-922.

Preparation, Structure and Mechanical Properties of Novel Nylon 56 Fibers

Yassir A. Eltahir[1,2], **Haroon A. M. Saeed**[1,2], **Yumin Xia**[1], **He Yong**[1] and **Wang Yimin**[1,*]

[1]*State Key Laboratory for Modification of Chemical Fibers and Polymer Materials, College of Materials Science and Engineering, Donghua University, 2999 North Renmin Road, Songjiang, Shanghai 201620, P.R. China;*

[2]*Faculty of Textiles, University of Gezira, Wad-Medani, P.O. Box 20, Sudan*

E-mail: yassir_eltahir@yahoo.com

The aim of this paper is the preparation and characterization of novel nylon 56 fibers. Nylon 56 (PA56) and nylon 6 (PA6) fibers were prepared by melt spinning process and drawn by a thermal drawing at different draw ratios (DRs). PA56 and PA6 fibers were characterized by means of differential scanning calorimeter (DSC), wide-angle x-ray diffraction (WAXD), sonic velocity, and tensile testing measurements. DSC results revealed that the melting temperature has no considerable variation, while the heat of fusion increased with increments in the draw ratio. This could be due to the existence of crystals with lower perfection and sizes. The equatorial WAXD revealed that PA56 as-spun fibers exhibit α-like crystal structure, while PA6 as-spun fibers observed γ-form structure. The WAXD results showed that the crystallinity and crystal size of PA56 and PA6 fibers were directly proportional to the draw ratio. This could be attributed to the molecular chain and adjacent crystals and small crystallite in the amorphous region transforming from a random state and becoming densely packed. The orientation factor was increased as expected upon drawing. However, the tenacity and Young's modulus were found to be significantly increased, while the elongation at break decreased as the draw ratio increased. The improved mechanical properties were attributed to the improvement of molecular orientation along the fiber axis, the development of more ordered crystalline structure, and the crystallinity. Compared with PA6, PA56 fibers show a little lower mechanical properties at different draw ratios (DRs). Nylon 56 spun fibers presumably a competitive fiber and the above results can help to optimize their manufacturing.

Keywords: Nylon 56; Melt spinning; Fibers; Tenacity; Modulus.

3D Printed Continuous Flow Reactor for Inline Reaction Tracking

Marcel Fassbender, Martin Wildemann, Werner Pauer*,
Gerrit A. Luinstra, Hans-Ulrich Moritz
Institute for Technical and Macromolecular Chemistry, University of Hamburg,
Bundesstraße 45, 20146 Hamburg, Germany
E-mail: werner.pauer@chemie.uni-hamburg.de

Fused Filament Fabrication (FFF) is a simple, competitive and effective additive manufacturing (AM) for producing three-dimensional objects of any favored shape. In FFF a thermoplastic material is extruded and in its semi molten state deposed onto a stage layer by layer. Currently only a small range of thermoplastic polymers could be used for 3D printing despite the large number of available polymeric materials. Polymers must have a low glass transition temperature and melting viscosity. With these properties, those polymers are suitable for extruding. Powder formed materials and a few photopolymers could be used as well for 3D printing.[1] However, a magnification of these limited materials is of necessity because industrial applications are currently very constricted. Composites of the given polymers enhance their performances. Nanomaterials can confer multifunctional properties to the printing of pure materials where they can serve as structural, sensing, heating, magnetic and conductive elements. The resulting nanocomposites, possessing unique properties, could expand the utilization of 3D printing in various areas. Synthesis of polyamide nanocomposites with single wall carbon nanotubes (SWCNT) gives us a conductive material, which satisfy the demand for low cost and easy to use analytical devices. Bringing the polyamide nanocomposites into a 3D printable filament, gives us the opportunity to print devices replicable reliable. Those devices are easily designed and rapidly deposed with 3D printing.

The trend in chemical industry of changing to continuous operation systems is ongoing. Modularization contributes to the flexibility of continuous manufacturing processes and allows a high product diversity (product by process). Continuous microfluidics systems make it possible to tailor the product characteristics. Deploying such analytical devices inline the existing continuous microfluidic systems, gives us the possibility for online measurements at any given point in the microfluidic system. This leads to new research opportunities in synthetic chemistry, including polymerization reactions. [2]

First, we will show the difference of material properties in 3D printed objects e.g. rigidity and elastic modulus between pure polyamide and polyamide composites with SWCNT. Pursuing experiments with 3D printed measurement cells in a continuous flow reactor is an aspired goal. Furthermore, to determination the residence time distributions (RTD) with a higher precision due to the possibility of measurements at any point in the reactor is an overall aim. The number of measurement cells could have an arbitrary number.

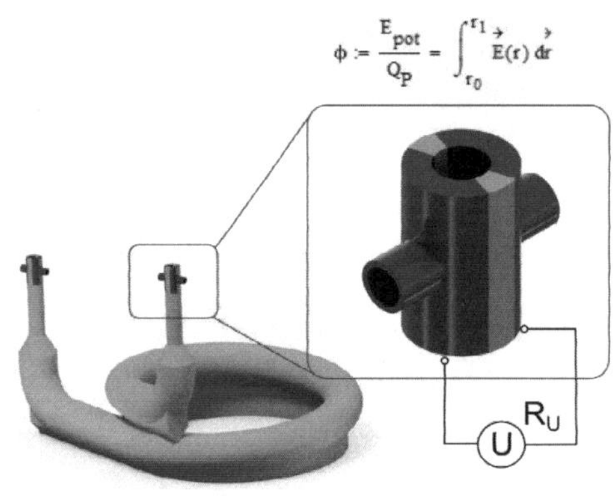

Figure 1. 3D printed reactor with tracking cells at the inlet and outlet.

References

[1] R. D. Farahani, M. Dubé, D. Therriault, *Advanced Materials* **2016**, 28, 5794.

[2] N. Kockmann, *Chemie Ingenieur Technik* **2015**, 87.

Regio-Selective Substitution at the 1,3- and 6,8-Positions of Pyrene for the Construction of Small Dipolar molecules

Xing Feng[a*], **Chuanzeng Wang**[b], **Takehiko Yamato**[b]

[a]*Beijing Institute of Graphic Communication, 1 Xinghua Avenue (Band Two), Daxing, Beijing, 102600, P.R.China;*

[b]*Department of Applied Chemistry, Faculty of Science and Engineering, Saga University, Honjo-machi 1, Saga 840-8502 Japan*

E-mail: hyxhn@sina.com

Pyrene derivatives, dipolar molecules, donor-acceptor system, intramolecular charge transfer

The construction of dipolar molecules with donor-acceptor (D-A) type structures is of interest given their potential application in organic optoelectronic devices.1 Such dipolar architectures can, via a suitable choice of the D/A units, fine-tune the electron redistribution, facilitate simultaneous manipulation of the HOMO-LUMO energy gap and the emission color by intramolecular charge transfer (ICT),2 control crystallinity,3 and be used to self-assemble molecular morphologies.4 this poster have presented a novel Dipolar molecules, which was prepared via efficient regio-selective functionalization along Z-axis of pyrene at both the 1,3- and 6,8-positions. Three asymmetrically substituted 1,3-diphenyl-6,8-R-disubsituted pyrenes were fully characterized by X-ray crystallography, electrochemistry, and DFT calculations. X-ray analysis has confirmed the novel asymmetric substitution of the pyrene core. The small dipolar molecules pyrene-based compound possess fine ICT state between D-A units in the ground state. This article presents a revolutionary methodology for the functionalization of the pyrene core and has potential application in organic photonics.

References

[1] X. Feng, J.-Y. Hu, L. Yi, N. Seto, Z. Tao, C. Redshaw, M. R. J. Elsegood, T. Yamato, *Chem.–Asian J.* **2012**, *7*, 2854–2836.

[2] X. Feng, J.-Y. Hu, F. Iwanaga, N. Seto, C. Redshaw, M. R. J. Elsegood, *Org. Lett.* **2013**, *15*, 1318–1321.

[3] X. Feng, H. Tomiyasu, J-Y. Hu, X. Wei, C. Redshaw, M. R. J. Elsegood, L. Horsburgh, S. J. Teat, T. Yamato, *J. Org. Chem.* **2015**, *80*, 10973–10978.

[4] X. Feng, J-Y. Hu, C. Redshaw, T. Yamato, *Chem. Eur.J.* **2016**, 22,11898-11916

Preparation and Characterization of Graphene Modified Poly (ethylene glycol) based Novel Chemical Polyurethane Phase Change Fiber

Ziwen Gan, **Jing Guo**

Department of High Polymer, College of Material Science and Engineering,

Dalian Polytechnic University, Dalian, 201620, China

E-mail:245723925 @qq.com

Cross-linked phase change polyurethane gets many problems, such as poor conductivity, poor thermal conductivity, low cross-linking probability and cross-linking agent is not environmental friendly. To solve these problems we have conducted some research. Firstly, the prepolymer (APEG) with active isocyanate groups is prepared by the reaction of PEG (polyethylene glycol) and MDI (diphenylmethane diisocyanate). Secondly, the target product is prepared by the reaction of hexahydroxyl xylitol and APEG. In the second step, the molarity of isocyanate group is measured in real time by measuring the weight loss of the sample and saturated sodium bicarbonate, at the same time, the molarity of isocyanate group is certificated by FTIR(Fourier transform infrared spectroscopy). On the other hand, in order to improve the compatibility of grapheme, multilayered graphene is added in the final stage of cross-linking and chain extension reaction. Finally, the phase change fibers are obtained and the dent of the fiber surface is found after post-processing, the ability to absorb sound waves has been obtained because of the dent.

The FTIR, POM, XRD, ^{1H}NMR and DSC analysis showed that the latent heat value of PU-PEG1500 was the highest when the molecular structure was successfully obtained. The thermal conductivity and electrical conductivity of the material has been improved because of multilayered grapheme. The ability to absorb sound waves has been obtained because of the dent of fiber surface.

Keywords: Phase change polyurethane; Multilayered graphene; Xylitol; MDI

In Situ Cross-Linking of Potato Starch Films and Control of its Enzymatic Degradability

Ran Gao, Jing Li, Xiaohuan Ji, Jun Li, Xiaoze Jiang, Bin Sun*, Meifang Zhu

State Key Laboratory for Modification of Chemical Fibers and Polymer Materials, College of Material Science and Engineering, Donghua University, Shanghai, 201620, China

E-mail: sunbin@dhu.edu.cn

As a non-toxic, good biocompatibility, biodegradable natural polymer, starch was widely used in biomedical fields. However, starch-based materials are easy to be degraded by enzyme system, which limits its application.

In the present work, cross-linked potato starch film was prepared by flow casting method and in situ crosslinking technology. The Sodium trimetaphosphate (STMP) were used as crosslinking agent and potato starch as raw material. And the influence of STMP content to the equilibrium swelling ratio(ESR) were investigated. Results showed that the ESR of starch film decreased with the increasing of STMP content when crosslinking agent content was below 15 wt%. In another word, the higher the STMP content was, the higher the degree of crosslinking was.

Meanwhile, based on the pH of 7.4 phosphate buffer solution containing alpha amylase simulated human body environment, the degradability of the different crosslinking degree of starch film was studied. The content of polysaccharides in degradation liquid was determined by 3,5-dinitrosalicylic acid colorimetry, which could quantitatively reflect the degradation rate of cross-linked starch film. Besides, the dynamic changes in the morphology and mechanical strength was characterized. Results showed that when the crosslinking agent content was below 15 wt%, the degradation rate of starch film decreased as the degree of crosslinking increased. The degradation of pure starch film within 7 d at a rate of 55.0%, while the degradation rate of crosslinked membrane is only 1.6%. The SEM and TG results indicate: starch film began to degrade from the surface which has loose structure.

Keywords: potato starch film; sodium trimetaphosphate; mechanical properties; enzymolysis properties

References

[1] Mali S, Grossmann M V E, Garcia M A, et al, *Food Hydrocolloids*, **2005**, 19(1), 157-164.

[2] Rodney J, Sahari J, Kamal M S M, et al, *E-Polymers*, **2015**, 15(6), 401-409.

[3] Ai Y, Jane J-L, *Starch-Starke*, **2015**, 67(3-4), 213-224.

[4] Wang S, Copeland L, *Food Chemistry*, **2012**, 135(3), 1635-1642.

[5] De Miranda C S, Ferreira M S, Magalhaes M T, et al, *Materials Research-Ibero-American Journal of Materials*, **2015**, 18, 260-264.

[6] Isotton F S, Bernardo G L, Baldasso C, et al, *Industrial Crops and Products*, **2015**, 76, 717-724.

[7] Sang Y, Prakash O, Seib P A, *Carbohydrate Polymers*, **2007**, 67(2), 201-212.

[8] Mello El Halal S L, Colussi R, Deon V G, et al, *Carbohydrate Polymers*, **2015**, 133, 644-653.

[9] Jeong W-H, Harada K, Yamada T, et al, *Breeding Science*, **2010**, 60(2), 160-163.

Shear Induced Crystallization of Bimodal and Unimodal High Density Polyethylene

Sumin Gao[1], Yatao Liu[1,†], Benjamin Hsiao[2], Huili Shao[1], Yaopeng Zhang[1,*]

[1]*State Key Laboratory for Modification of Chemical Fibers and Polymer Materials, College of Materials Science and Engineering, Donghua University, Shanghai 201620, P. R. China;*

[2] *Department of Chemistry, Stony Brook University, NY 11790, USA*

[*]E-mail: zyp@dhu.edu.cn

High-density polyethylene (HDPE) pipes are light, strong, extremely tough and very durable, and therefore have widespread applications in water pipeline to replace aging pipeline. To produce high quality HDPE pipeline, it is necessary to understand the crystallization behavior of bimodal and unimodal HDPE under shear field. Shish-kebab structure is considered as the fundamental form in crystalline polymers induced by flow fields (elongation, shear, mixed). Previous studies aforementioned provide valuable knowledge about shear-induced crystallization behavior of crystalline polymers. However, detailed investigations about how molecular weight distributions affect the crystallization behavior are rare. In this paper, synchrotron radiation X-ray diffraction and scattering techniques were applied to intensively study crystallinity, lamellae orientation, long period and orientation of HDPE. The effect of polymer molecular structure on the structure and properties of HDPE were analyzed. In situ shear experiments were performed to study the shear-induced crystallization of bimodal and unimodal HDPE samples. The present study indicates that the bimodal HDPE has medium molecular weight (MW), broad molecular weight distributions (MWD) and short chain branching (SCB) at high MW regions. One of the unimodal sample has high MW, narrower MWD and SCB at medium-high MW regions, and another unimodal sample has high MW, medium narrower MWD and high SCB at low MW regions. It is found that discrete primary lamellae and secondary lamellae are formed in all samples. The secondary lamellae grow faster in bimodal HDPE than in unimodal HDPE. Shear generally results in oriented lamellae along the shear direction and enhances the crystallization. Upon shear, the bimodal HDPE has higher crystalline orientation and lamellar orientation than the unimodal HDPE samples, whereas the bimodal HDPE has relatively higher crystallinity. In the case of the two unimodal HDPE samples, the one which has high MW, narrower MWD and SCB at medium-high MW regions has higher crystallinity and crystalline orientation. It is revealed that large fractions of medium-high molecular weight (MMW) chains with SCB are effective to generate shear-induced nuclei, enabling higher crystalline orientation.

Keywords: HDPE, Synchrotron radiation technology, Bimodal, Unimodal

[†] Present address: Institute of New Energy Material Chemistry, School of Materials Science and Engineering, National Institute for Advanced Materials, Nankai University, Tianjin 300350, P.R. China

Preparation and Characterization of Carboxymethyl Chitosan Base Superabsorbent Composite Sponge

Xing Ge, Jing Li, Jiaming Lou, Zhe Zhou, Bin Sun*, Meifang Zhu

State Key Laboratory for Modification of Chemical Fibers and Polymer Materials, College of Material Science and Engineering, Donghua University, Shanghai, 201620, China

E-mail: sunbin@dhu.edu.cn

In recent years, the sponge has been widely used in wound dressings and other fields. How to prepare biomedical sponge via the natural biopolymer is one of the hottest research topics. Chitosan materials have many advantages, such as abundant raw materials, good biocompatibility, antibacterial activity and prevent infection, so they have broad prospects in the field of biomedicine. Carboxymethyl chitosan (CMCS) as the product of chitosan carboxylation, has a wide range of applications due to its special properties with carboxyl ($-CH_2COO-$) functional groups. CMCS not only keeps the excellent hemostatic properties of chitosawn, and the water-soluble improve greatly, it is the ideal medical hemostatic material, but pure CMCS sponge has low water absorption, only 800%, and no water resistance. Carboxymethyl starch sodium (CMS), as a water-soluble starch, has swelling and high water absorption, but it was found that the pure CMS sponge was low in strength and fragile, and starch granules were falling off, which seriously affected its application.

In this paper, a solution or hydrogel is formed by combining CMCS and CMS or other gelatinized starch, CMCS superabsorbent composite sponge was prepared by vacuum freeze-drying method. The preparation of the best morphology of CMCS superabsorbent composite sponge, optimum technological parameters of freeze-drying and the related properties of hemostasis were investigated. The composite sponges were characterized by infrared spectroscopy (IR) and scanning electron microscope (SEM), and their structure shape and mechanical properties and water absorption speed were measured.

Its result indicated: CMCS superabsorbent compound sponge, not only enhanced the sponge formation performance, and had fast absorbing water as the CMCS composite sponge. It was found that the compressive modulus of the sponge was doubled after crosslinking, and its water absorption rate can reach more than 2800%, which is the foundation for the study of rapid hemostasis in related fields.

Keywords: Superabsorbent sponge; Carboxymethyl chitosan; Freeze drying;

References

[1] Wedmore I, et al. *Journal of Trauma*, **2006**, 60(3): 655-8.
[2] Park Y, et al. *Biomaterials*, **2000**, 153-159.
[3] Jiang, H., et al. *Applied Materials & Interfaces*, **2013**,12036-12044.
[4] Mi F L, et al. *Biomaterials*, **2001**, 165-173.
[5] Park Y J, et al. *Journal of Controlled Release*, **2000**, 385-394.
[6] Yan T, et al. *Carbohydrate Polymers*, **2017**, 170:271.
[7] Jayakumar R, et al. *Biotechnology Advances*, **2011**, 322-337.
[8] Zamani A. *Ind.eng.chem.res,* **2010**, 8094-8099.
[9] Guo M, et al. China Pharmacist, **2012**.
[10] Liu Y, et al. *Bioresources*, **2015**, 10(4).
[11] Hu, Z., et al. Materials Science and Engineering,**2017**,012045.
[12] Kamel N A, et al. *Materials Science & Engineering C*, **2017**, 72:543.

Controllable Micro-nano Assembly of Bio-functionalized Poly(3,4-ethylenedioxythiophene) and their Applications in Tumor Cell Capturing

Zhi Geng , Qichao Pan, Zhengwei You, Bo Zhu

State Key Lab for Modification of Chemical Fibers and Polymer Materials, College of Materials Science and Engineering, Donghua University, 2999 North Renmin Road, Songjiang, Shanghai, 201600, China

E-mail: gengzhi@dhu.edu.cn

Conducting polymers had been widely used in the fields of cell-related diagnostic and therapeutic platforms. Among them, researchers showed great interests on poly(3,4-ethylenedioxythiophene) (PEDOT) due to its excellent properties of low structure defects, high aqueous stability and biocompatibility. Meanwhile, properly nanostructures could extremely enhance the tumor cell-capturing efficiency of bio-functional PEDOTs[1]. Several approaches have been made to fabricate PEDOT materials with nanostructures via template method, seeding, interfacial polymerization, vapor polymerization, and self-assembly approach. However, most of them require hard-templates or surfactants, which would complicate the preparation process and reduce the quality of the obtained materials. Therefore, we have proposed the approach for fabricating template-free bio-functional poly(3,4-ethylenedioxythiophene) with controllable nanostructures via electrochemical deposition. By tuning the experimental condition, controllable switching micro-nano dots and micro-nano tubes structure of PEDOT, PEDOT-OH and PEDOT-COOH could be fabricated. Meanwhile, changing of electric charges and physical structures of micro-nano dots and micro-nano tubes PEDOT-OH films were studied to explain the morphology transformation. Besides, the cell-capturing efficiency was characterized of the micro-nano dots and micro-nano tubes structure EDOT-COOH films after bio-conjugated. We hope this current study would provide a simple and effective method for fabricating high efficiency tumor cell-capturing materials.

References

[1] Sekine J., Luo S-C, Wang S.,Zhu B., Tseng H-R and Yu H-H, *Adv. Mater.* **2011**, *23*, 4788–4792.

Starch-*graft*-Polyacrylonitrile Nanofibers by Electrospinning

Yumei Gong, Zhaomei Sun, and Jing Guo

School of Textile and Material Engineering, Dalian Polytechnic University, 116034, Dalian, China

E-mail: ymgong@dlpu.edu.cn

As the second largest natural renewable materials on earth every year, starch is widely used in a wide range in industries[1-5]. However, the applications of native starch are restricted because of its inherent defects such as insoluble in cooled water and a tendency to degrade in heated state. To overcome these defects and achieve innovation in the utilization of starch, researchers paid their attentions to chemical modification via introduction of various functional groups such as carboxyl, acetyl and hydroxypropyl onto the backbone of starch[2-4]. Studies on chemical modification of starch have been performed in early 1940s[5]. In lots kinds of chemical modifications, grafting method is the simplest performance and can remain the native chemical structure of the starch.

At the same time, the application of nanofibers has attracted increasing attention[6]. And electrospinning has been recognized as an efficient and concise method to fabricate nanofibers in the range of few micrometers to hundred of nanometers[7-8]. Moreover, the properties of nanofibers are affected by some parameters such as the applied voltage, temperature, concentration and flow rate of the polymer solution or melt, the collector distance[9-12].

Here, we prepare starch nanofibers by modifying native starch through grafting polyacrylonitrile (PAN) and electrospinning. The starch–*graft*–PAN copolymer with grafting ratio ~ 49% was homosynthesized by free radical polymerization of acrylonitrile from starch initiated by ammonium ceric nitrate. And then the copolymer is dissolved in DMSO and electrospun to nanofibers under controlled applied voltage, polymer flow rate and so on. Fig 1 shows ^{13}C-NMR spectra of the native starch and the graft copolymer. The degree of substitution defined as the average amount of grafted acrylonitriles from a glucose unit of starch in the copolymer and the mass fraction of grafted acrylonitriles from the copolymer estimated from Fig 1 are 13.0 and 196%, respectively. And Fig 2 displays the scan electronic microscopy (SEM) images of the nanofibers prepared from the collector distance 15 cm, flow rate 0.5 mL/h, temperature 25 °C, and controlled applied voltages. It shows that the diameter of the nanofibers is 250 ~ 400 nm and decreases with increasing of the applied voltage. Fig 3 displays the SEM images of the nanofibers prepared from the collector distance 15 cm, applied voltage 15 kV, temperature 25 °C, and different flow rates. And it suggests that the diameter of the nanofibers is 300 ~ 510 nm and increases with the increasing of the flow rate.

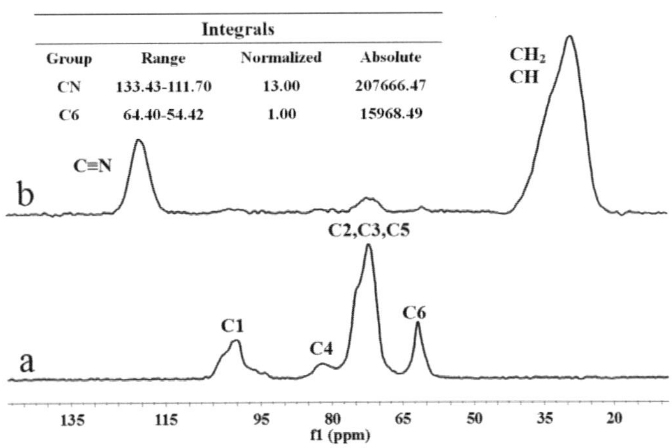

Fig 1. ^{13}C-NMR of native starch (a) and the synthesized starch–*graft*–polyacrylonitrile copolymer (b).

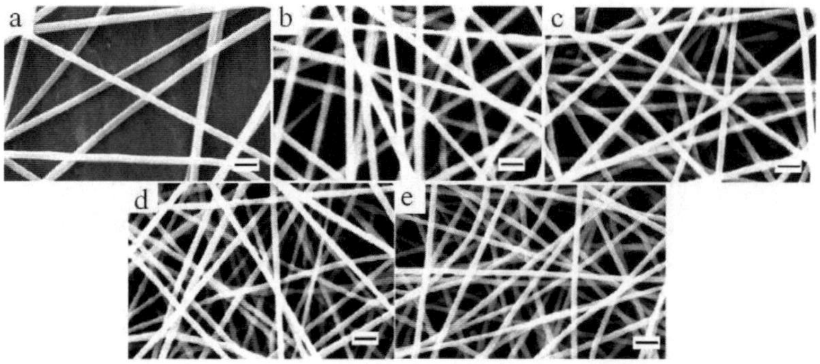

Fig 2. Typical SEM images of the graft copolymer nanofibers under various applied voltages.
(a) 5 kV, (b) 10 kV, (c) 15 kV, (d) 20 kV, (e) 25 kV.
(collector distance: 15 cm, flow rate: 0.5 mL/h, temperature 25 °C). The scale bar is 1 μm.

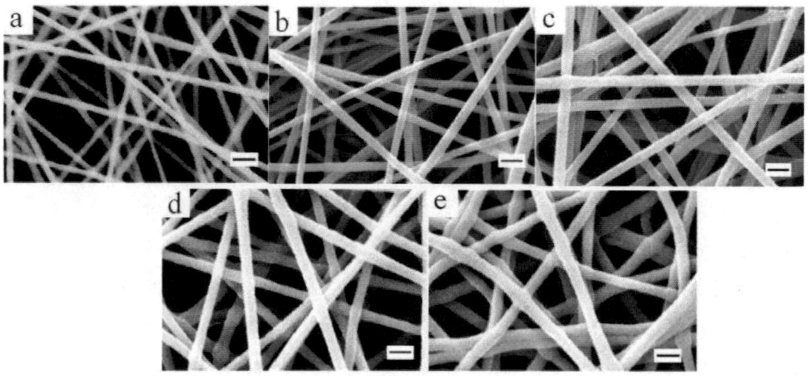

Fig 3. Typical SEM images of the graft copolymer nanofibers under various flow rates.
(a) 0.5 mL/h, (b) 1 mL/h, (c) 1.5 mL/h, (d) 2.0 mL/h, (e) 2.5 mL/h.
(collector distance: 15 cm, applied voltage: 15 kV, temperature 25 °C). The scale bar is 1 μm.

References

[1] S. Doi, J. H. Clark, D. J. Macquarrie, K. Milkowski, *Chemical Communications*, **2002**, 34, 2632-2633.

[2] M. A. Misman, A. R. Azura, Z. A. A. Hamid, *Carbohydrate Polymers*, **2015**, 128, 1-10.

[3] H. Song, D. Wu, R. Q. Zhang, L. Y. Qiao, S. H. Zhang, S. Lin, J. Ye, *Carbohydrate Polymers*, **2009**, 78, 253-257.

[4] K. N. Waliszewski, M. A. Aparicio, L. A. Bello, J. A. Monroy, *Carbohydrate Polymers*, **2003**, 52, 237-242.

[5] Q. Chen, H. Yu, L. Wang, Z. U. Abdin, Y. Chen, J. Wang, W. Zhou, X. Yang, R. U. Khan, H. Zhang, X. Chen, *Rsc Advances*, **2015**, 5, 67459-67474.

[6] S. Tang, Z. Zhao, G. Chen, Y. Su, L. Lu, B. Li, D. Liang, R. Jin, *Journal of Sol-Gel Science and Technology*, **2016**, 77, 594-603.

[7] H. Wang, W. Wang, S. Jiang, S. Jiang, L. Zhai, Q. Jiang, *Iranian Polymer Journal*, **2011**, 20, 551-558.

[8] J. Sunthornvarabhas, P. Chatakanonda, K. Piyachomkwan, K. Sriroth, *Materials Letters*, **2011**, 65, 985-987.

[9] C. J. Thompson, G. G. Chase, A. L. Yarin, D. H. Reneker, *Polymer*, **2007**, 48, 6913-6922.

[10] H. Fong, I. Chun, D. H. Reneker, *Polymer*, **1999**, 40, 4585-4592.

[11] S. D. Vrieze, T. V. Camp, A. Nelvig, B. Hagström, P. Westbroek, K. D. Clerck, *Journal of Materials Science*, **2009**, 44, 1357-1362.

[12] A. Rezaei, A. Nasirpour, M. Fathi, *Comprehensive Reviews in Food Science & Food Safety*, **2015**, 14, 269-284.

Functionalized Polymer Matrices for Metal Catalysis by Electrospinning

Franziska Griesing[1], Dr. N. Steinfeldt[2], Prof. Dr. H.-U. Moritz[1], Prof. Dr. M. Zhu[3], Dr. W. Pauer[1]

[1]*Institute for Technical and Macromolecular Chemistry University of Hamburg, Germany, Bundesstraße 45, 20146 Hamburg;*

[2]*Leibniz-Institute for Catalysis, Albert-Einstein-Straße 29a, 18059 Rostock;*

[3]*Donghua University, College of Materials Sciense & Engineering, Shanghai, 2999 North Renmin Road, Songjiang, Shanghai, 201620 China.*

E-mail: werner.pauer@chemie.uni-hamburg.de

Commonly commercial available carrier materials for catalysis are made of oxides from aluminum, silica, activated carbon or functionalized crosslinked polymeric particles. Over the last decade, the interest in functionalized polymer carrier materials has increased continuously, especially for pharmaceutical products, fine chemicals as well as in catalysis. The large surface to mass ratio as well as the opportunity to design the carrier material tailor-made with different functional groups opens an attractive research field.

Fibers with a functional surface, such as alkyne and azide could be produced by the means of electrospinning of functionalized polystyrene. To sustain the functionalities, the polymer for the spinning solution was synthesized through dispersion polymerization which guaranteed a good reproducibility. Depending on the ratio of the functionalized groups the particle size varies. In different solvents, depending on the functional groups, functionalized particles were dissolved for at least 12 h. The fibers were spun and collected on a plate or a collector and analyzed via SEM, IR and TGA. Azide and alkyne functional groups could be identified via IR-spectroscopy in the fiber matts. The SEM-images show different structures of the fibers depending on the relative humidity as well as the solvent. A decomposition temperature in the range of 460 °C for the fibers could be obtained through TGA. It was possible to crosslink the azide functionalized fibers through UV-light which made them stable in most solvents afterwards.

The investigation of the catalytic activity is pending and will be conducted at the Leibniz Institute for Catalysis (LIKAT). The results will be compared to previous investigations of functionalized microspheres .

Synthesis and Texture Transformation of Cellulose Side Chain Liquid Crystals

Yangyang Han

Key Laboratory of Tianjin City Advanced Fiber and Energy Storage, Department of Materials Science and Engineering, Tianjin Polytechnic University, Tianjin 300387, China

E-mail: sjhb2000@163.com

A liquid crystal unit containing methyl cyanide and methyl imidazole at one end was synthesized and the cellulose liquid crystal solution was synthesized by the hydrogen bond interaction between the two groups. Then the cellulose liquid crystal solution was put into a certain magnetic field, so that the liquid crystal solution was changed from the smectic phase to nematic spherulites. The images of the nematic spherulites were obtained by POM observation under different conditions. The DSC results show that the phase transition temperature range of the smectic phase to the nematic phase was 70 °C to 80 °C. The test of Magnetic field strength on the phase transition shows that phase transformation could be achieved at 70(A/m)-80(A/m). Tensile test results show that the strength of liquid crystal fiber was up to 3.15cN/dTex, which was increased by 40% compared to pure cellulose fiber.

Keywords: liquid crystal; Crystal phase transformation; magnetic field

Effect of Graphene Oxide on Interface of Carbon Fiber/Polyetheretherketone Composites

Elwathig. A. M. Hassan, Zhu Shu, Yu Muhuo*

[1]State Key Laboratory for Modification of Chemical Fibers and Polymer Materials, Donghua University, North Renmin Rd. 2999, Songjiang District, Shanghai 201620, China;

[2]Key Laboratory of Shanghai City for lightweight composites, Center for civil aviation composites, Donghua University, Shanghai 201620, China

E-mail: elwathig2009@yahoo.com

Carbon fiber (CF) reinforced polyetheretherketone (PEEK) composite is one of the most promising thermoplastic composite materials for aerospace applications [1, 2]. However, the interfacial adhesions between CF and PEEK matrix are relatively weak, thus mechanical performances of these composites prepared without surface treatments on CF are rather poor. In this work, graphene oxide (GO) dispersed in Polyetherimide as media was introduced as an interface phase to enhance the interfacial bonding between CF and PEEK. Adding GO sheet in sizing agent had an obvious effect on the external fiber surface and grooves of the fiber have been covered with PEI/GO layer. Fourier transform infrared (FT-IR) spectrometer were used to evaluate the surface characterizations of treated CF. The results demonstrated that modified GO could interact with PEI sizing agent to form hydrogen bonds at the interfaces. The interlaminar shear strength (ILSS) increased by 41 % for modified composite compared with that without treatment. Moreover, the storage modulus of modified composite was showed higher value and treated carbon fiber lead to an increase in T_g temperature, which indicated that carbon fiber/matrix interaction improved. SEM observations on the fractural surfaces showed that each CF was covered with a substantial PEEK matrix and the boundary gaps disappeared, indicating efficient modifications on the adhesion between CF and PEEK. Scheme and some results of surface modifications for CF are shown in **Figure 1**.

Keywords: Carbon fiber, polyetheretherketone, interfacial modification, graphene oxide.

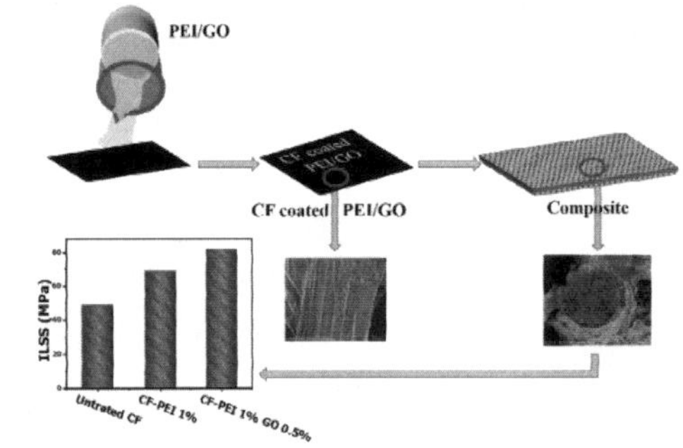

Figure 1. Show scheme of surface modification of carbon fibers, SEM and ILSS values

References

[1] Giraud. I, *Journal of Applied Polymer Science*, **2015**, 132, 42550.

[2] Yapıcı. U, *Applied Mechanics and Materials*, **2014**, 598 , 66-72.

Acknowledgment: This project is supported by National Natural Science Foundation of China (No. 21404023), funding from State Key Laboratory for Modification of Chemical Fibers and Polymer Materials, Donghua University (No. LK1417) and the Fundamental Research Funds for the Central Universities (2232015D-10).

Interlayer-Functionalization of α-ZrP with Vinyl Groups and its Thoil-ene Click for improving Grafting Ratio

Ahmed Hossamaldin, **Ya Du, XiaoHuan Ji, Bin Sun, MeiFang Zhu**

State Key Laboratory for Modification of Chemical Fibers and Polymer Materials, College of Materials Science and Engineering, Donghua University, 201620, Shanghai, P. R. China

*Corresponding author: sunbin@dhu.edu.cn, zmf@dhu.edu.cn

P-OH groups existing on surface and interlayers spaces of α-ZrP impart organic functionality to α-ZrP. However, numerous reports for the intercalation of the different groups into α-ZrP show a small grafting ratio. This work demonstrates a new strategy to functionalize the α-ZrP interlayers via thiol–ene chemistry application which can attain a high grafting ratio. Interlayer-functionalized α-ZrP with vinyl groups were synthesized using a combination of ion exchange having amino-propane and silylation having vinyl-trimethoxy-silane (C5H12O3Si) (VTMS). The preintercalation and intercalation layers of α-ZrP were probed using X-ray Powder Diffraction (XRD) and Fourier Transform Infrared Spectroscopy (FT-IR). To quantify the grafting ratio of VTMS in the interlayer spacing of ZrP, Thermo-Gravimetric Analysis (TGA) was used. XRD and TGA showed the d-spacing of ZrP/VMTS increase to 10.06 Å, 11.87 Å and 17.04 Å with grafting ratio 9.81%, 10.45% and 17.98% respectively, The grafting ratio increased with increasing the interlayer distance of α-ZrP. The novel process incorporated 2-Mercaptobenzothiazoles (2MBT) into ZrP/VTMS under UV irradiation, thereby the thiol-group (-SH) in 2MBT interacts with vinyl group (-CH=CH-) in ZrP/VMTS by thiol-ene click chemistry. Carbon Mass Nuclear Magnetic Resonance (^{13}C NMR) and UV-Vis spectroscopy confirmed that 2MBT successfully reacted with vinyl groups in functionalized α-ZrP and the grafting ratio increased to 18.72 wt%, the evidence is shown by the disappearance of the vinyl peaks. Absorption bands at 240 nm and 327 nm in UV-vis spectroscopy correspond to the interaction between (-CH=CH-) in ZrP/VTMS and (-SH) in 2MBT. The click-chemistry of α-ZrP lead to multi-functional and Uv protection with high effective and grafting ratio.

Acknowledgements: This research is financially supported by the Program of Shanghai Academic/Technology Research Leader (16XD1424500), and the Program for Changjiang Scholars and Innovative Research Team in University (T2011079, IRT1221)

Preparation of Porous Ultrafine SF Fibers via Selective Dissolution of Electrospun SF/PLA Blend Fibers

Jiwei Huang[1,2,3], Wane Ning[1,2,3], Haitao Lin[1,2,3], Xinlong Ling[1,2,3], Xinxia Yue[1,2,3], Fang Jiang[1,2,3]

[1]*College of Biological and Chemical Engineering, Guangxi University of Science and Technology, Liuzhou, 545006, China;*

[2]*Guangxi Key Laboratory of Green Processing of Sugar Resources, Guangxi University of Science and Technology, Liuzhou, 545006, China;*

[3]*Key Laboratory for Processing of Sugar Resources of Guangxi Higher Education Institutes, Guangxi University of Science and Technology, Liuzhou, 545006, China*

E-mail: jiweihuang@yeah.net

A Porous and continuous ultra-fine fibers were prepared via selective dissolution of electrospun Bombyx mori silk fibroin/poly(L-lactic acid) (SF/PLA) fibers. SF and PLA were partially miscible and mainly phase separation processed rapidly during the electrospinning process. Most of the PLA in SF/PLA blend fibers was dissolved in chloroform within 1h, while SF remained undissolved. After extraction of the PLA, the resulting SF fibers had three-dimensionally interconnected pores with a circular shape and the pore size distribution was narrow.

Keywords: Bombyx mori silk fibroin (SF), Poly(L-lactic acid) (PLA), Electrospinning, Selective dissolution, Porous ultrafine fibers.

Amine(imine)diphosphine Iron Catalysts for Asymmetric Transfer Hydrogenation of Ketones and Imines

Shangfei Huo, Jie Li, Qingquan Xue, Di Wang, Kunda, Weiwei Zuo*

State Key Laboratory for Modification of Chemical Fibers and Polymer Materials, College of Materials Science and Engineering, Donghua University, 2999 North Renmin Road, Shanghai 201620, P. R. China

E-mail: zuoweiwei@dhu.edu.cn

A rational approach is needed to design hydrogenation catalysts that make use of Earth- abundant elements to replace the rare elements such as ruthenium, rhodium, and palladium that are traditionally used. Here, we validate a prior mechanistic hypothesis that partially saturated amine(imine)diphosphine ligands (P-NH-N-P) activate iron to catalyze the asymmetric reduction of the polar bonds of ketones and imines to valuable enantiopure alcohols and amines. We present a direct synthetic approach to enantiopure ligands of this type that takes advantage of the iron(II) ion as a template. The catalytic mechanism is elucidated by the spectroscopic detection of iron hydride and amide intermediates.

Keywords: diphosphine, hydrogenation, iron catalysts

Synthesis of Monodisperse and Porous Ag Nanoparticles/Polystyrene Microcomposite Particles by Seeded Suspension Polymerization

Xiaohuan Ji,[1] Franziska Griesing,[2] Bin Sun,[1] Werner Pauer,[2] Meifang Zhu,[1] Yushan Sun[1,3] and Hans-Ulrich Mortiz[2]

[1] *State Key Lab for Modification of Chemical Fibers and Polymer Materials, College of Material Science and Engineering, Donghua University, Shanghai, 201620, China;*

[2] *Institute for Technical and Macromolecular Chemistry, University of Hamburg, Bundesstraße 45, 20146 Hamburg, Germany;*

[3] *State Key Lab of Biobased Fiber Manufacturing Technology, China Textile Academy, Beijing, 100025, China*

ContactAuthor: sunbin@dhu.edu.cn; pauer@chemie.uni-hamburg.de; Speaker: jsjtjxh@gmail.com

In recent years, silver nanoparticles (AgNPs) have attracted much interest in the application of catalysis. However, single AgNPs used as catalytic materials are very easy to be aggregated due to their high surface energy that usually leads to a remarkable reduction in their catalytic activities. In addition, as an efficient catalyst, it should be easily recycled and reused. To solve these problems, AgNPs /polymer composites have been widely used. Especially the porous polymers are now widely used because of its high specific surface area, tunable pore morphology, strong adsorption capacity, excellent physicochemical stability and stable mechanical properties.

In our research, monodisperse and porous polystyrene (PS) microspheres with homogeneous and uniform dispersed AgNPs embedded were obtained in a two-step seed swelling polymerization. This procedure is involving precipitation polymerization synthesis of PS seed particles, swelling of seed particles with monomers, cross-linker, AgNPs, plasticizer, initiator and porogen, and polymerization of those substances inside the swelled PS seed particles. In order to keep high dispersity of AgNPs in the porous PS matrix and to have strong combination between the AgNPs and the matrix, an organic-soluble AgNPs capped with double bond were synthesized previously and used during the polymerization. The AgNPs can be well dissolved in the monomer, cross-linker and plasticizer, so they can be easily swelled into the PS seed together with monomers during the swelling step. In addition, the double bond on the surface of AgNPs can be reacted with styrene monomers and divinylbenzene cross-linkers to firmly trap AgNPs into the PS network by covalent attachment. In the experiments, the reaction temperature, the amounts of AgNPs, the ratio between the monomer and the crosslinker, the kinds of porogen were changed to investigate their effects on the structure and morphology of AgNPs/PS porous composites. FTIR, SEM, XRD, TEM, BET and TGA methods were used to characterize the composite, structure, morphology, specific surface area and AgNPs loading amount, respectively.

Herein, a method to synthesize monodisperse and porous AgNPs/PS microcomposite particles is studied. In view of the high specific surface area of porous PS and excellent catalytic activity of AgNPs, the catalytic properties of this kind of composite will be further explored in the future. As AgNPs/PS microcomposite particles can be a functional particles, it is possible and meaningful to produce crosslinked and highly porous AgNPs/PS composite fibers in the next step.

Keywords: silver nanoparticles, polystyrene, porous, composite

Interaction between Carbon Black and NMMO in the Preparation of Dope Dyed Lyocell Fibers

Hong Jin[1], Yining Liu[2], Si Chen[2], Lejun Wang[2], Yumei Zhang[1*], Huaping Wang[1]

[1] *College of Materials Science and Engineering, Donghua University, Laboratory of fiber material modification, Shanghai, 201620, China;*

[2] *Hi Tech Fiber Group Co.Ltd, Beijing, 100020, China*

E-mail: 18817320642@163.com

Compared with the traditional dyeing of Lyocell, dope dyeing method has advantages of low energy consumption, simple process, high dyeing rate and color fastness. The problem in dope dyeing process of Lyocell is whether the additives such as carbon black, affect the recovery of Lyocell solvent NMMO or not. The spinning process with the change of the filling amount of the carbon black was analyzed. The relationship between structure and properties was analyzed by X-ray diffraction method. Infrared spectrum and UV spectrum were used to analyze the influence of filling agent on NMMO, and also the migration quantity of carbon black in spinning process was studied. The results show that carbon black does not migrate during the spinning process, and the addition of carbon black does not affect the recovery of solvent. The dope dyed Lyocell has a high color fastness, but the crstallinity of the fiber and the fiberstrength will be decreased slightly with the increase of the carbon black, which can still meet the requirement of taking.

Key Words: Lyocell, dope dyeing, carbon black, NMMO

Fabrication of Nano-sized Attapulgite-based Aerogels as Anode Material for Lithium Ion Batteries

Ye Lan, Dajun Chen*

State Key Laboratory for Modification of Chemical Fiber and Polymer Materials, College of Material Science and Engineering, Donghua University, Shanghai, 201620, China

E-mail: 770575785@qq.com

Lithium-ion batteries (LIBs) can store considerable green and clean energies. Meanwhile, as environment-friendly second batteries, LIBs have high energy density and power density [1]. However, graphite, a kind of commercial anode materials [2], can't meet the demand of high capacity, high safety and long lifespan due to its poor rate performance and low specific capacity (372 mAh g^{-1}) [3]. To fulfill these requirements, Li-alloying materials [4], such as Si, Sn and Ge, have been attracted much attention. Among these materials, SiO_2 reacts with metal lithium generating Li_2O, Li_4SiO_4 and Si [5] in the process of initial insertion. The first two products are inert materials, which can not only effectively impede electrochemical aggregation of Si, but also alleviate the volume expansion of Si during charging/discharging process [6]. Although it has high theoretical capacity (1965 mAh g^{-1}) [7], SiO_2 particles due to its low conductivity and poor mechanical property can't be widely applied in the field of LIBs. To solve above problems, carbon modification has been elucidated to effectively enhance the electrochemical performance of SiO_2-based anodes [8], such as SiO_2-carbon composite , SiO_2@C hollow sphere, SiO_2-mesoporous carbon composite, etc. However, there are still some issues of silica–carbon composites which obstruct their practical application. For example, the complicated preparation processes of silica–carbon composites are generally with high cost and low production. Hence, it is necessary to develop a kind of cost-effective silica sources and convenient synthesis method for preparing silica–carbon anode materials with excellent electrochemical performance.

Attapulgite (AT) is a natural crystalline hydrated magnesium aluminum silicate material with excellent chemical and thermal stability. Its uniform fiber-like porous structure [9] can absorb more electrolytes into active materials and shorten lithium ions diffusion pathway, which can obtain good rate capability. We adopted a facile method to prepare AT/polyacrylonitrile (PAN) aerogel precursor and then the precursor was pre-oxidated and carbonized, seen in Figure 1. We expected AT-based aerogels contributed to high specific capacity and PAN provided high structural integrity and electric conductivity. The morphology of AT-based composite aerogel was investigated, seen in Figure 2. The porous anode material displayed superior initial specific capacity of 1628.4 mAh g^{-1} and average discharge capacity of 534.6 mAh g^{-1} at 100 mA g^{-1} after 50 cycles. Moreover, coulombic efficiency reached over 97%, seen in Figure 3. Therefore, our work presented a practicable and potential way to prepare AT-based composite aerogel with excellent electrochemical performance of anode materials for LIBs.

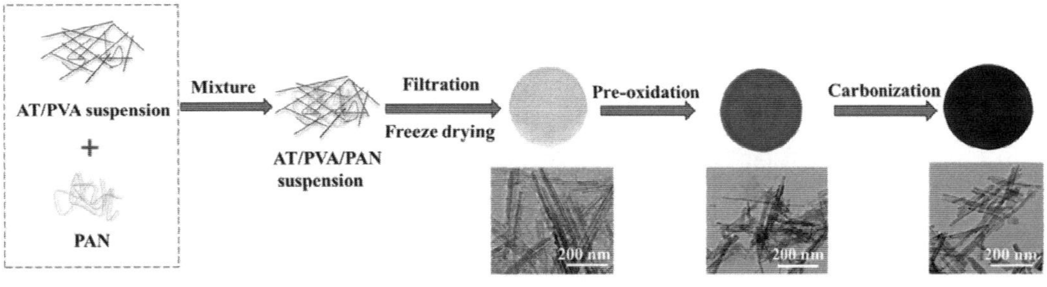

Figure 1. Schematic process of generating AT-based anode materials

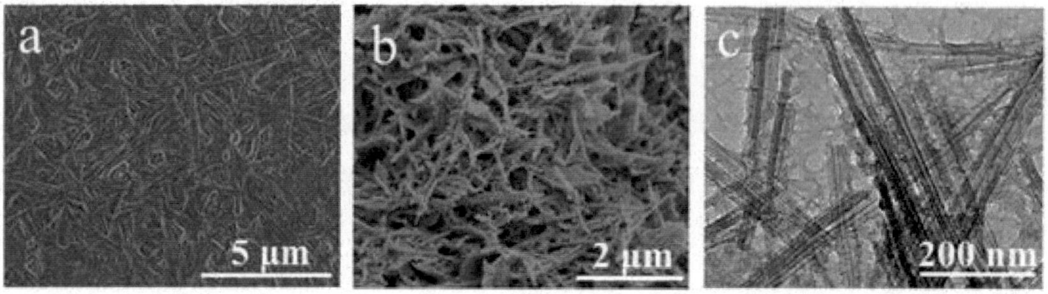

Figure 2. SEM images of AT-based precursor (a) surface, (b) cross-section and (c) TEM image of AT-based precursor

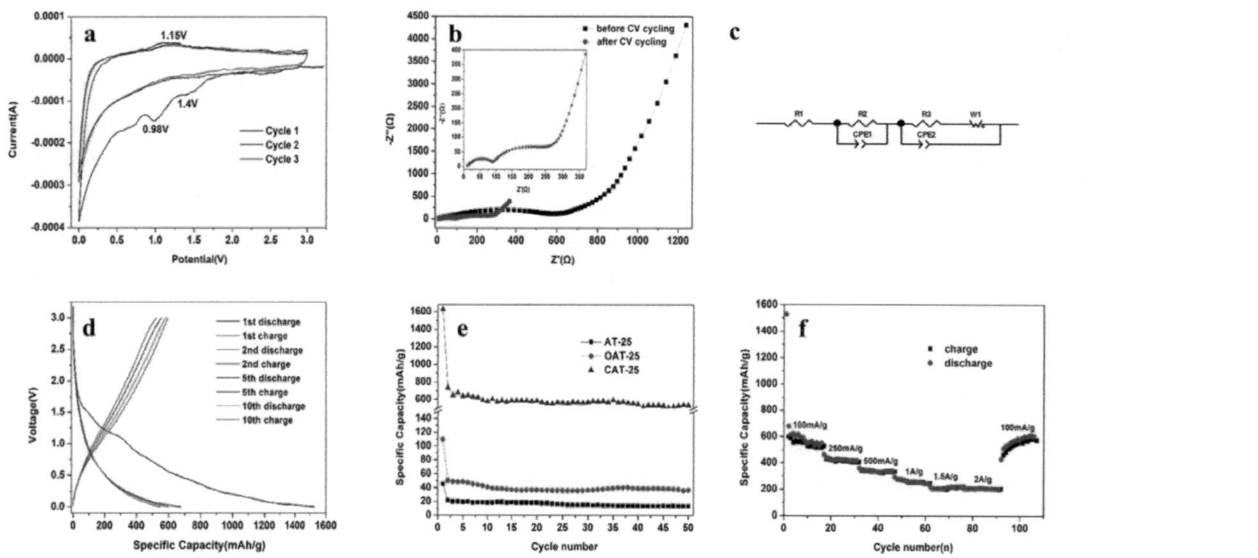

Figure 3. (a) CV curve of carbonized aerogel. (b) The EIS plot of carbonized aerogel before CV cycling and after CV cycling. The inset is magnified image of the EIS plot of carbonized aerogel before CV cycling. (c) The equivalent circuit for quantitative simulation of the Nyquist plots. (d) Galvanostatic charge-discharge profile of carbonized aerogel. (e) Cyclic behavior of AT-based precursor, pre-oxidated aerogel and carbonized aerogel. (f) Rating performance diagram of carbonized aerogel

References

[1] Ryu J, Hong D, Choi S, Park S, *ACS Nano*, **2016**, 10, 2843-2851.

[2] Andreev YG, Bruce PG, *Journal of Applied Crystallography*, **2016**, 49, 24-30.

[3] Park HY, Kim MS, Bae TS, Yuan J, Yu JS, *Langmuir*, **2016**, 32, 4415-4423.

[4] Chang WS, Park CM, Kim JH, Kim YU, Jeong G, Sohn HJ, *Energy & Environmental Science*, **2012**, 5, 6895-6899.

[5] Liu X, Chen Y, Liu H, Liu ZQ, *Journal of Materials Science & Technology*, **2017**, 33, 239-245.

[6] Lv P, Zhao H, Wang J, Liu X, Zhang T, Xia Q, *Journal of Power Sources*, **2013**, 237, 291-294.

[7] Yan N, Wang F, Zhong H, Li Y, Wang Y, Hu L, Chen Q, *Scientific Reports*, **2013**, 3, 1568-1573.

[8] He W, Tian H, Zhang S, Ying H, Meng Z, Han W, *Journal of Power Sources*, **2017**, 353, 270-276.

[9] Xie Q, Zheng A, Xie C, Fu W, Wu S, Zhang Y, Zhao P, *Microporous and Mesoporous Materials*, **2016**, 224, 239-244.

The Interaction between Graphene Oxide and Polyacrylonitrile in Spinning Solutions

Fengmei Li, Biao Wang*

State Key Laboratory for Modification of Chemical Fiber and Polymer Materials, College of Material Science and Engineering, Donghua University, Shanghai, 201620, China

E-mail: wbiao2000@dhu.edu.cn

The homogeneous graphene oxide (GO)/polyacrylonitrile (PAN) spinning solutions containing different amount of GO were prepared by solution blending method in NaSCN aqueous solution and DMSO solvent, respectively. And two kinds of PAN with different electronegativity were used in this study, PAN1 with sulfonic groups had negative charges and PAN2 with tertiary amine groups positively charged in the solutions. The interaction between GO and PAN in spinning solutions were investigated through both steady-state and dynamic rheological measurements. The parameters such as apparent viscosity (η_α), storage modulus (G'), loss modulus (G'') and mechanical loss factor (tanδ) were illustrated the rheological behaviors of these solutions. The results showed that the apparent viscosity, storage modulus and loss modulus of GO/PAN1 spinning solutions decreased with adding appropriate amount of GO, while the mechanical loss factor of GO/PAN1 spinning solutions increased. However, unlike the GO/PAN1 spinning solutions, there was shown different rheological behaviors of GO/PAN2 spinning solutions. There was an increase of ηα, G' and G'' when adding appropriate amount of GO in PAN2 solutions at the same temperature, while tanδ of GO/PAN2 spinning solutions decreased. According to the experimental phenomenon, we think that the interaction between GO and PAN in spinning solutions were forced by electrostatic interactions. When GO contacted with PAN in solutions, the negatively charged GO might be interact with PAN chains, no matter in what kind of solvents. The positively charged PAN2 interacted with GO by electrostatic attraction, resulting that the motion of PAN2 chains was obstructed evidently by GO sheets and the GO/PAN2 solutions exhibited a thickening behavior under the shear stress. However, PAN1 had negative charges in solutions and interacted with GO by electrostatic repulsive. Therefore GO/PAN1 solutions exhibited a thinning behavior under the shear stress.

Keywords: polyacrylonitrile (PAN); graphene oxide; GO/PAN spinning solutions; rheological properties; interaction

Preparation a Novel Fluorescent Emulsion: a Possible Way to Solve the ACQ Effect of Traditional Fluorescent Molecules

Weili Li[1], Jun Wang[1], Zhenyu Qiu[1], Wei Yao[1], Liming Zhou[1], Yixiao Xie[1], Liwei Zhang[1], Shengyuan Yang[2], Meifang Zhu[2]

[1]*School of Material Science and Engineering, National Demonstration Center for Experimental Materials Science and Engineering Education, Jiangsu University of Science and Technology, Zhenjiang 212003, China;*

[2]*State Key Laboratory for Modification of Chemical Fibers and Polymer Materials, Donghua University, Shanghai, 201620, China*

E-mail: just_liweili@163.com

As the commercialized fluorescent organic molecules, rhodamine B, emits intensive fluorescent light in solution state, which is now applied in various areas, such as cellular fluorescence stain, tinted glass, featured fireworks, firecrackers, and so on. However, its fluorescence behaviors can be inhibited in the aggregation state due to ACQ (aggregation caused quenching) effect. In addition, it may be restricted due to its carcinogenicity in the state of small organic molecules. To overcome these flaws and broaden its application fields, in this paper, a novel fluorescent emulsion based on rhodamine B was synthesized and characterized. With the electrostatic attraction force, the fluorescent molecule, was absorbed and dispersed around the latex particles to formed electric double layer by the emulsifying agent, which will avoid its fluorescence quenching. According to this mechanism, the properties of fluorescent emulsion were studied systematically. Firstly, the emulsion polymerization process was studied with real-time monitoring by using of fluorescence spectrum. Comparing the two types of fluorescent emulsion, that is, cationic fluorescent emulsion and anionic fluorescent emulsion, the latter is more stable during the reaction, which might due to the formed more stable electric double layer between anionic emulsifier and the dissolved rhodamine B in neutral or slightly acidic solution solvent. Secondly, the factors that affect the luminescence behavior of fluorescent emulsion were studied systematically by varying the dispersed state of rhodamine B, the concentration of fluorescent molecule & emulsion, and the size of the latex particles. Thirdly, the relationship between fluorescence property of the cured fluorescent emulsion and the testing temperature was studied. As polymer segmental motion behavior depends on the ambient temperature, and it tends to move freely when the ambient temperature exceeds its glass transition temperature, which will influence the fluorescent luminescence behavior of the joint rhodamine B. This relationship may make the fluorescence materials be applied in test. The research findings give a systematic result to study the fluorescent organic molecules and the related polymer materials, and pave a new way to develop the fluorescent materials, such as fluorescent fiber, fluorescent coatings, and so on.

Preparation of Nanofibers by Electrospinning and its Application in Filter Masks

You Li[1,2]

[1]Key Laboratory of Human Factors Engineering, China Astronaut Research and Training Center, Beijing, 100094, China;

[2]Chongqing Sino science and Technology Co., Ltd., China Astronaut Research and training center

E-mail: 76819024@qq.com

In view of the problem of low yield in electrospinning process from laboratory to industrialization, multi batch spinning and no needle spinning were adopted to improve production efficiency. The basic principle and implementation method of electrospinning are briefly described. The situation and characteristics of electrospun polymer fibers, inorganic fibers, coaxial and hollow fibers are introduced. A method for preparing non-woven spinning non-woven fabric by controlling the distribution of high voltage electric field is proposed. The diameter of the nanofiber prepared by this method is about 20~120nm, and the diameter of the three-dimensional nano film is less than 0.3 microns (protection precision: PM0.3), and the filtration efficiency is higher than 99%. In addition, the three-dimensional nanoporous filtration material has small pores and high porosity, and the air resistance is only 100Pa, and the porosity is more than 80%. The nanofiber has been successfully used as an anti fog and haze nanofiber PM0.3 mask as an efficient and low resistance air filter.

Effects of Poloxamer 188 on the Structure and Properties of PLA-CS Nerve Guidance Conduits

Gandong Liu, **Jing Guo, Shuang Chen, Yiyong Ma**

Department of Polymer materials and engineering, College of Material Science and Engineering, Dalian Polytechnic University, Dalian, 116034, China

E-mail: 452328195@qq.com

Currently, tissue engineered nerve conduits instead of autograft nerve to address the limited sources of autologous nerve grafts produced in the clinical repair of peripheral nerve injury, resulting in donor neuroma, scarring and infection, and many other issues. Based on the traditional PLA-CS nerve guidance conduits, poloxamer 188, which has protective effect on cells, is introduced to improve the cell compatibility of PLA-CS nerve guidance conduits and to enhance cell adhesion on nerve guidance conduits to improve neuronal cells Growth in the interior of the nerve tube. The morphology and structure of the neural tube were analyzed by Fourier transform infrared spectroscopy (FT-IR), scanning electron microscopy (SEM), thermogravimetric analysis (TG) and differential scanning calorimetry (DSC), and the cell compatibility of the catheter was tested by co-culture experiments of Schwann cells. The results showed that the addition of poloxamer 188 increases the number of adherents in the Schwann cells within the neural tube, and the cell growth is better(the absence of poloxamer 188 nerve guidance conduits as shown in Figure 1-(a), the addition of poloxamer 188 nerve guidance conduits as shown in Figure 1-(b)).

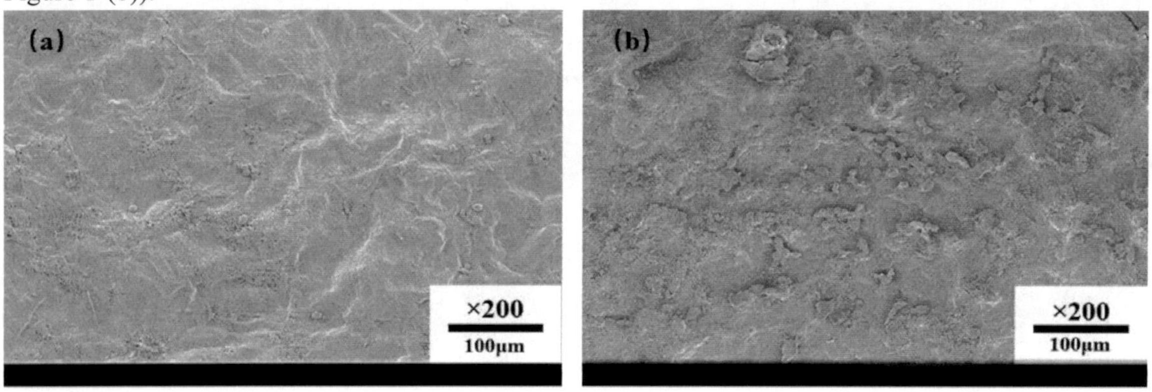

Figure 1. The growth of Schwann cells in the absence of poloxamer 188 nerve catheters(a) and the addition of poloxamer 188 nerve catheters(b).

References

[1] Panseri S, Cunha C, Lowery J, et al. Electrospun micro- and nanofiber tubes for functional nervous regeneration in sciatic nerve transections[J]. *Bmc Biotechnology*, **2008**, 8(1):39.

[2] Serbest G, Horwitz J, Jost M, et al. Mechanisms of cell death and neuroprotection by poloxamer 188 after mechanical trauma.[J]. *Faseb Journal Official Publication of the Federation of American Societies for Experimental Biology*, **2006**, 20(2):308-10.

[3] Chwalek K, Dening Y, Hinüber C, et al. Providing the right cues in nerve guidance conduits: Biofunctionalization versus fiber profile to facilitate oriented neuronal outgrowth.[J]. *Materials Science & Engineering C*, **2016**, 61(4):466-472.

[4] Geng X, Kwon O H, Jang J. Electrospinning of chitosan dissolved in concentrated acetic acid solution.[J]. *Biomaterials*, **2005**, 26(27):5427-5432.

Preparation and Characterization of Halogen-Free Flame Retardant Polyamide 6 Fiber

Ke Liu[1], Yuanyuan Li[1], Jidong Zhang[2], Ru Xiao[1]

[1]*State Key Laboratory for Modification of Chemical Fibers and Polymer Materials, College of Materials Science and Engineering, Donghua University, Shanghai, 201620, P. R. China;*

[2]*State Key Laboratory of Polymer Physics and Chemistry, Changchun Institute of Applied Chemistry, Chinese Academy of Sciences, Changchun, 130022, P.R. China*

E-mail: xiaoru@dhu.edu.cn

Polyamide 6 (PA6) is widely used in the fields of engineering thermoplastics and textiles due to its low density and high mechanical properties. However, PA6 burns easily and the fire caused by the burning of fibers and fabrics has brought enormous losses in recent years. To meet the strict fire safety requirements in some fields, it is necessary to improve the flame retardancy of PA6. Conventionally, flame retardants are introduced to the PA6 via melt blending which is convenient and simple. However, the preparation of fiber is difficult due to large loading of flame retardant for melt blending. In this paper, flame retardant polyamide 6 (FRPA6) resins were prepared by introducing the phosphorus-containing reactive flame retardant (P-FR) and molybdenum compounds during the polymerization of PA6. Then the FRPA6 fibers were obtained by melt spinning. Differential scanning calorimetry, thermogravimetric analysis, vertical burning test, limiting oxygen index test and tensile test were utilized to investigate the properties of FRPA6 samples. The results showed that the P-FR destroyed the structure regularity of FRPA6 and decreased its crystallization. The incorporation of P-FR significantly improved the yield of char residue. The FRPA6 can reach the V-0 rating according to the vertical burning test and the LOI value is over 30%. The tenacity at break of FRPA6 fiber is over 3.0 $cN \cdot dtex^{-1}$ which meets the requirements for textile.

References

[1] Richard Horrocks, Ahilan Sitpalan, Chen Zhou, Baljinder K. Kandola, *Polymers*, **2016**, 8, 288-304

[2] A. R. Horrocks, B. K. Kandola, P. J. Davies, S. Zhang,S. A. Padbury. *Polymer Degradation and Stability* **2005**, 88, 3-12

[3] Zheng Wang, Wei Wu, Yuhua Zhong, Mingzhu Ruan, Lin Lin Hui, *Journal of Applied Polymer Science*, **2015**, 132, 41545-41554

[4] Yuanyuan Li, Yunzhi Lin, Kai Sha, Ru Xiao, *Textile Research Journal,* **2017**, 87, 561-569

A Novel Approach to Design A Nanoporous Polyethylene/Polyester Composite Fabric for Human Body Cooling

Rong Liu[a,b], Xinwei Wang[b,c], JunrongYu[a,*], Yan Wang[a], Jing Zhu[a], Zuming Hu[a]

[a] *State Key Laboratory for Modification of Chemical Fibers and Polymer Materials, College of Materials Science and Engineering, Donghua University, Shanghai 201620, China;*

[b] *State Key Laboratory of Polyolefins and Catalysis, Shanghai Research Institute of Chemical Industry, Shanghai 200031, China;*

[c] *Shanghai Key Laboratory of Catalysis Technology for Polyolefins, Shanghai Research Institute of Chemical Industry, Shanghai 200031, China*

E-mail: yjr@dhu.edu.cn

Currently, some infrared-transparent visible-opaque materials based on nanoporous polyethylene have been developed to improve the personal thermal management. These materials made the comfortable, economical, practical and wearable cloth became reality. In this study, a facile and novel approach was proposed to design a newly ultra-high molecular weight polyethylene (UHMWPE)/polyester composite fabrics with nanoporous structure *via* thermally induced phase separation (TIPS) method. In order to promote the ability of water wicking, mPEG-NH2/PDA (mPPDAPs) particles was used to modify the hydrophilcity of the composite fabric. MPPDAPs/UHMWPE/liquid paraffin (LP) solution was pressed into the holes among the polyester fabric and wrapped the fabric yarns under pressure at a proper temperature, and then the nanoporous structure was formed in UHMWPE phase after cooling down and removing LP by extraction. The newly mPPDAPs/UHMWPE/polyester composite fabric has connective nanoporous structure in the pore size range of 50 to 100 nm. It is an effective, economical and practical textile with infrared-transparent visible-opaque property, block ultraviolet (UV) radiation, sufficient moisture wicking rate, air permeability, and mechanical strength for wearability.

Keywords: UHMWPE/polyester composite; Thermally induced phase separation; Infrared-transparent; Human body cooling

Bladder Acellular Matrix Hydrogel Dually-Reinforced with Silk Fibroin Mats and Biocompatible Cross Linker

Wenjing Liu[1], Jianwen Huang[2], Lujie Song[2], Huili Shao[1], Yaopeng Zhang[1,*]
[1]State Key Laboratory for Modification of Chemical Fibers and Polymer Materials, College of Materials Science and Engineering, Donghua University, Shanghai 201620, P. R. China;
[2]Department of Urology, Shanghai Jiao Tong University Affiliated Sixth People's Hospital, Shanghai 200233, P. R. China
E-mail: zyp@dhu.edu.cn

Complex urethral repair and reconstruction using tissue engineering technology is one of the most promising clinical application researches. Meanwhile, biomaterials composed of extracellular matrix (ECM) have been widely used to repair and reconstruct variety of tissues and organs. As a promising material in clinical and pre-clinical researches, bladder acellular matrix(BAM) hydrogel not only mimics the native extracellular matrix environment, but also releases angiogenic growth factors. However, the poor mechanical properties of BAM hydrogel limited its further application. In this study, silk fibroin (SF) mats were added in BAM hydrogel precursor as a macroscopic reinforcement network of BAM hydrogel. After the gelation of BAM, genipin, a natural plant extract, was applied to cross link the BAM hydrogel in molecular level. Scanning electron microscope (SEM) analyses revealed that pristine BAM hydrogel, genipin cross-linked BAM hydrogel, and SF mats reinforced BAMG hydrogel all have porous structures. Furthermore, the morphologies of the SF mats reinforced composite hydrogel show that SF fibrous mats were evenly embedded in the BAM hydrogel. Fourier-transform infrared spectroscopy (FTIR) analyses show that BAM and SF have strong interaction in the composite hydrogel. Rheological properties suggest that both SF mats reinforced composite hydrogel and genipin cross-linked BAM hydrogel have a higher elasticity modulus than pure BAM hydrogel. The addition of SF mats in the BAMG gel precursor and the immersion of BAM gel in genipin significantly improve the compression strength of BAM hydrogel. Moreover, the BAM hydrogel dually-reinforced with SF mats and genipin exhibits more obvious mechanical improvement than individually reinforced hydrogel. Both BAM and SF in the hydrogel were cross linked by genipin. The reinforced BAM hydrogel may have great potential in further tissue engineering application.

Keywords: Hydrogel, Bladder acellular matrix, Silk fibroin mats, Genipin, Crosslink

Preparation of Paclitaxel/Polysaccharides Micro-nano Fibers Based on Non-Covalent Interactions Assembly

Yongjia Liu[a,b], Tongtong Zhou[a,b], Bangshang Zhu[a*]

[a] *Instrumental Analysis Center, Shanghai Jiao Tong University, 200240 Shanghai, China;*

[b] *School of Chemistry and Chemical Engineering, Shanghai Jiao Tong University, 200240 Shanghai, China*

E-mail: liuyongjia@sjtu.edu.cn; tongtz@sjtu.edu.cn; bshzhu@sjtu.edu.cn

Paclitaxel (PTX) is a well-known drug for cancer therapy, while its application has been restricted by hydrophobility, instability, burst release and low drug loading efficiency. In order to increase the water-solubility and decrease the toxicity, the improved drug delivery systems have been developed for paclitaxel in last decades, including the modified- PTX, liposome, micelles and nanoparticles. Here, we reported a method to establish a paclitaxel-loaded system via co-assembly of PTX and polysaccharides (PSC). The two model systems chosen in this study were paclitaxel/dextran and paclitaxel/chotisan. Dextran (DEX) possesses a large amount of hydroxyl groups and highly water soluble, which was mainly used as blood expander, anticoagulant and imaging agent. Chitosan (CS) is another polysaccharide employed in this research, which is one of most popular materials used in drug delivery system. The diameter of PTX/PSC fibers was less 1000 nm and relying on the specific polysaccharide added. The result of FRIT analysis also demonstrated that polysaccharides have involved in formation of fiber. Furthermore, the drug loading efficiency and drug release in vitro were investigated. The results suggested that these co-assembly fibers have high PTX loading efficiency and the maximum value reached 82% for PTX/DEX and 72% for PTX/CS respectively. The drug release profile of PTX/PSC fibers was sustained and the PTX released amount in fibers were less than 50% over 14 days. So the fibers prepared via the co-assembly of PTX and polysaccharides have a great potential for PTX drug delivery.

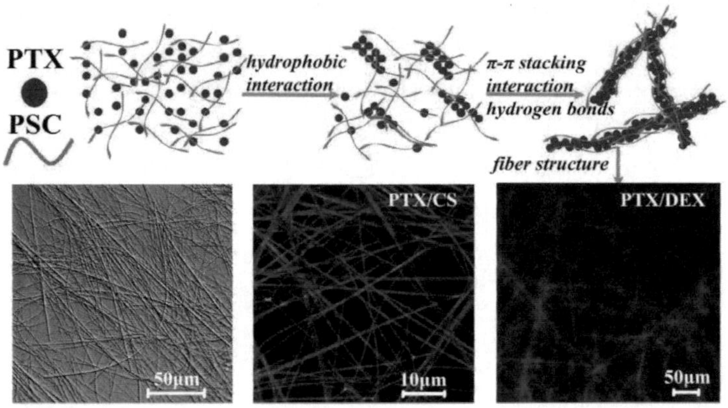

References

[1] De Weger, V.A., Beijnen, J.H. and Schellens, J.H.M., *Anti-cancer drugs*, **2014**, 25, 488-494.

[2] Li, Y., Chen, N., Palmisano, M. and Zhou, S., *Pharmaceutics*, **2015**, 12, 1308-1317.

[3] Li, S., Wang, X., Li, W., Yuan, G., Pan, Y. and Chen, H., *Carbohydrate Polymers*, **2016**, 146, 52-57.

[4] Marupudi, N.I., Han, J.E., Li, K.W., Renard, V.M., Tyler, B.M. and Brem, H., *Expert opinion on drug safety*, **2007**, 6, 609-621.

Chemical Crosslinked EVOH Nanofiber Aerogels with Superelasticity and Superoleophilicity for Removal of Oils and Organic Solvents

Jianwei Lu, Shan Yan, Ru Xiao*

State Key Laboratory for Modification of Chemical Fiber and Polymer Materials, College of Material Science and Engineering, Donghua University, Shanghai 201620, China

E-mail: xiaoru@dhu.edu.cn

Chemical crosslinked thermoplastic polymer–poly(vinyl alcohol-co-ethylene) (EVOH) nanofiber aerogel (NFA) was fabricated employing an economical and facile freeze-drying process, which is an effective material for the purification of water through removing oils and organic solvents. The chemical crosslinking of NFA was characterized by scanning electron microscopy, attenuated total reflection-Fourier transform infrared spectrometer, and X-ray diffraction. The resulting aerogels showed high porosity, superior elasticity, elastic durability, high hydrophobicity (water contact angle >144°), and superoleophilicity. The crosslinked EVOH NFAs exhibited excellent absorption capacity (ranging from 45 to 102 times their own weight) for various oils and organic solvents. Consequently, it is envisaged that the crosslinked EVOH NFA would play an important role in many fields of pollution removal.

Keywords: crosslinked; EVOH NFA; elasticity; superoleophilicity

Effect of Aging Environment on the Properties of Basalt Fiber

Lida Luo[1,2,3], Jiwen Xiao[1], Qingwei Wang[2,3*], Wei Ning[2,3], Jie Gao[1,3]

[1]*College of Materials Science and Engineering of Donghua University, Shanghai, 201620, China;*

[2] *Glass & Enamel Institute of Donghua University, Shanghai, 201620, China;*

[3]*Engineering Research Center of Advanced Glass Manufacturing Technology, Ministry of Education, Donghua University, Shanghai, 201620, China*

E-mail: wqwq888@dhu.edu.cn

Due to their high thermal stability, chemical and mechanical resistance and low cost, basalt fibers were extensively employed in various kinds of substrate materials which are generally kept in alkaline and acidic environment. As iron-rich aluminosilicate glass network, the aging mechanism of the basalt fiber would be studied in various outer circumstances.

Basalt fibers were treated with saturated solution of calcium hydroxide, certain concentrations of Sulfuric acid solution and 95% humidity respectively for different temperature and different periods of time. Both the weight loss rate and the tensile strength of fibers were examined after the treatment. The fiber surfaces were characterized by scanning electron microscopy and energy dispersive X-ray spectroscopy (SEM/EDS). The microstructure was analyzed by Fourier transform infrared spectroscopy (FTIR). Besides inductive coupled plasma emission spectrometer (ICP) was used to test the ion concentration of the solutions.

The results showed that tensile strength of basalt fibers aged in humidity condition declined by 16%, whereas it reduced respectively with 50% and 60% when aged in acid and alkali environment. Helical crack morphology was observed on the surface of basalt fiber which aged in H_2SO_4 after 84 days. The fracture layers seemed to drop off when aged in the $Ca(OH)_2$ after 84 days. However, some punctate particles were observed on the surface when aged in the humid circumstance after 28 days.

Figure 1. SEM images of basalt fibers treated in H_2SO_4 solution after 84 days.

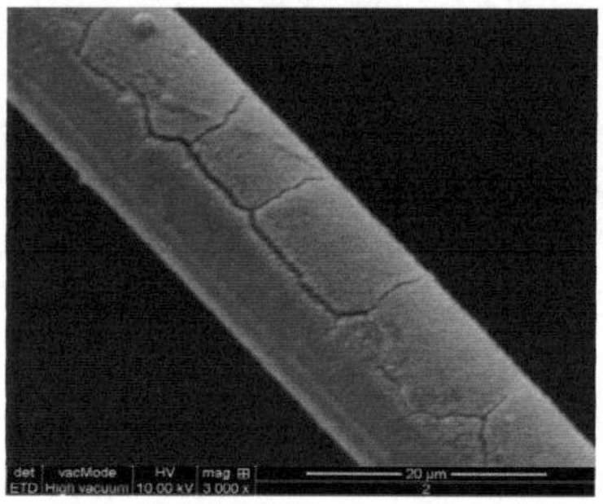

Figure 2. SEM images of basalt fibers treated in $Ca(OH)_2$ solution after 84 days.

Keywords: basalt fiber; corrosion; aging; filament strength

References

[1] V. Dhand, G. Mittal, K.Y. Rhee, S. J. Park, D. Hui, A short review on basalt fiber reinforced polymer composites[J]. *Composites: Part B*, **2015**,73:166-180.

[2] Y.V. Lipatov, S.I. Gutnikov, M.S. Manylov, E.S. Zhukovskaya, B.I. Lazoryak, High alkali-resistant basalt fiber for reinforcing concrete[J]. *Materials and Design*, **2015**,73:60-66

[3] Q. W. Wang, R. K. Brow, H. Li, E. A. Ronchetto, Effect of aging on the failure characteristics of E-glass fibers[J]. *Journal of Materials Science*, **2016**, 51:2404-2410

Effect of Redox States of Iron in Basalt Fiber on its Mechanical Properties

Lida Luo[1,2,3], **Jiwen Xiao**[1], **Qingwei Wang**[2,3*], **Weizhong Jiang**[1,2,3], **Wei Ning**[2,3]

[1]*College of Materials Science and Engineering of Donghua University, Shanghai, 201620, China;*

[2] *Glass & Enamel Institute of Donghua University, Shanghai, 201620, China;*

[3]*Engineering Research Center of Advanced Glass Manufacturing Technology, Ministry of Education, Donghua University, Shanghai, 201620, China*

E-mail: wqwq888@dhu.edu.cn

Basalt glasses belong to the iron-rich aluminosilicate systems. The iron content, Fe^{2+}/Fe^{3+} ratio and the coordination influence the viscosity, density, mechanical and crystallization properties of basalt fiber. Basalt fibers with different iron redox index (Fe^{2+}/Fe_{total}) were prepared by adding different amount of oxidizing and reducing agent to the basalt mineral. The content of iron oxide and ferrous oxide was measured by o-phenanthroline spectrophotometry. X-ray fluorescence spectrometer, X-ray diffractometer, scanning electron microscope, Fourier transform infrared spectroscopy were used to test and characterize the basalt mineral or fiber samples. The crystallization behavior of basalt fiber will not be caused by redox treatment, which were proved in XRD results. After redox treatment, the basalt fiber minimum and maximum iron redox index is 0.262 and 0.804. The iron redox index of untreated basalt fiber is 0.405.The mechanical properties results showed that the tensile strength and breaking strain of monofilament is inversely proportional to the iron redox index. It was inferred that the fiber structure could be strengthened by the ferric oxide as tetrahedrally coordinated network-former. However the ferrous oxide as a free network body could weaken the structure.

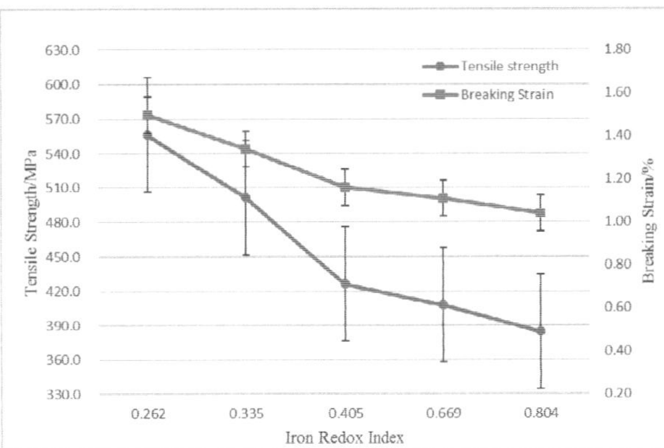

Figure 1. Tensile strength and breaking strain of basalt fibers with different iron redox index.

Keywords: Basalt fiber; Iron redox index; o-phenanthroline spectrophotometry; Tensile strength;

References

[1] S. I. Gutnikov, M. S. Manylov, Y. V. Lipatov, B. I. Lazoryak, K. V. Pokholok, Effect of the reduction treatment on the basalt continuous fiber crystallization properties [J]. *Journal of Non-Crystalline Solids*, **2013**, 368:45-50.

[2] M. S. Manylov, S. I. Gutnikov, Y. V. Lipatov, A. P. Malakho, B. I. Lazoryak, Effect of deferrization on continuous basalt fiber properties [J]. *Mendeleev Communications*, **2015**,25:386-388.

[3] V. Fiore, T. Scalici, G. Di Bella, A. Valenza, A review on basalt fibre and its composites [J]. *Composites Part B*, **2015**, 74:74-94.

Facile Synthesis of Polyaniline-Nanorods/MnO_2 Nanocomposites for Supercapacitor Electrodes with Improved Electrochemical Properties

Mebrahtu Melake Mezgebe, Hongyao Xu, Shanyi Guang

State Key Lab for Modification of Chemical Fibers and Polymer Materials,
College of Materials Science and Engineering, Donghua University,
2999 North Renmin Road, Songjiang, Shanghai, 201600, China
E-mail: josyboy15@yahoo.com

The best electrode candidates for supercapacitors are carbon based nanomaterials, conducting polymers and metal oxides due to their high charge storage characteristics. A binary composite of Polyaniline nanorods and manganese dioxide (PANI-nr/ MnO_2) was designed via hydrothermal method to upgrade the specific capacitance of PANI and improve its cycle stability. The structure and performances were characterized by scanning electron microscope (SEM), Fourier transform infrared spectroscopy (FTIR) and ultraviolet-visible absorption spectroscopy (UV-Vis). Careful study of the electrochemical performance of the electrode has revealed that synergy has been achieved due to the uniform deposition of MnO_2 nanoparticles on the surface of the PANI-nr by strong interaction. The uniform distribution of MnO_2 nanoparticles on PANI-nr increased porosity and surface area of the composite. Specific capacitance of PANI-nr/MnO_2 composite (573.75 F g^{-1}) showed significant increase compared to the individual components like PANI-nr (413.75 F g^{-1}) and MnO_2 nanoparticles (41.25 F g^{-1}) at 1 A/g current density attributed to the synergistic effect of the components. More importantly, retention ratio of the composite electrode could still retain 82% after 1500 cycles compared to the 48% for PANI-nr.

Keywords: Polyaniline; Manganese dioxide; Composite; Supercapacitor

References

[1] Sun X, Gan M, Ma L, et al. Fabrication of PANI-coated honeycomb-like MnO_2 nanospheres with enhanced electrochemical performance for energy storage. *Electrochimica Acta* **2015**, 180: 977-982

[2] Chen W, Tao X, Li Y, et al. Hydrothermal synthesis of graphene-MnO_2-polyaniline composite and its electrochemical performance. Journal of Materials Science: *Materials in Electronics* **2016**, 27: 6816-6822.

[3] Mezgebe M M, Yan Z, Wei G, et al. 3D graphene-Fe_3O_4-polyaniline, a novel ternary composite for supercapacitor electrodes with improved electrochemical properties. *Materials Today Energy* **2017**, 5: 164-172.

[4] Toufiq A M, Wang F, Javed Q U A, et al. Hydrothermal synthesis of MnO_2 nanowires: structural characterizations, optical and magnetic properties. *Applied Physics* A **2014**, 116: 1127-1132.

Synthesis and Controllable Time Reaction of Crosslink Organic-Inorganic Hybrids with Near Infrared Absorption

Nahla Omer, and Hongyao Xu*

College of Material Science and Engineering &State Key Laboratory for Modification of Chemical Fibers and Polymer Materials, Donghua University, Shanghai 201620, China

Email: nahala61@gmail.com

Semisquaraine and squaraine dyes have been the subject of many recent investigations owing to their excellent ultra-strong absorption properties ($\varepsilon \geq 10^5 L \cdot mo^{l-}1 \cdot cm^{-1}$), molecular design and low cost. But these kinds of materials suffering from poor thermal and optical stability. In this work, octavinyl-polyhedral oligomeric silsesquioxane OV-POSS have used to improve the properties of the design. A novel structure of organic-inorganic hybrids optical material, synthesis by OV-POSS with 6-Bromoquanaldine and Squaric acid (semisquaraine, squaraine) at different time. Hence, fixed the best time of theses reaction with high yield see **figure 1**. So the system was characterized by Fourier Transform Infrared Spectroscopy FTIR, UV-Vis-Spectra,^{1}HNMR Spectra. When hybrid applied as sensitizers in solar cell DSSCs exhibited good performance.

Keywords: OV- POSS hybrids, near-IR dyes, semi-squaraines, squaraines, Time reaction.

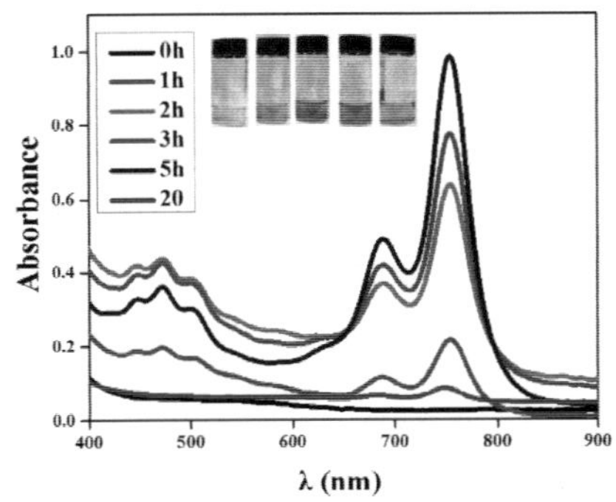

Figuer1 UV-vis absorbance spectra of OV-POSS with 6-Bromoquanaldine and Squaric acid change in absorption spectra at various time intervals. 0h,1h,2h,3h,5h,and 20h in DMF.

References

[1] Guo, Wenxi, Xu, Zijie, Zhang, Fayin, Xie, Shuyao, Xu, Hongyao, Liu, Xiang Yang, *Advanced Functional Materials*, **2016**, 26, 8855~8884

[2] Naibo Lin, Fan Hu, Yilin Sun,Chenxu Wu, Hongyao Xu, Xiangyang Liu. *Adv. Funct. Mater*, **2014**, 24, 5284–5290.

[3] Yan, Z.; Xu, H.*; Guang, S.; Zhao, X.; Fan, W.; Liu, X. Y., *Advanced Functional Materials*, **2012**, 22, 345–352.

Acknowledge: This research was financially supported by the National Natural Science Fund of China (Grant Nos. 21671037, 21471030, 21771036).

Study on the Interfacial Tension of Immiscible Polystyrene/Polypropylene Blend with Improved Deformed Drop Retraction Method

Dan Pan, Houkang He, He Ren, Min Cao, Long Chen*, Zongyi Qin*

College of Material Science and Engineering, State Key Laboratory for Modification of Chemical Fibers and Polymer Materials, Donghua University, Shanghai, 201620, China

E-mail: chnspan@163.com

Interfacial tension is one of the major factors governing the compatibility of polymer blends and plays a major role in controlling the morphological structure and the final state properties of polymer blends. Therefore, it is necessary to quantify the effect of process condition and raw material property, such as the temperature, pressure, molecular weight and molecular weight distribution, on the interfacial tension and the morphological structure of polymer blends during the development of novel polymer materials. Especially in melt spinning of blend fibers, there is an obvious non uniform distribution of temperature along the spinning line and radial direction, which result in a non uniform morphological structure within blend fibers, as showed in our previous researches.[1,2] So the better understanding of the quantitative relation between the temperature, molecular weight and interfacial tension of polymers will give us a deep insight into the formation of special morphological structure, and help us precisely control the morphological structure and the final state properties of polymer blends. The measurement of interfacial tension of polymer pairs has been studied by many researchers and the methods are summarized in review.[3]

In this paper, the effect of temperature and molecular weight on the interfacial tension of PS/PP, using four different molecular weights PS and one single PP was evaluated with IDDRM. Series of methods were used to improve the accuracy of measurement. The typical retraction process of imbedded PS fiber in PP matrix is presented in Figure 1.

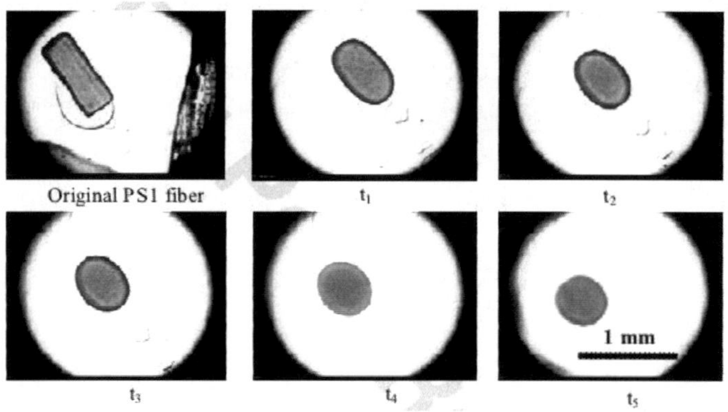

Figure 1. The shape evolution of PS fiber imbedded in PP at 220 °C.

A quantitative relation between interfacial tension, temperature and molecular weight was established based on mathematic fitting, and would be used to simulate the evolution of morphology of polymer blends during the process, as shown in Figure 2. Even though we cannot obtain a general equation for all PS/PP systems. This also gives us a possibility to quantify the effect of interfacial tension on the morphology during the process of polymer blend in nonisothermal flow field.

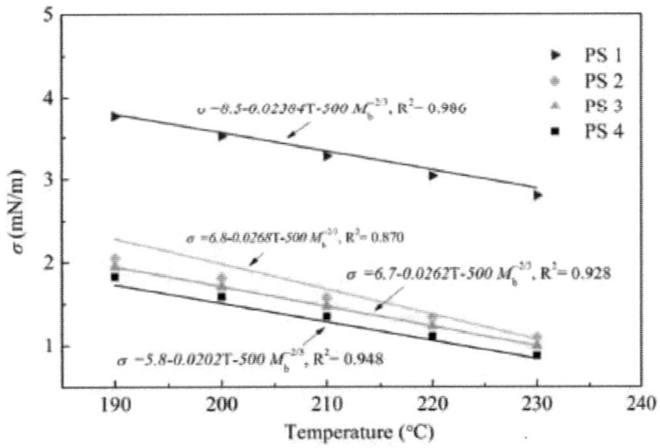

Figure 2. Simulation of interfacial tension for PS/PP blends

Reference

[1] Pan ZJ, Zhu MF, Chen YM, et al. *Fiber Polym,* **2010**; 11: 249-257.

[2] Xing Q, Zhu M, Wang Y, et al. *Polymer,* **2005**; 46: 5406-5416.

[3] Demarquette NR, De Souza AMC, Palmer G, et al. *Polym Eng Sci,* **2003**; 43: 670-683.

A Knittable Fiber-shaped Supercapacitor based on Commercial Polyacrylonitrile Thread for Wearable Electronics

Wei Pan,[a,b]* **Yuerong Ba,**[a,b] **Dong Liu**[a]

[a]*School of Materials and Chemical Engineering, Zhongyuan University of Technology, Zhengzhou, 451191, China;*

[b]*Collaborative Innovation Center of Textile and garment industry, Henan Province, Zhengzhou, 451191, China*

E-mail: doctorpan0152@163.com

In recent years, with the rapidly increasing demands for wearable and portable consumer electronics, wearable technology has attracted great interest. Innovatively, the smart garments, which is a high-level pattern for functional wearable electronics integrating into daily clothing, are potentially used into multifunctional sports wears and military equipments.[1,2,3] Fibriform supercapacitor (FSC), which is a new type of flexible energy storage device, has received great attention due to its advantages including tiny volume and lightweight. More attractively, this kind of supercapacitors can be knitted into textile without sacrificing the original softness of textile fabrics [4, 5]. Accordingly, FSC has rapidly developed and turned into an important type of energy storage system. The core objective in designing FSC is to endow it intelligent versatility while maintaining its comfortability. In this regard, fabricating functional fiber by using existing textile is rather promising.[6,7]

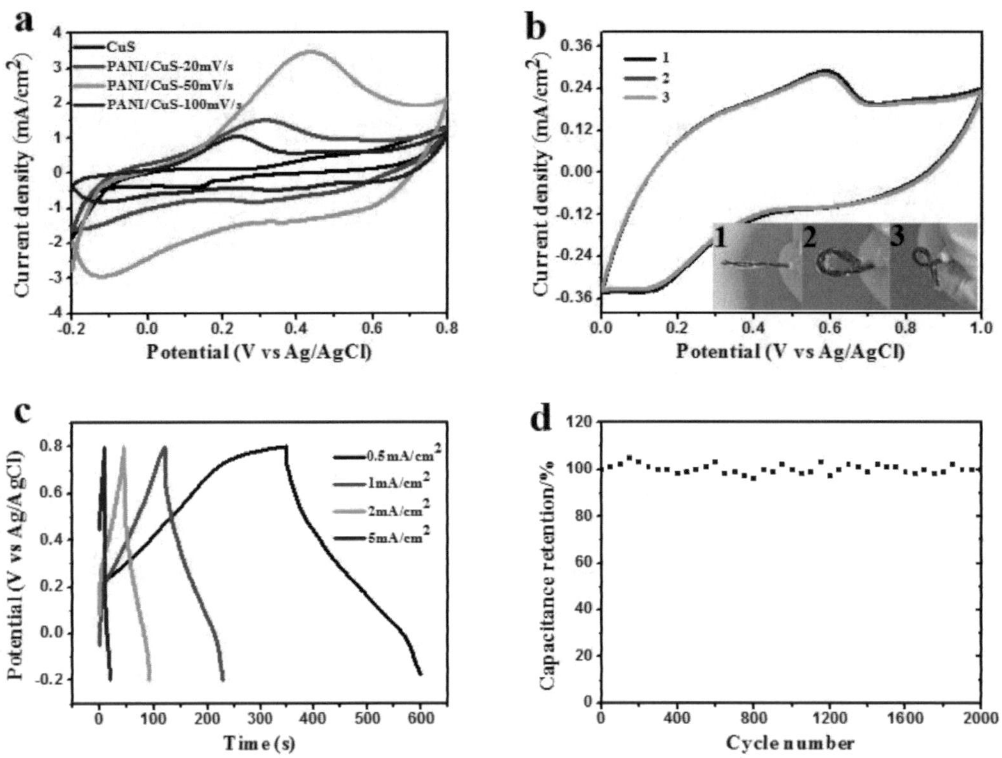

Figure 1. (a) CV curves of CuS-PAN thread, PANI/CuS-PAN thread-20mV/s, PANI/CuS-PAN thread-50mV/s and PANI/CuS-PAN thread-100mV/s at a constant scan rate of 10mV/s; (b) CV curves of the PANI/CuS-PAN thread-50mV/s symmetric supercapacitor under different bending angles at 10 mV/s; (c) GCD curves of PANI/CuS-PAN thread-50mV/s at different current densities; (d) cycling stability of the PANI/CuS-PAN thread-50mV/s symmetric supercapacitor at current density of $0.5mA/cm^2$.

In this work, we successfully converted commercial polyacrylonitrile(PAN) threads into electrically conductive and electrochemically active threads by introducing copper sulfide (CuS) and polyaniline(PANI) via a simple chemical bath deposition and electrochemical deposition.The prepared PANI/CuS-PAN thread derived from commercial polyacrylonitrile thread is flexibility and mechanically strong.The supercapacitive behavior of the electrodes was investigated using cyclic voltammetry, galvanostatic charge-discharge and electrochemical impedance spectroscopy techniques. Under the current density of 0.5 mA/cm^2, the PANI/CuS-PAN electrode exhibited a specific capacitance of 136 mF/cm^2 and retains 98% when current density is increased 10-fold to 5mA/cm^2.The interconnected network structure of PANI can increase the surface area as well as the electrode/electrolyte interface area, thus resulting in improved electrochemical performance.

Fiber-shaped supercapacitor based on the prepared fiber electrodes and polyvinyl alcohol-H_3PO_4 gel electrolyte was fabricated. The flexible supercapacitor showed a specific capacitance of 19.1 mF/cm^2 when the current density was 0.5mA/cm^2 and a high stability retained almost 100% of the initial capacitance after 2000 cycles of charge-discharge at a current density of 0.5mA/cm^2. Due to its good mechanical and electrochemical properties, the PANI/CuS-PAN thread based all-solid fiber-shaped supercapacitor can be directly knitted into fabrics and maintain its original capacitive performance. Such a low-cost textile thread based versatile energy storage device may hold great potential for future wearable electronics applications.

References

[1] R.F. Service, *Science*, **2003**, 301, 909-911.

[2] K. Jost, G. Dion, Y. Gogotsi, *J. Mater. Chem. A*, **2014**, 2, 10776-10787.

[3] X. Wang , X. Lu , B. Liu , D. Chen , Y. Tong , G. Shen , *Adv. Mater.* , **2014**, 26, 4763-4771.

[4] J.A. Lee, M.K. Shin, S.H. Kim, H.U. Cho, G.M. Spinks, G.G. Wallace, M.D. Lima, X. Lepro, M.E.Kozlov, R.H. Baughman, S.J. Kim, *Nat. Commun.* , **2013**, 4, 1970-1978.

[5] J. Bae, M.K. Song, Y.J. Park, J.M. Kim, M. Liu, Z.L. Wang, *Angew. Chem. Int. Ed.*, 2011, 50, 1683-1687.

[6] W.Weng, P. Chen, S. He, X. Sun, H. Peng, *Angew. Chem. Int. Ed.*, **2016**, 55, 6140-6169

[7] L.Dong, C. Xu, Y. Li, Z.-H. Huang,F. Kang, Q.-H.Yang, X. Zhao, *J. Mater. Chem.* A, **2016**, 4, 4659- 4685.

Ring-Opening Metathesis Polymerization of Norbornene Derivatives with Long Flexible Functionalized Side Group

Haiyan Pei [a*], Yingliang Liu [b], Dongfang Wang [b], Jin Wang[b], Chunjiang Wang[b], Shaokui Cao [b]

[a] *School of Materials and Chemical Engineering, Zhongyuan University of Technology, Zhengzhou 450002 China;*

[b] *School of Materials Science and Engineering, Zhengzhou University, Zhengzhou 4500052 China*

*E-mail: peihaiyan2008@126.com

Fully functionalized photorefractive polymers have attracted much attention in recent years[1-3]. While the synthesis of chromophore-containing functionalized monomers is not a trivial task, and the polymerization of functionalized monomer normally gives a low molecular weight owing to both the monomer bulkiness and the side effect of functional group on polymerization, which is not beneficial to the film-forming property[4-6]. Recent development on ring-opening metathesis polymerization (ROMP) of cyclic olefinic derivatives provides an excellent approach, especially the ROMP of norbornene functional monomers[7-9]. With the aim of fully functionalized photorefractive polymers bearing a non-conjugated main chain (NCMC), the carbazole-based functionalized monomers were designed and synthesized by introduction of the norbornene groups.

In this paper, three norbornene monomers with long flexible functionalied side group were designed and synthesized. One was without chromophores and the other two was with azobenzene chromophore and methine chromophore respectively. Corresponding polymers were prepared via ring-opening metathesis polymerization (ROMP) using Grubbs 2 as a catalyst (Scheme 1).

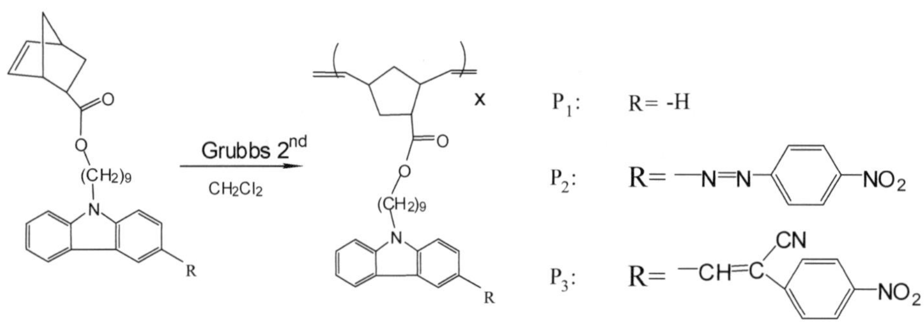

Scheme 1. Synthetic route for polymers.

These three polymers were characterized at the first stage by using IR, UV-Vis, GPC and DSC etc. Molecular weight ($\overline{M}_n$) of the polymers are 13.1×10^4, 5.0×10^4, and 5.4×10^4 respectively as measured by GPC. The glass transition temperature (T_g) of the polymers are 18 °C, 43 °C and 80 °C respectively (Figure 1). The thermogravimetrical analysis shows that the polymers of P_2 and P_3 start to decompose from 260 °C, which is stable enough for normal PR applications. The characterization of photorefractive performance of P_2 and P_3 is in due course.

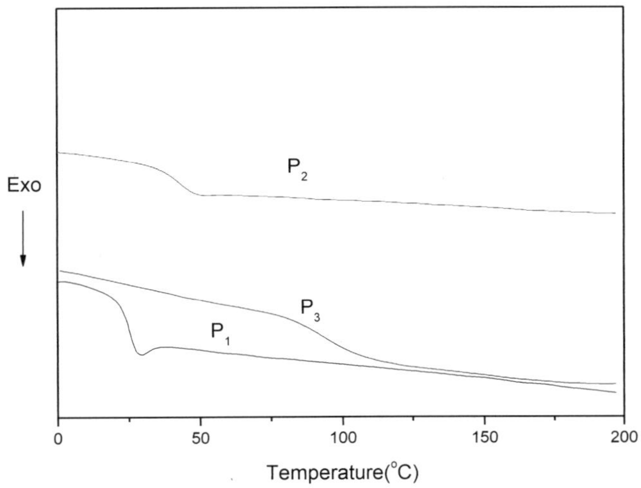

Figure 1. DSC curves of the polymers P_1, P_2 and P_3.

Keywords: ROMP, chromophore, norbornene

References

[1] S. Ducharme, J. C. Scott, R. J. Twieg, W. E. Moerner, *Phys Rev Lett*, **1991,** 66, 1846-1849.

[2] O. Ostroverkhova, W. E. Moerner, *Chem Rev*, **2004,** 104, 3267-3314.

[3] O. A. Scherman, H. M. Kim, R. H. Grubbs, *Macromolecules,* **2002,** 35, 5366–5371.

[4] Shi J, Huang MM, Chen ZJ, Gong QH, Cao SK. *J Mater Sci,* **2004,** 39, 3783–3785.

[5] Zhang L, Shi J, Huang MM, Jiang ZW, Yang Z, Chen ZJ, Gong QH, Cao SK. *Polymer,* **2008,** 49, 2107-2114.

[6] Zhang L, Huang MM, Jiang ZW, Yang Z, Chen ZJ, Gong QH, Cao SK. *React. Funct Polym ,* **2006,** 66, 1404-1410.

[7] O. A. Scherman, H. M. Kim, R. H. Grubbs, *Macromolecules,* **2002,** 35, 5366–5371.

[8] D. J. Liaw, K. L. Wang, K. R. Lee, J. Y. Lai, *J Polym Sci: Part A: Polym Chem,* **2007,** 45, 3022–3031.

[9] Christopher W. Bielawskia, Robert H. Grubbs, *Prog Polym Sci,* **2007,** 32, 1–29.

Functionalization of Activated Carbon Fiber with Nano Titanium Dioxide as Efficiency Filter Material for Formaldehyde Removal

Yidai Ying[1], Shirui Cui[1], Zhaorong Zhang[2], Kemei Pei[1,*]

[1]*Department of Chemistry, Zhejiang Sci-Tech University, Hangzhou 310018, China;*

[2]*Hangzhou Benniao science and Technology Co Ltd, Hangzhou 310051, China*

E-mail: peikemei@zstu.edu.cn

Activated carbon fiber (ACF) is a common material in air or water purification[1]. As we know, the most common activated carbon fiber is more proleophilic than hydrophobic, therefore its removal ability for formaldehyde is weak. The superior photocatalytic performance of TiO_2 has been known for almost 40 years, and TiO_2 is widely used to removal organic pollutants[2]. As we know, TiO_2 is difficult to control for it's a fine powder, and developing the carriers to load TiO_2 is a hot issue for research. In this work, the functionalization of activated carbon fiber by nano TiO_2 was reported as a formaldehyde removal material for water or air purification, and similar behaviour can be expected for activated carbon. Dopamine(DA) polymer was used as adhesive for ACF and TiO_2. This work supplies a method for development of carbon-supported hybrid catalyst for removal pollutants in air or water.

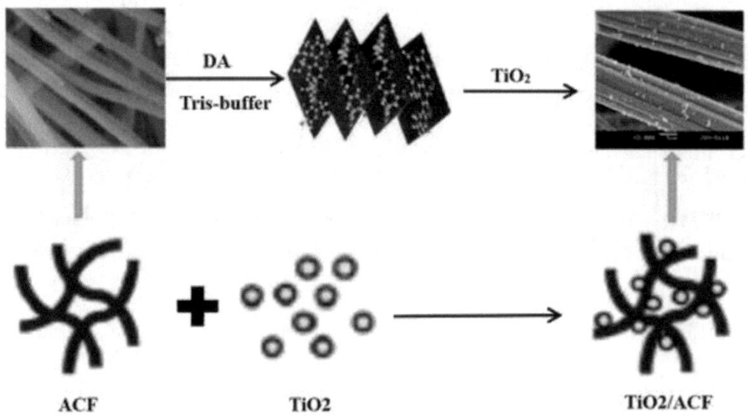

Figure 1. Schematic illustration of the preparation procedure of TiO_2/ACF.

References

[1] T. Salthammer, *Angewandte Chemie International Edition*,**2013**,52,3320-3327.

[2] Weiyi Cui, Dan Xue, Xiaoling Yuan, Bin Zheng, Mingjun Jia, Wenxiang Zhang, *Applied Surface Science*, **2017**, 411, 105-112.

An Ultra-sensitive and Stretchable Nanofiber-based Flexible Sensor for Electronic Skin Applications

Kun Qi[1,2], **Jianxin He**[2], **Hongbo Wang**[1], **Yuman Zhou**[1,2], **Xiaolu You**[2], **Nan Nan**[2]

[1]*School of Textile and Clothing, Jiangnan University, Wuxi, 214122, China;*

[2]*Collaborative Innovation Center of Textile and Garment Industry, Zhengzhou 450007, China*

E-mail: qikuntwins@163.com

An ultra-sensitive and stretchable sensor based on composite nanofibers is successfully fabricated through the electrospinning and in-situ polymerization process. This sensor which exhibited an incorporation of ultrahigh sensitivity to tensile strain deformation and wide maximal sensing range was constructed using an electrospun polyurethane (PU) nanofiber membrane as the flexible substrate and in situ polymerization of conductive polymer poly(3,4-ethylenedioxythiophene) (PEDOT) on the oxygen plasma-treated nanofibers. The results show PEDOT is successfully synthesized on the surface of PU nanofiber and PEDOT/PU composite nanofibers possess skin-core structure. In addition, this sensor showed highly sensitive response towards tensile strain with low working voltage (1 V) and wide sensing range (up to 200% strain), it also exhibited excellent long-term stability under 50,000 cycles and can detect both large- and small scale human motions. For this study, wearable electronics are fabricated with this sensor that can monitor diverse human motions, including joint movement, phonation, swallowing, and breathing.

Keywords: Electrospinning ; Polyurethane ; PEDOT; Nanofiber ; Stretchable senor

Preparation of Conductive Textile Electrodes and Their Properties

Haiming Qin, **Wanwan Zhou, Junrong Li, Liying Qian***

School of Light Industry and Engineering, South China University of Technology, Guangzhou, 510641, China

E-mail: 794376242@qq.com

In recent years, the Marathon attracts more and more people to join in. Sudden death in the Marathon is a potential event risk. Electrocardiogram (ECG) includes meaningful information about the person's heart and performance and is a standard procedure in current cardiac medicine. There has been a particular interest in wearable electronics for electrocardiogram recording recently. The discovery of new applications in ECG devices that provide sportsman with the caution at the right time and will help to avoid or reduce the possibility of sudden death of Marathon. The electrocardiograph was a bulky and main-powered instrument with multiple cables and electrodes originally. A special electrode as interface between the body and the device is needed to catch biosignals. Wearable electrodes based fabrics can monitor the ECG signal easily and anywhere-anytime and contact with the body without skin irritation. In this paper, first, silver-coated chitosan fibers with excellent conductivity was successfully obtained by electroless coating using glucose as a reducing agent. The morphology of chitosan/Ag nanocomposite fabrics were examined by X-ray diffraction (XRD) and scanning electron microscopy (SEM). Chemical composition on fiber surface was examined by energy dispersive spectroscopy (EDS). The hydrophobicity of surface, thermal stability and mechanical property of the chitosan/Ag nanocomposite fabrics were investigated. SEM-EDS and XRD indicated that the silver particles on the chitosan fiber surface are in the metallic state and plated compactly and completely. The morphology of chitosan fiber before and after silver coating indicated a strong interaction between the silver layer and the chitosan fiber. Electroless coating showed slight positive effect on the thermal stability of chitosan fiber. The four-point probe resistivity meter was used to study the electrical resistivity of the silver-coated chitosan fibers. The treated fabrics displayed a very low surface resisitance of 20.23mΩ/square. After 40 washing times, the surface resisitance was still below 1Ω/square. Then, to evaluate the performance of the conductive textile, we apply conductive textile as our ECG electrode by embeddable in a sportswear and use it to measure ECG signals. The result revealed the capability of the proposed electrodes of silver-coated chitosan fibers to accurately capture the ECG signal morphology.

Preparation and Properties of Polyamide 6 Fiber Nibs

He Ren, Junfen Sun, Dan Pan, Long Chen*

College of Material Science and Engineering, State Key Laboratory for Modification of Chemical Fibers and Polymer Materials, Donghua University, Shanghai, 201620, China

E-mail: rhdhuedu@163.com

Fiber nib is a kind of composite material formed by fiber and polyurethane adhesive[1]. According to the character, fineness and quantity of fiber and the types and usage of adhesives, it can be made into various fiber nibs for whiteboard pen, fluorescent pen, watercolor pen, brush and cosmetic pen. At present, there are three main kinds of fibers used in fiber nibs, polyester fiber, polyamide 6 fiber and acrylic fiber. Polyamide 6 fiber has been widely used in making fiber nibs for brush and cosmetic pens because of its high strength, high toughness, good abrasion resistance and excellent hydrophilicity. Generally, we use hardness, water absorption and water delivery time to evaluate the quality of a fiber nib. So far, the quantitative relationship between fiber structure and the properties of fiber nibs has not been established. We know that water absorption is related to the number of pores in the fiber nibs and the water delivery time is related to the pore size of the fiber nibs. According to the formula of capillarity: h=, the fiber nibs with smaller pore size have the stronger capillary attraction [2,4]. However, the influence of fiber cross section shape and fineness on the pore number and pore size of fiber nibs needs to be explored.

In this paper, the polyamide 6 fiber with different fineness and different cross section shapes and the polyurethane adhesives were used to prepare five kinds of fiber nibs. The effect of fiber cross-section shape and fineness on the hardness, pore size distribution, water absorption and water delivery time of fiber nibs were discussed.

The fiber nibs which made of profiled fiber have smaller pore size and higher water absorption than the fiber nibs which made of round fiber, as shown in Figure 1. Among of them, the fiber nibs which made of cross fiber have the highest water absorption rate, reaching 61.86%, and the shortest water delivery time, only 40.58s, as shown in Figure 2.

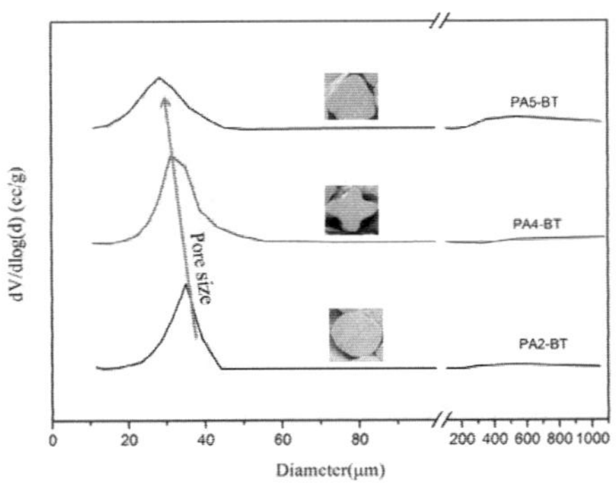

Figure 1. Pore size distribution curve of fiber nibs with fiber which has different cross section shape.

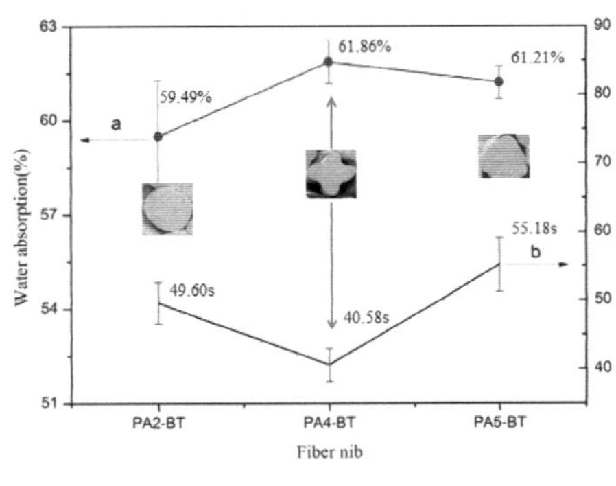

Figure 2. Water absorption curve and water delivery time curve of fiber nibs with fiber which has different cross section shape.

For the fiber with round section shape, with decrease of fiber fineness, the pore size and water delivery time of the fiber nibs are decreased gradually, but the water absorption of the fiber nibs are gradually increased. The fiber nibs whose fiber fineness is 1.16dtex have the highest water absorption rate, reaching 60.43%, and the shortest water delivery time, only 46s. In addition, the fiber with 1.16 dtex fineness is too small to fill the fiber nibs, there are larger pores in the middle of the fiber nib which is difficult to produce.

References

[1] Black D A., *Journal of Criminal Law Criminology & Police Studies*, **1967,** 57(4):521-525.
[2] Hsieh Y L, Yu B, Hartzell M M., *Textile Research Journal*, **1992**, 62(12):697-704.
[3] Yanılmaz M, Kalaoğlu F., *Textile Research Journal*, **2012**, 82(8):820-831.
[4] Crow R M, Osczevski R J., *Textile Research Journal*, **1998**, 68(4):280-288.

Growth of C_3N_4 Nanosheets on Carbon-fiber Cloth as Flexible and Macroscale Filter-membrane-shaped Photocatalyst for Degrading the Flowing Wastewater

Xiaofeng Shen,[1] Zhigang Chen,[2] Jianshe Liu,[1] Lisha Zhang[*,1]

[1]*State Key Laboratory for Modification of Chemical Fibers and Polymer Materials, College of Environmental Science and Engineering, Donghua University, Shanghai 201620, China;*

[2]*College of Materials Science and Engineering, Donghua University, Shanghai, 201620, China.*

E-mail: lszhang@dhu.edu.cn

Photocatalytic technology has been demonstrated to be an efficient and cheap method to degrade organic contamination[1]. The filtration technology has been widely used in the purification of flowing water[2]. If the filtration technology is combined with photocatalysis process, undoubtedly the pollutants can be blocked, concentrated and then degraded. Herein, we report the design and preparation of C_3N_4 nanosheets on carbon-fiber (CF) cloth as filter-membrane-shaped photocatalyst[3]. The growth of C_3N_4 nanosheets has been realized by using a dip-coating and thermal condensation method with carbon-fiber cloth as the substrate. The resulting cloth (area: 4×4 cm^2) is composed of carbon fiber (diameter: 15 μm) bunches which are decorated with C_3N_4 nanosheets with the thickness of 30 nm. CF/C_3N_4 cloth exhibits excellent flexibility and strong visible-light absorption at ~ 450 nm. When CF/C_3N_4 cloth (area: 4×4 cm^2) is floated on the polluted water, it can degrade 98% Rhodamine B (RhB) in 60 min and 99.3% colorless parachlorophenol (4-CP) after 120 min of visible-light irradiation. Interestingly, when CF/C_3N_4 cloth as the filter-membrane is used to construct a new photocatalytic setup for degrading the flowing wastewater (rate: 1.5 L h^{-1}) under visible-light irradiation, the degradation efficiency of RhB goes up from 18% to 92% with the increase of filtering/degrading grade from 1 to 7. Therefore, CF/C_3N_4 cloth has great potential to be used as an efficient, stable, macroscale filter-membrane-shaped photocatalyst for degrading the flowing wastewater.

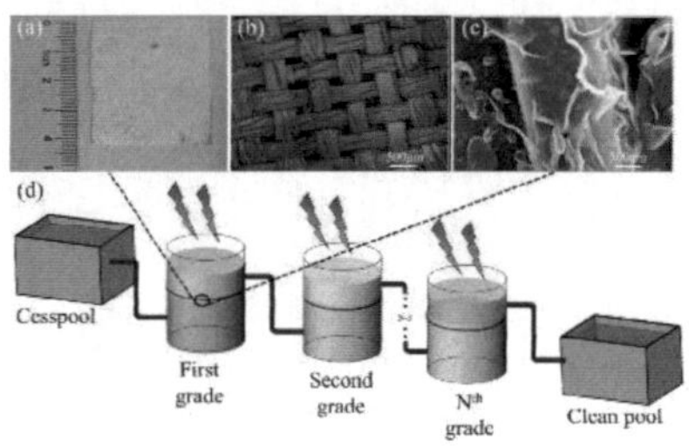

Figure 1. The photo (a) and SEM images (b, c) of CF/C_3N_4 cloth, (d) Schematic diagram of the seven photocatalytic reactors for RhB removal in series

References

[1] H. Wang, L. Zhang, Z. Chen, J. Hu, S. Li, Z. Wang, J. Liu and X. Wang, *Chem. Soc. Rev.*, **2014**, 43, 5234-5244.

[2] J. Zheng, J. Ma, Z. Wang, S. Xu, T.D. Waite, Z. Wu, *Environ. Sci. Technol.* **2017**, 5, 12757-2765.

[3] X. Shen, T. Zhang, P. Xu, L. Zhang, J. Liu and Z. Chen, *Appl. Catal. B: Environ.*, **2017**, 219, 425-431.

Core-shell Nanocapsules Containing Essential Oil for Textile Application

Cong Liu[1], Baijun Liang[1], Yuhai Wang[1*], Yingxian Li[1], Guang Shi[1]

School of Chemistry and Environment, South China Normal University, Guangzhou, 510006, China

E-mail: shiguang@scnu.edu.cn

A novel polysiloxane-modified aromatic nanocapsule with hard core and soft shell has been prepared by two-stage emulsion polymerization method. The core of the nanocapsule is made up of styrene/methyl methacrylate (St/MMA) copolymer, while the shell is polybutyl acrylate (PBA) modified by octamethylcyclotetrasiloxane (D4). Scanning electron microscopy (SEM) and transmission electron microscope (TEM) images reveal well-defined core-shell morphology of the particles, and the average particle size is about 100 nm. UV-Vis analysis shows that the essential oil can be well encapsulated by the nanocapsules and the encapsulation efficiency and loading capacity are 78.76% and 26.25%, respectively. The residual rate of essential oil is about 25.3% after being washed 15 times, which indicates that the nanocapsules have excellent washing durability. These core-shell aromatic nanocapsules show great potential application in fabrics treatment.

Key words: nanocapsule; core-shell structure; essential oil; properties and characterization; textiles

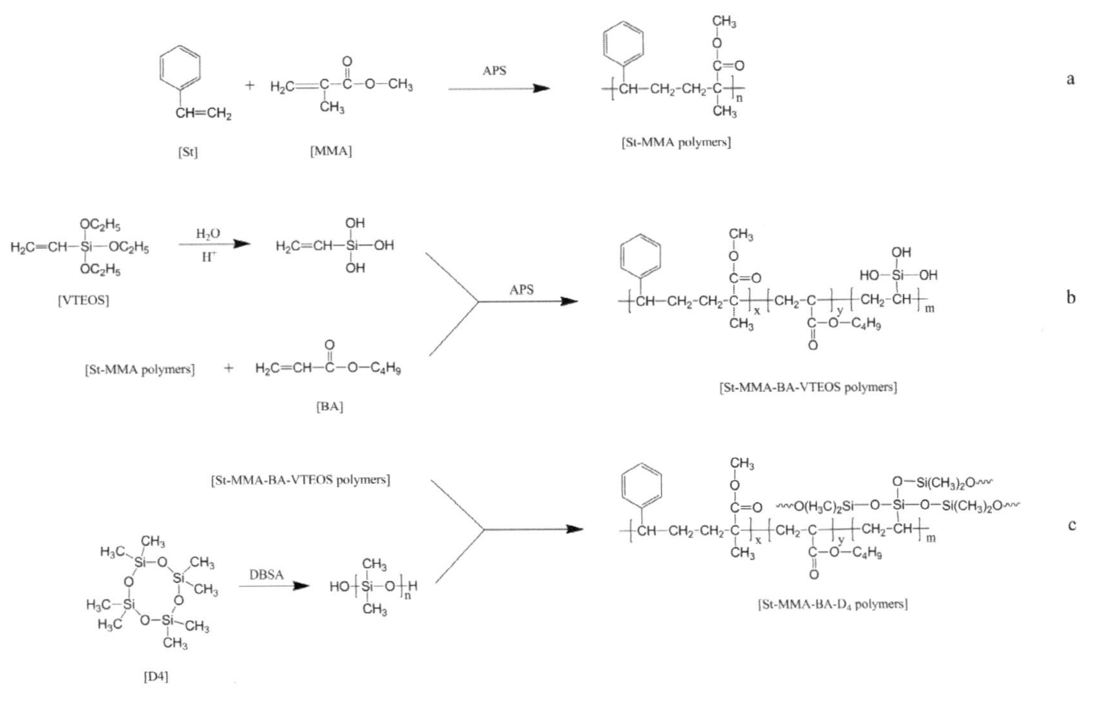

Figure 1. Figure 1 Schematic presentation of possible polymer formation and the synthetic design of P(St-MMA)/PBA core-shell nanocapsule.

a The hard core of P(St-MMA) obtained by copolymerization of St and MMA.b P(St-MMA-BA-VTEOS) obtained by copolymerization of P(MMA-St), BA and silane coupling agent VTEOS. c P(St-MMA)/PBA modified by PDMS formed by octamethylcyclotetrasiloxane (D4) ring-opening and VTEOS as coupling agent.

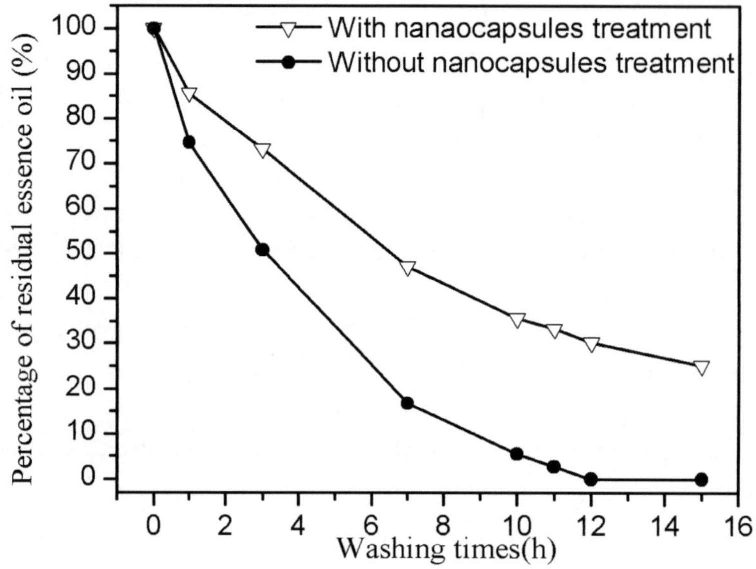

Figure 9. Essential oil content of the fabrics treated with and without nanocapsules after different washing times.

References

[1] Liu C.; Liang B.J.; Shi G.; Li Z.C.; Zheng X.B.; Huang Y.; L. Lin. **Flavour Frag. J. 2015**, 30, 295.

[2] Xiao Z.; Tian T.; Hu J.; Wang M.; Zhou R. **Flavour Fragr. J. 2014**, 29, 22.

An Ultra-high Performance Fiber Catalyst for Biomass Conversion

Xianlei Shi, **Qianqian Hu**, **Yongju Chen**

Henan Polytechnic University, Jiaozuo, Henan 454003, China

E-mail: shixl@tju.edu.cn

A commercially available ultra-high performance fiber, capable of acting as a novel support for heterogeneous Brønsted acid catalyst for the conversion of biomass components to methyl levulinate, is presented. The catalyst was prepared facilely from polyphenylene sulfide fiber according to a simple one-step procedure by sulfonation, and characterized detailedly with elemental analysis, FTIR spectroscopy, and SEM during the modification and utilization processes. Moreover, the fiber catalyst exhibited prominent activity for the catalytic conversion of biomass components such as microcrystalline cellulose, starch, inulin, sucrose, glucose, fructose, 5-hydroxymethylfurfural and furfuryl alcohol to methyl levulinate (51.9%-95.7%), and the ability to be very easily recycled with superior recyclability and stability over 5 runs, as well as the performance of effective large-scale process which in turn is very attractive for the industrial application. This is the first report about the utilization of the polyphenylene sulfide fiber as a new support for heterogeneous catalysis in the practice of biomass molecules conversion research.

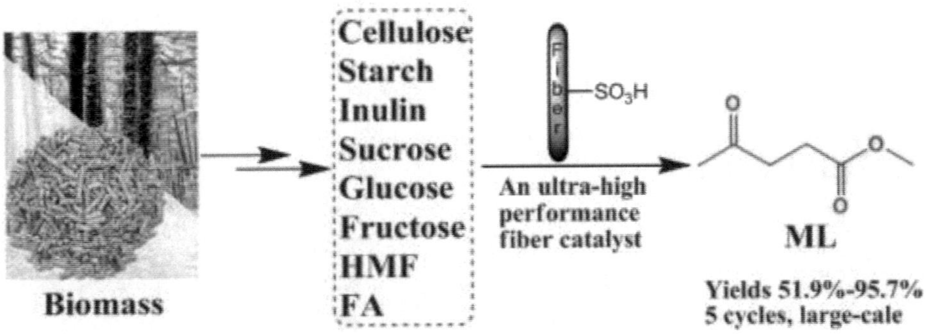

Figure 1. The diagrammatic sketch.

References

[1] L. Shi, Q. Hu, F. Wang, W. Zhang, P. Duan, *J. Catal.*, **2016**, 337, 233-239.

[2] X.-L. Shi, M. Zhang, Y. Li, W. Zhang, *Green Chem.*, **2013**, 15, 3438-3445.

Preparation and Fluorescence Property of PET Luminescence Fibers with Different Rare Earth Complexes

Dengkun Shu, Peng Xi*, Shuwang Li, Bowen Cheng*

Tianjin Key Laboratory of Advanced Fibers and Energy Storage, Tianjin Polytechnic University, 399 Bin Shui West Road, 300387, Tianjin, P. R. China

E-mail: dengkunshu@sina.com

PET was selected as the fiber-forming polymer in this paper, and the solvent was hexafluoroisopropanol (HFIP). Different rare earth luminescence fibers were prepared by means of adding different rare earth complexes to system of spinning solution by electrospinning technique, and the rare earth complexes included p-methylbenzoic acid-Tb rare earth complex, para-chlorobenzoic-acid glyceride-Eu rare earth complex, 1,3,5-trimesic acid glycollic acid ester-Eu rare earth complex, para toluic acid glycollic acid ester-Tb rare earth complex and p-phthalic acid benzilate-Tb rare earth complex. The morphology of the PET luminescence fibers was characterized by field emission scanning electron microscope; The composition structure of PET luminescence fibers was characterized by fourier transform infrared (FTIR); The fluorescence property of PET luminescence fibers was characterized by fluorescence spectrophotometer; The mechanical property of PET luminescence fibers was characterized by electronic single fiber strength tester. The spinnability of PET luminescence fibers under the condition of different rare earth complexes was explored, and fluorescence property and mechanical property of PET luminescence fibers with different rare earth complexes were compared.

Keywords: PET; electrospinning; rare earth complex; luminescence fibers

1.The morphology of the fibers

Figure 1. Effect of spinning solution concentration on fiber morphology.

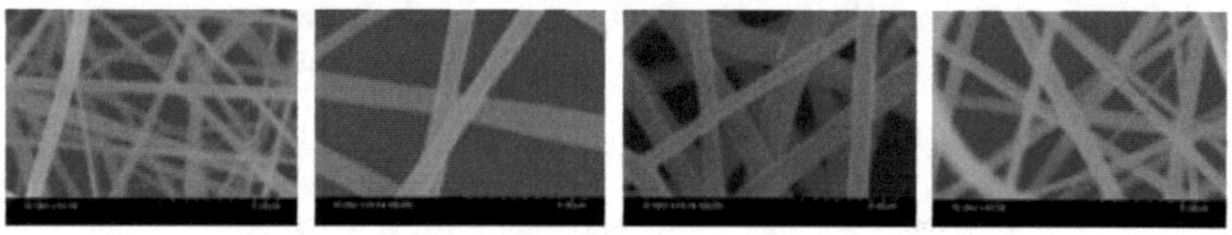

Figure 2. Effect of spinning voltage on fiber morphology.

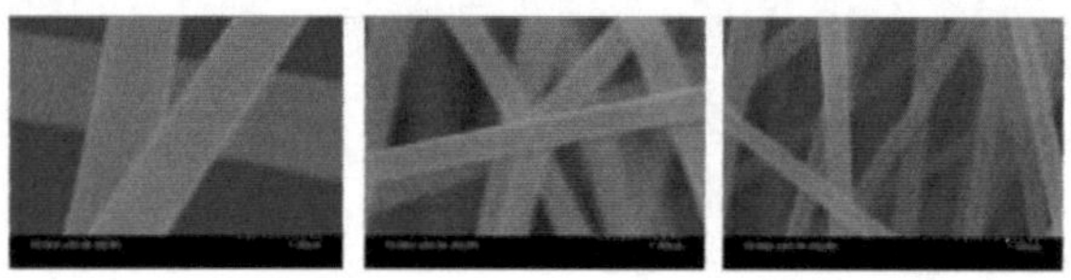

Figure 3. Effect of the receiving distance on fiber morphology.

Figure 4. Effect of different rare earth complex on fiber morphology.

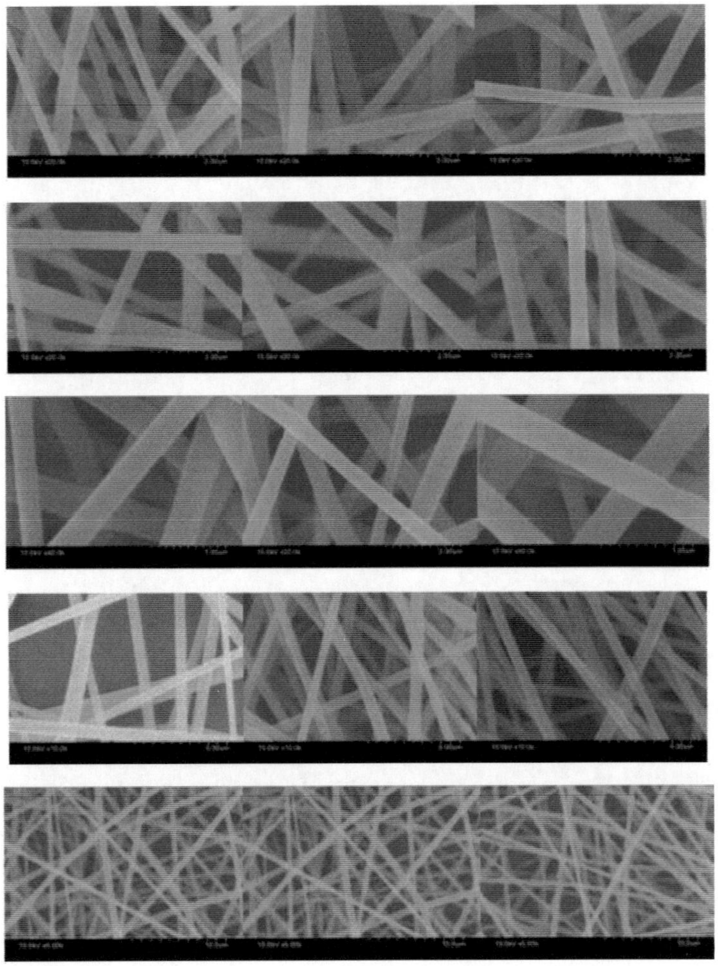

Figure 5. Effect of different addition of rare earth complex on fiber morphology.

2.Mechanical property

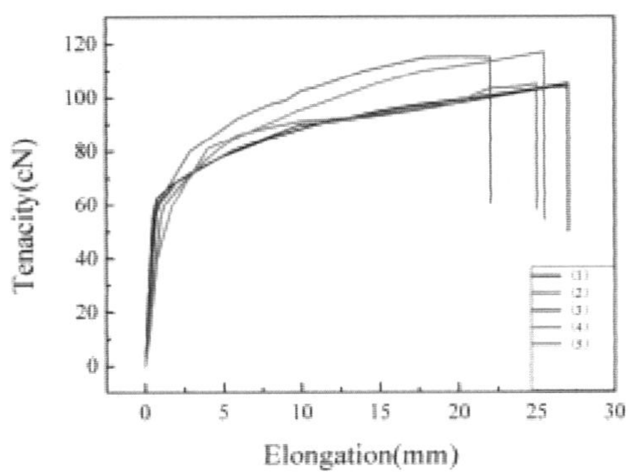

Figure 6. Tensile strength of the PET nanofibers.

3.Fluorescence property

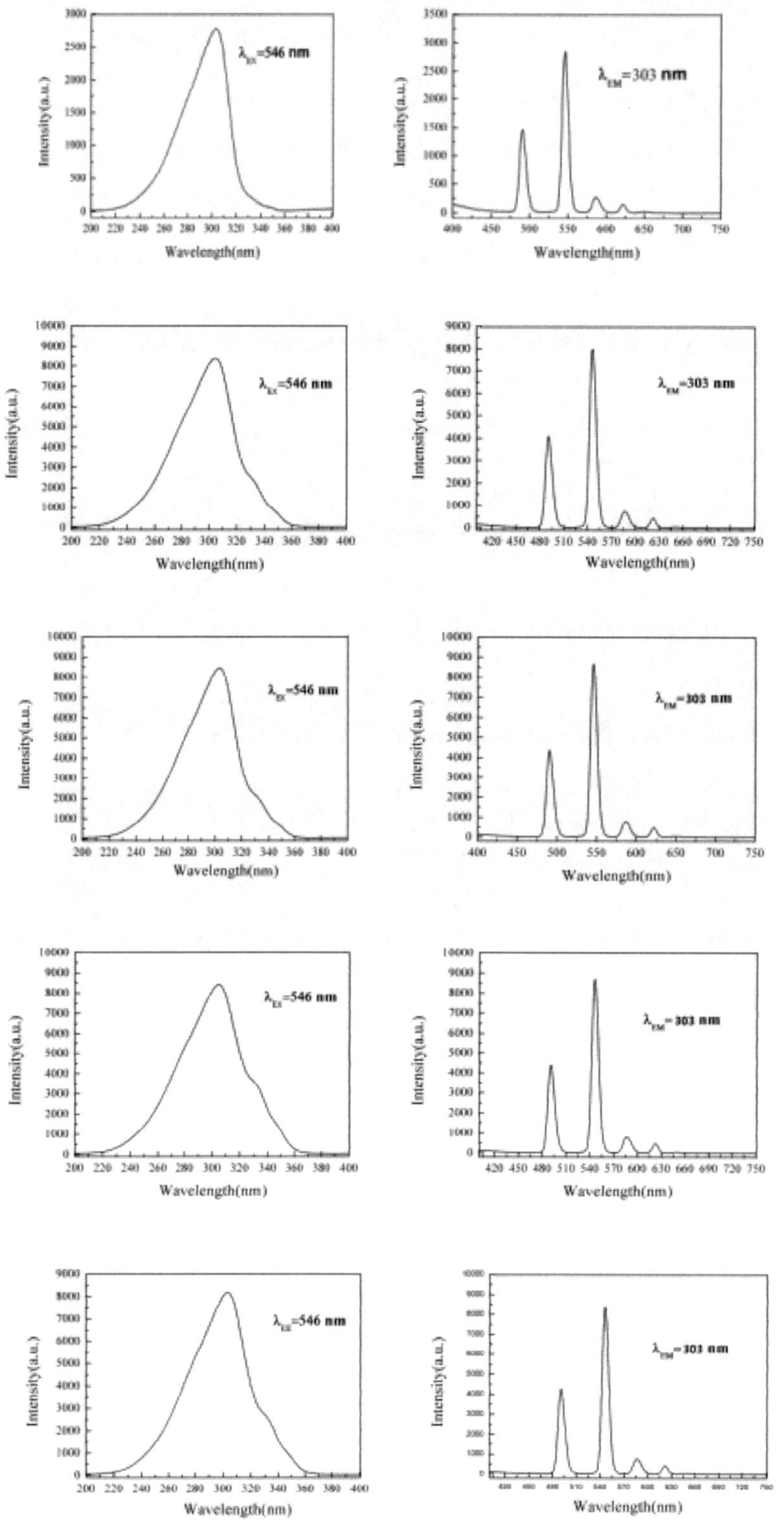
λEX=546 nm
λEM=303 nm
Intensity(a.u.)
Wavelength(nm)

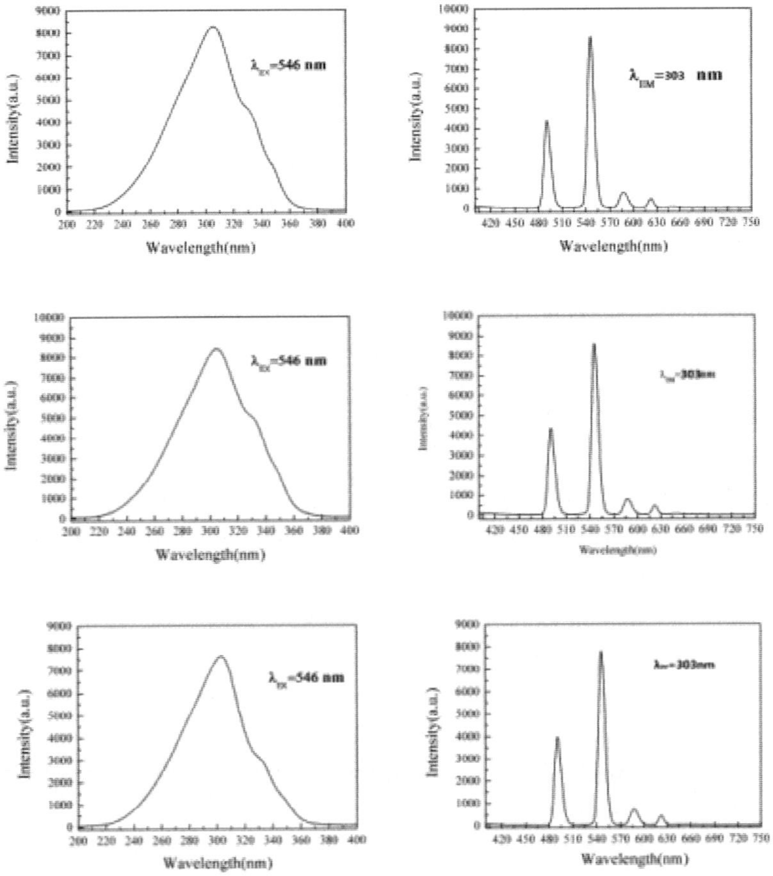

Figure 7. Effect of different addition of rare earth complex on fluorescence intensity.

4. The FTIR results

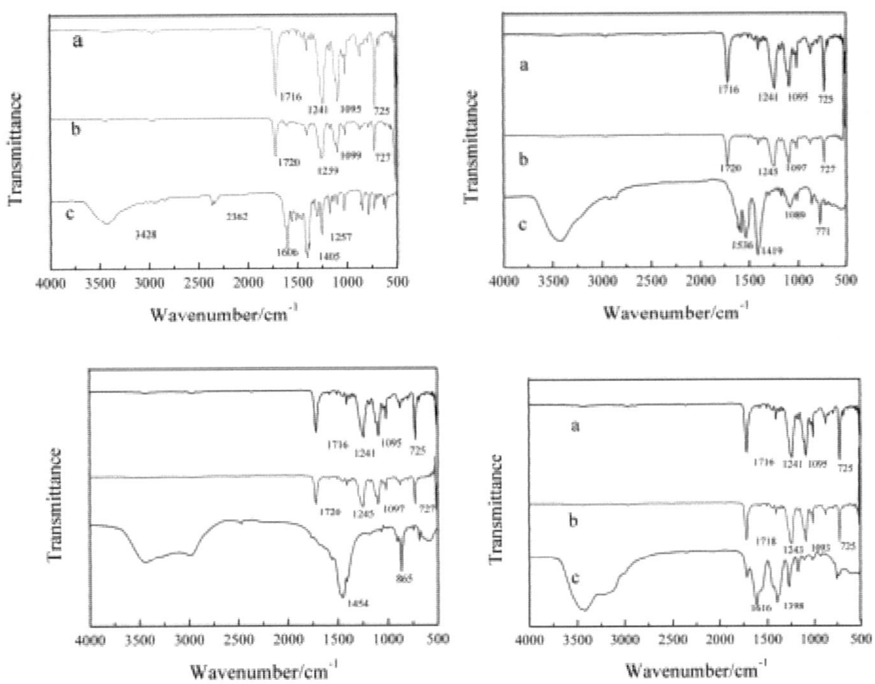

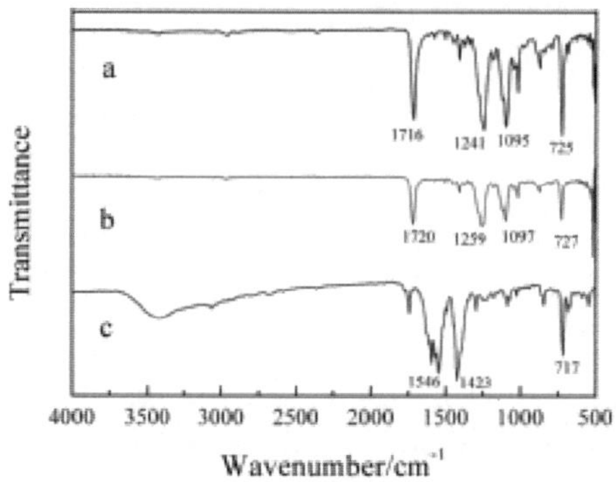

Figure 7. Infrared spectrum.

References

[1] Veleirinho B, Rei M F, Lopes-Da-Silva J A. Solvent and concentration effects on the properties of electrospun poly(ethylene terephthalate) nanofiber mats[J]. *Journal of Polymer Science Part B Polymer Physics*, **2008**, 46(5):460–471.

[2] Reneker D H, Yarin A L, Zussman E, et al. Electrospinning of Nanofibers from Polymer Solutions and Melts[J]. *Advances in Applied Mechanics*, **2007**, 41(07):43-195.

[3] Hou Z, Cheng Z, Li G, et al. Electrospinning-derived Tb2(WO4)3:Eu(3+) nanowires: energy transfer and tunable luminescence properties.[J]. *Nanoscale*, **2011**, 3(4):1568-74.

[4] Veleirinho B, Rei M F, Lopes-DA-Silva J A. Solvent and concentration effects on the properties of electrospun poly (ethylene terephthalate) nanofiber mats[J]. *Journal of Polymer Science Part B: Polymer Physics*, **2008**, 46(5): 460-471.

[5] Ge M, Guo X, Yan Y. Preparation and study on the structure and properties of rare-earth luminescent fiber[J]. *Textile Research Journal*, **2012**, 82(7): 677-684.

[6] Zander N E, Gillan M, Sweetser D. Recycled PET nanofibers for water filtration applications[J]. *Materials*, **2016**, 9(4): 247.

[7] Li C, Quan Z, Yang J, et al. Highly uniform and monodisperse β-NaYF4: Ln^{3+} (Ln= Eu, Tb, Yb/Er, and Yb/Tm) hexagonal microprism crystals: hydrothermal synthesis and luminescent properties[J]. *Inorganic chemistry*, **2007**, 46(16): 6329-6337.

[8] Wang M, Jin H J, Kaplan D L, et al. Mechanical properties of electrospun silk fibers[J]. *Macromolecules*, **2004**, 37(18): 6856-6864.

Preparation and Characterization of Ternary Poly(arylene ether sulfone) Copolymer Nanofibrous Membranes via Soultion-Blowing

Guocheng Song, Jing Li, Jing Zhu, Junrong Yu, Yan Wang, Zuming Hu*
College of Material Science and Engineering, Donghua University, Shanghai, 201620, China
E-mail: sgc_34@163.com (Guocheng Song)
Corresponding author: Pro. Zuming Hu, hzm@dhu.edu.cn (Zuming Hu)

In this paper we synthesized a series of linear high molecular weight modified poly(arylene ether sulfone) with Cardo groups, and then different content of the copolymers were dissolved in DMAc to obtained varying concentrations of spinning solution. The corresponding nanofibers of the polymers were fabricated via solution-blowing technology. The spinning process showed the co-PAES with the concentration of 25wt% was suitable to form the ideal and near uniform nanofibers which the mean diameter of the fiber was about 0.5μm. Scanning electron microscopy (SEM), fourier transform infrared (FTIR), differential scanning calorimetry (DSC), thermogravimetric analysis (TGA) and water contact angles (WCA) were conducted to characterize the nanofibrous membranes. The test results indicated the nanofiber membranes had a higher glass transition temperature (Tg) and better thermal stability with the increase content of Cardo group. The Tg ranging from Cardo-PAES-1 to Cardo-PAES-3 was about 215℃, 226℃ and 246℃ respectively, and the 5% weight loss temperature of nanofibers was 427.3℃, 444.5℃ and 481.9℃ respectively. Along with the improvement of these properties, its potential application could be extended in heat microfiltration membranes field.

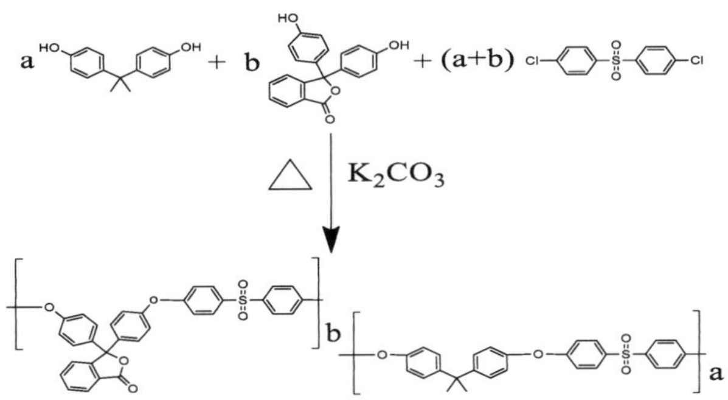

Scheme 1. Synthesis of the copolymers

copolymer	n(DCDPS): n(PHT): n(BPA)	[η] (dL/g)
Cardo-PAES-1	4 : 1 : 3	0.65
Cardo-PAES-2	4 : 2 : 2	0.66
Cardo-PAES-3	4 : 3 : 1	0.64

Table 1. Characterization of the copolymers

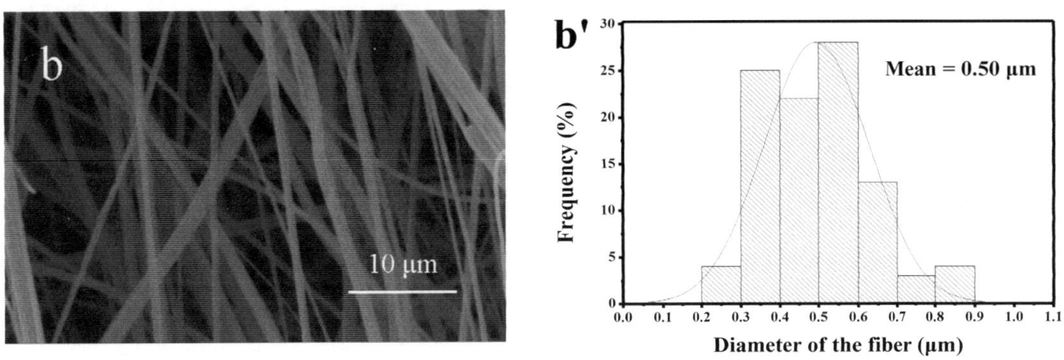

Fig 1. SEM images and corresponding fiber diameters of Cardo-PAES-2 with 25 wt%

References

[1] Zhang CH,Chen T L, Yuan Y G. *CN:85108751.A*, **1985**-11-25.

[2] B. Liu, G. Wang, W. Hu, Y. Jin, C. Chen, Z. Jiang, W. Zhang, Z.Wu and Y. Wei. *J.Polym.Chem.Ed.,* **2002,** 40, 3392-3398.

[3] Y. Zhang, X. Sun, R. Xu, Y. Niu, G. Wang and Z. Jiang. *Mater. Chem.Phys.,***2006**, 99, 465-469.

[4] V.L. Rao, P.U. Sabeena, A. Saxena, C. Gopalakrishnan, K. Krishnan, P.V. Ravindran, and K.N. Ninan.*Eur. Polym. J.,* **2004**, 40, 2645-2651.

[5] Bin Huang, Minghong Zhou, Xiaojun Zhu, Mingzhong Cai. *Polymer Engineering and Science,* **2015,** 10, 2140-2147.

[6] Mengzhu Liu, Yongpeng Wang, Zhiqiang Cheng,Mingyue Zhang, Meijuan Hu and Junfeng Li. *High Performance Polymers,* **2015**, 27, 939–949.

[7] Gang Wang, Hao Zhang, Bingqing Qian, Jinyan Wang,Xigao Jian, Jieshan Qiu. *Applied Surface Science,* **2015,** 351, 169–174.

[8] Hang Wang, Xupin Zhuang, Xiaojie Li, Wei Wang, Yannan Wang, Bowen Cheng. *Journal of Applied Polymer Science,* **2015,** 10, 42572(1-7).

[9] Jin Xiaoming, Wang Fosong, Liu Kejing. *Acta Polymerica Sinica,* **1991,** 6, 646-652.

[10] E.N. Bolbasov, K.S. Stankevich, E.A.Sudarev, V.M. Bouznik, V.L.Kudryavtseva, L.V. Antonova,V.G. Matveeva, Y.G. Anissimov, S.I. Tverdokhlebov. *Materials Chemistry and Physics,***2016,** 182, 338-346.

Preparation and Mechanical Properties of Cellulose /Antarctic Krill Protein Composite Fiber

Jingxing Song, Jing Guo, Qian Tan, Sen Zhang

Department of heigh polymer, College of Material Science and Engineering, Dalian Polytechnic University, Dalian, 116034, China

E-mail: 1509372911@qq.com

An cellulose /antarctic krill protein (C/AKP) composite fiber was prepared by wet spinning process. The raw material is waste cellulose and antarctic krill protein. As shown in Figure 1, a mixture of aqueous solution was prepared by mixing water, NaOH and urea, with H_2O: NaOH: urea of 81:7:12 by weight[1-2]. Cellulose was immersed immediately in the cooled solvent and stirred quickly for about 3 min at -12 °C. Then the AKP was added into this mixture under mechanical agitation. The solution was spinning with wet spinning machine. And the fiber were dispersed in the coagulation bath for 5 min. Cellulose and antarctic krill proteins present large amounts of hydrogen bond receptors, the broad peak 3000-3800cm^{-1}, getting from FTIR, was smoothed using the Savitzky–Golay method by 15 points in Origin 8.0 under six times infrared spectrum fitting operation[3] shown in Figure 2(a). Then the peak area and the content of hydrogen bond were counted in the Figure 2(b). It can be seen from Figure 2(b) that the hydrogen bond contribution in the intermolecular hydrogen bonds is in the order of OH…π>OH…O>OH…N. The mechanical properties of different temperature was tested, the Figure 3 shown that with the tensile temperature increases, the fracture strength increases. From Figure 4 we can see that the surface of the fiber has a lot of grooves, which is caused by double diffusion.

Keywords: cellulose; antarctic krill protein; hydrogen bond; mechanical properties

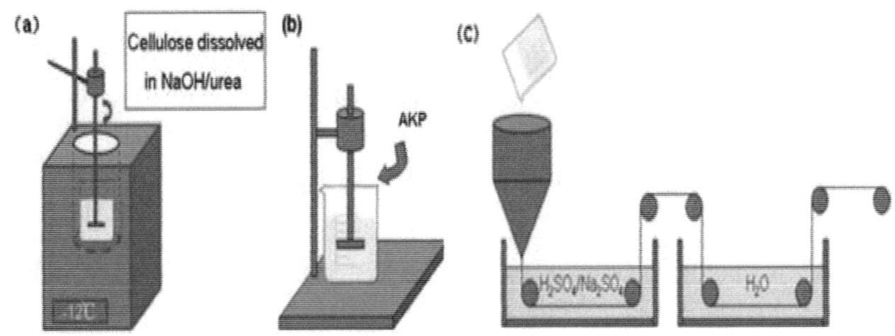

Figure 1. The preparation of C/AKP composites fiber.

(a)

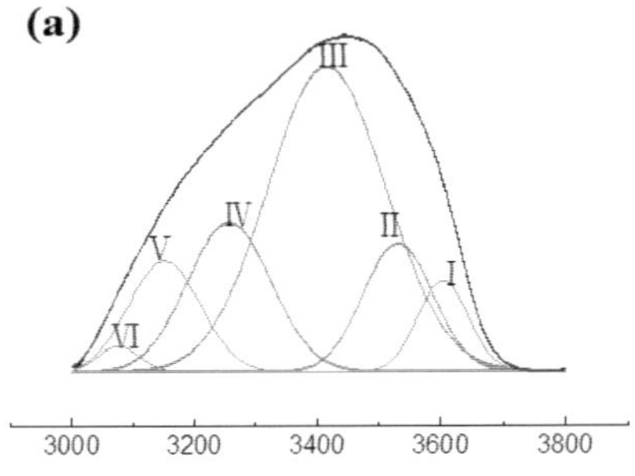

(b)

	Abbreviations	Wave number /cm^{-1}	Average peak area	Standard deviation s/ %
I	—OH	3605	103.62	6.07
III	OH. . .OH	3418	218.80	53.46
V	Annular polymer	3149	143.64	9.05
II	OH. . .π	3536	153.64	14.72
VI	OH. . .ether O	3256	164.66	15.87
VI	OH. . .N	3079	76.41	0.82

Figure 2. (a) Gauss curve fitting of composite material by different types of hydrogen bonds, (b) the fitting results of various kinds of hydrogen bonds.

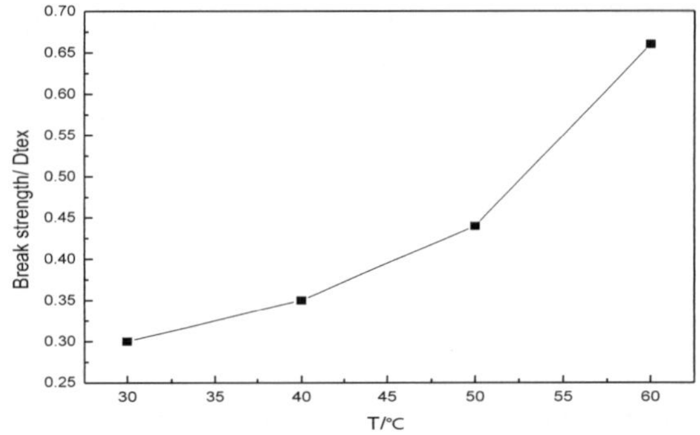

Figure 3. The Relationship between the Breaking Strength of C / AKP Composite Fibers and Molding Temperature.

Figure 4. The SEM images of SA/AKP composite fiber

References

[1] Cai J, Zhang L. Rapid dissolution of cellulose in LiOH/urea and NaOH/urea aqueous solutions[J]. *Macromolecular Bioscience*, **2005**, 5(6):539–548.

[2] Wang W, Li F, Yu J, et al. Effects of coagulation conditions on structure and properties of cellulose-based fibers from aqueous NaOH solvent[J]. *Carbohydrate Polymers*, **2017**, 164:118.

[3] Yang L, Jing G, Yue Y, et al. Hydrogen bonds of sodium alginate/Antarctic krill protein composite material[J]. *Carbohydrate Polymers*, **2016**, 142:275-281.

The Synthesis of PEDOT: PSS and its Application in Supercapacitor Electrode Materials

Weili Shao, Xianqiang Sun, Shuyuan Zhao, Qian Wang, Jianxin He

School of Textiles, Zhongyuan University of Technology, Zhengzhou 450007,China

E-mail: 1796167229@qq.com

In this paper, a novel yarn-type supercapacitor electrode is designed by the method of electrospinning and the addition of PEDOT: PSS. Nitrogen desorption and Fourier transform infrared spectroscopy and a series of characterization were carried out. Eventually, the capacitor electrode material has a mass specific capacity of up to 28.75 F/g. In addition, it also shows a good cycle performance and rate capability. Because of its excellent electrochemical properties, this supercapacitor electrode material, made of all solid capacitors, can be further woven into flexible, wearable functional textiles.

Keywords: supercapacitor; nanofiber core yarn; nickel-plated cotton yarn; polythiophene

Novel and Reusable Porous Linear Polyimide Electrospun Nanofibers for Hot Oil Adsorption

Lidong Tian [a], **Chongyin Zhang** [b], **Xiaowei He** [a], **Yongqiang Guo** [a], **Mingtao Qiao** [a], **Junwei Gu** [a,*], **Qiuyu Zhang** [a,*]

[a] *Shaanxi Key Laboratory of Macromolecular Science and Technology, School of Science, Northwestern Polytechnical University, Xi'an, Shaanxi, 710072, PR China;*

[b] *Shanghai Academy of Spaceflight Technology, Shanghai, 201109, PR China.*

E-mail: nwpugjw@163.com & qyzhang@nwpu.edu.cn

Eco-friendly, efficient and easily scaled-up processes are needed to fabricate oil adsorption materials which are very useful to deal with the oil pollution in heat environment, to deal with the oil spill pollution. Herein, highly efficient hot oil adsorption of PI fibers with excellent thermal stability was obtained by a facile electrospinning method followed by post-treatment. Morphologies, porous structures and oil adsorption properties of PLPI fibers at different temperatures were characterized. And the relationship between morphologies and oil adsorption properties of PLPI fibers was analyzed. The results showed that PLPI fibers possessed a stable morphology and pore structure at 300°C. The oil adsorption capacity of porous PI fiber for hot motor oil (300°C) was approximate to 58.9 g g^{-1}, which was higher than the PI fibers with non-porous structure for the motor oil at room temperature. Even after ten adsorption cycles, the oil retention was only 5.1%. Besides, the PLPI fibers also possessed favorable flexibility and excellent mechanical property. The obtained porous PI fibers possessed excellent hot oil adsorption capacity, reusability and recoverability, matching well with the requirements for treating the oil spills pollution and presenting potential applications in hot oil adsorption and separation.

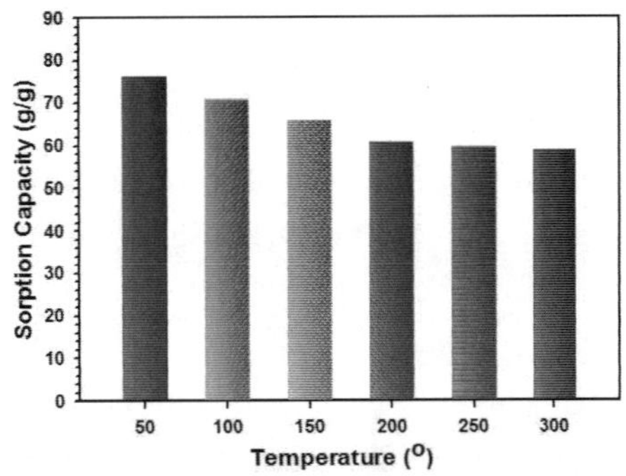

Figure 1. Maximum sorption capacities of PLPI fibers for motor oil at different temperature

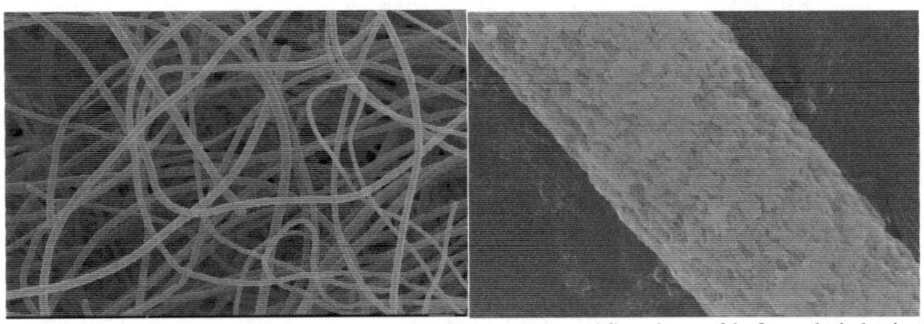

Figure 2. FE-SEM images of the porous PI fibers. Magnification of left and right is 1000x and 20000x, respectively.

References

[1] J. Gu*, Z. Lv, Y. Wu, Y. Guo, L. Tian, H. Qiu, W. Li, Q. Zhang*, *Compos. Part A*, **2017**, 94, 209-216.
[2] L. Tian, J. Gu*, X. Lei, Z. Lv, M. Qiao, C. Yin, Q. Zhang*, *Macromol. Mater. Eng.* **2016**, 301, 625-635.
[3] J. Wu, N. Wang, Y. Zhao, L. Jiang*, *Nanoscale*, **2015**, 7, 2625-2632.

Electrospinning of Chitosan/pullulan Nanofiber Mats

Wajeeh Ullah, Tongna Shi, Yali Ji

College of Materials Science and Engineering, Donghua University, Shanghai 201620, China

E-mail: engrwajeeh95@outlook.com.

Electrospinning has been widely used as nanofiber fabrication method [1]. The nanofibers have a wide range of applications in the field of tissue engineering, wound dressing, drug delivery, catalysis, filtration, enzyme immobilization and sensors [2]. Chitosan, a copolymer of (1→4)-2-amino-2-deoxy-β-D-glucan and (1→4)-2-acetamido-2-deoxy-β-D-glucan, is well known for its nontoxic, biocompatible and biodegradable properties [3]. In addition, it has unique antimicrobial property. It can inhibit the growth of a wide variety of fungi yeasts, and bacteria, which is beneficial in the field of biomedical applications [4]. Electrospinning of chitosan has been investigated by several authors, who thought the manufacturing of pure chitosan nanofibers was extremely challenging [3]. The electrospinnabilty of chitosan is limited mainly by its polycationic nature in solution, rigid chemical structure and specific inter and intramolecular interactions, which makes chitosan highly viscous at low pH and room temperature [5]. Herein, we use pullulan as carrier polymer to electrospin chitosan nanofibers considering its excellent nontoxicity, biocompatibility and electrospinnability [6]. Owing to the present of pullulan, the low concentration of acetic acid can be used as spinning solvent, i.e. 3 to 30% acetic acid instead of commonly used 90% acetic acid. Thus pullulan increases the electrospinnabilty of chitosan and decreases the toxicity of spinning solution. In addition, dopamine is used as a crosslinking agent to improve the water stability of the nanofiber mats. Dopamine is added into spinning solution before electrospinning, and then the obtained electrospun nanofiber mats are cross-linked in the presence of NaOH vapor for 24h. The effects of dopamine content on the structure and properties of chitosan/pullulan nanofiber mats are investigated. The representative morphologies of nanofiber mats are shown in Figure 1. It is found the cross-linked nanofiber mats exhibited good mechanical properties and water stability after crosslinking with NaOH vapor. In conclusion, we can increase the electrospinnability of chitosan just by simply using natural polymer pullulan. The obtained nanofiber mats have potential use in the field wound dressing, tissue engineering and drug delivery carrier.

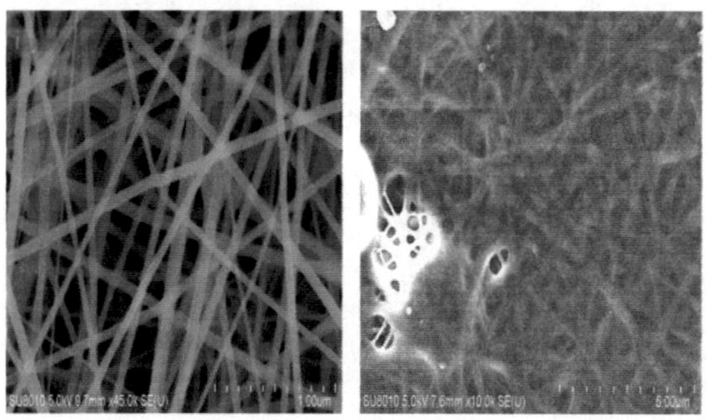

Figure 1. SEM image of chitosan/pullulan nanofiber mats before crosslinking (i) and after crosslinking (ii).

Keywords: Electrospinning, Chitosan, Pullulan, Dopamine, Crosslinking

References

[1] Liu, Minghuan, *Materials Science and Engineering,* **2017,** 76, 1413-1423.

[2] Homayoni, Homa," *Carbohydrate polymers,***2009**, 77,656-661.

[3] Kriegel, C., Kit, K. M., *Polymer,* **2009**, *50*(1), 189-200.

[4] Sun, K., & Li, Z. H., *Express Polymer Letters,* **2011**, *5*(4), 342-361.

[5] Tonda-Turo, C., Ruini, *Carbohydrate Polymers,* **2017**, 162, 82-92.

[6] Geng, Xinying, Oh-Hyeong Kwon, *Biomaterials,* **2005**, 26.27, 5427-5432.

Preparation and Filtration Properties of Porous PET Nano-fibrous Membrane

Lina Wang[1], Xia Xuelian[2], Zong Dingding[3]

[1]*School of Materials and Chemical Engineering, Henan University of Engineering, Zhengzhou, Henan province, 450007, China;*

[2]*School of Chemistry and Material Engineering, Henan University of Urban construction, Pingdingshan, Henan province, 467036, China;*

[3]*Material Science & Engineering, Donghua University, Shanghai, 201620, China*

Correspondence to: Wen-tao Liu (E-mail: wtliu@zzu.edu.cn)

Porous PET fibrous membranes were prepared by selective PEG dissolution from electrospun PET/PEG composite fibrous membrane. The PET/PEG composite fibers were fabricated as precursors by electrospinning PET/PEG solutions, and then the PEG was removed via a selective dissolution technique in hot water at 75 °C. Due to the incompatibility of PET and PEG, the co-continuous morphology was developed during the electrospinning process. The morphology was characterized by scanning electron microscopy (SEM) and fileld emission scanning eletcro microscopy (FESEM), and the images showed the porous structures were formed. Nitrogen adsorption based on BET (Brunauer-Emmett-Teller) principle was employed to measure the specific surface area and porous structure, which indicated that porous fibrous membranes possess mounts of mesoporous. The specific surface area increased about five times, and the pore volume increased nearly eight times for porous PET fibrous membranes compared to the PET/PEG composite fibrous membranes when the ratio of PET to PEG was 6/1. Both hydrophility and filtration properties were performed to further investigate the relationship of porous structure and filtration properties for the porous fibrous membranes. The static water contact angle (WCA) measurement was used to investigate the effect of porous structure on hydrophility, which indicated that the WCA firstly increased slightly and then dramatically decreased with the PEG content increasing, peaking at 6/1 PET/PEG, and the drop was due to forming of macropores, which caused the water more easily penetrated. The filtration efficiency and resistance were measured by TSI 8130 automated filter tester, and the results indicated that the filtration resistance and efficiency of porous membranes increased with the PEG content increasing. The equilibrium value of filtration efficiency is 99.2% for 6/1 PET/PEG, which proved the formation of microporous and mesoporous structure in the porous PET fibrous membranes.

Keywords: porous PET fibrous membranes, electrospinning, specific surface area, hydrophility, filtration properties

Study on Preparation of Rare Earth Fluorescent Fiber by Melt Spinning

Shiwei Wang, Sen Zhang, Zhichun Zhou

Department of High Polymer, College of Material Science and Engineering, Dalian Polytechnic University, Dalian, 116034, China

E-mail: 863885908@qq.com

Firstly, the fluorescent powder of rare earth nitrate was synthesised by water bath[1]. Then, the fluorescent powder and PA6 were mixed thoroughly. The fluorescent fibers with the different content of fluorescent powder were prepared by melt spinning respectively. Without adding any other modification agent, so some of problems such as degradation, the decrease of glass transition temperature could be fundamentally avoided. And in the melt spinning process, the phosphor to maintain good integrity, thus ensuring a good fiber fluorescence performance. In the experiment, we systematically studied the influence of phosphor content on the fluorescence intensity, the mechanical properties and the thermodynamic stability of composite fiber. Meanwhile, the surface, mechanical properties, and thermodynamic properties of fluorescent fibers were characterized. The results show that the fluorescence intensity of the fiber promotes linearly with the increase of phosphor content. In terms of mechanical properties, the strength of the fluorescent fiber shows the trend of increasing first and then decreasing. The influence of the addition of phosphor on the thermodynamic stability of the fiber was not very obvious, which only changed slightly and the stability increased first and then decreased. The surface of the fluorescent fiber was observed by scanning electron microscope (SEM), and it was found that the surface of the fiber was very smooth. It indicates the phosphor is well dispersedly in PA6 matrix, without nano agglomeration occuring.

Keywords: fluorescent fiber; fluorescent intensity; PA6; mechanical properties; thermodynamic stability

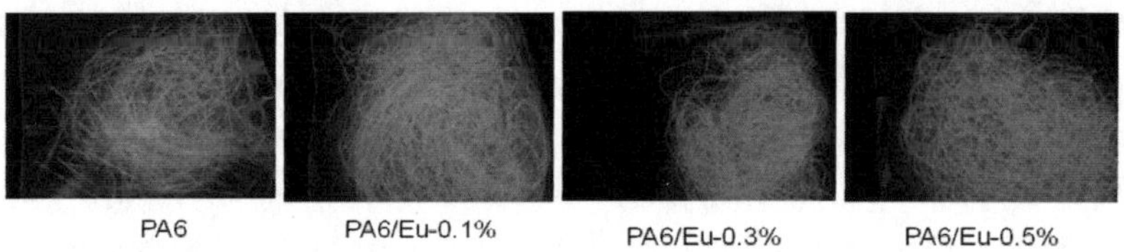

Figure 1. Fluorescent fibers photographs of different fluorescent powder contents of under ultraviolet light.

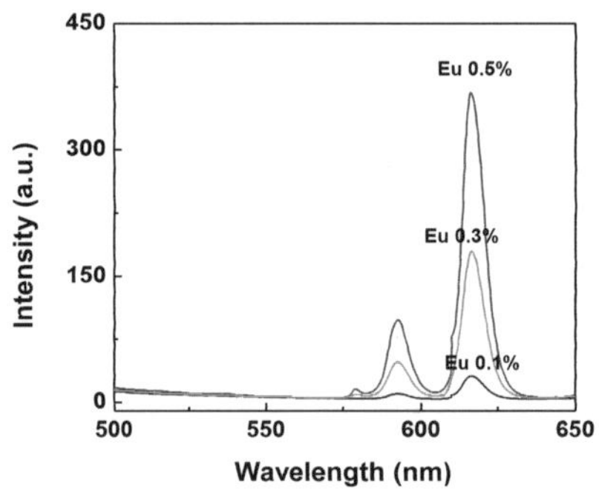

Figure 2. Fluorescence intensity of fibers with different fluorescent powder contents.
(EX Slit: 1.0 nm EM Slit: 1.0 nm)

References

[1] Jilin Zhang, Bowei Chen, Xuan Luo, Kai Dua, *Functional Polymer Science*, **2013**, 55(3), 158-163.

Efficient Visible-light Driven mpg-C_3N_4 /TiO_2 Composite Photocatalyst: Facile Preparation and Enhanced Catalytic Activity

Shuang Wang, Bo Yang, Yaping Zhao*, Zaisheng Cai, Junchao Ji

Department of Chemistry, Chemical Engineering and Biotechnology, Donghua University, Shanghai, 201620, China

E-mail: zhaoyping@dhu.edu.cn

The application of the titanium dioxide (TiO_2) as a photocatalyst is seriously hindered by its intrinsic large band gap leading to its effectiveness only under UV- light.[1] However, UV-light accounts for only 3-5% of solar radiation and enhancement of visible response of TiO_2 is in urgent need.[2] Recently, as a free-metal semiconductor with a wealth of attractive properties such as excellent chemical and thermal endurance and unique energy band structure, polymeric graphitic carbon nitride (g-C_3N_4) is becoming increasingly significant as a visible-light active photocatalyst in the sustainable utilization of solar energy.[3] But the photocatalytic performance of g-C_3N_4 is still limited due to quick recombination of the carriers and a very small surface area which cannot provide more possible reaction sites for the catalytic reaction.[4,5] Mesoporous carbon nitride (mpg-C_3N_4) provide more contact areas, [4] in principle enhancing the light harvesting ability and the reactant adsorption capability of the material.

Here, we reported the exploration of compositing mpg-C_3N_4 and TiO_2 toward the efficient visible-light-driven photocatalyst. A facile template approach has been developed to prepare mpg-C_3N_4 using SiO_2 nanoparticles as a template and cyanamide as a precursor. Based on the mpg-C_3N_4 and TiO_2 powder as raw materials, the solvent evaporation method was adopted to mix different mpg-C_3N_4 to TiO_2 ratios (marked as mpg-C_3N_4/TiO_2-2:1; mpg-C_3N_4/TiO_2-1:1; mpg-C_3N_4/TiO_2-2:3; mpg-C_3N_4/TiO_2-1:2; mpg-C_3N_4/TiO_2-1:4 ; mpg- C_3N_4/TiO_2-1:10). Then, mpg-C_3N_4/TiO_2 composite photocatalyst was prepared by high temperature annealing method. The samples have been characterized using Fourier transform infrared spectra (FT-IR), X-ray diffraction (XRD), scanning electron microscopy (SEM), the Brunauer-Emmett-Teller (BET), and ultraviolet-visible diffuse reflectance spectroscopy (UV-vis). The photocatalytic experimented indicated that the mpg-C_3N_4/TiO_2 composites showed a much higher photocatalytic activity than the pristine TiO_2 or mpg-C_3N_4 for the degradation of methylene blue (MB) under visible-light irradiation. The degradation of MB of mpg-C_3N_4/TiO_2-1:2 in visible-light was 1.6 times as efficient for that of pure g-C_3N_4 and 6.15 times as that of TiO_2.The mpg-C_3N_4/TiO_2 catalyst has a porous structure and large surface area, causing the increase of contact areas with degraded materials, which can be attributed to the heterojunction formed at the interface between mpg-C_3N_4 and TiO_2. Furthermore, based on systematic characterization and discussion, a possible photocatalytic mechanism for the excellent photocatalytic performance was proposed.

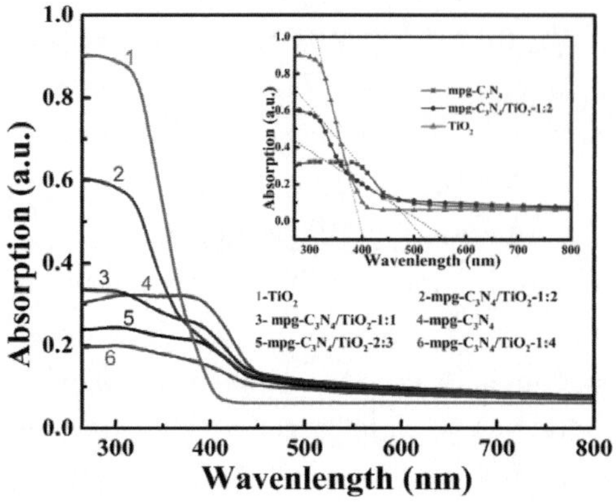

Figure 1. UV-Vis diffuse reflection spectra of various samples: mpg-C_3N_4, mpg-C_3N_4/TiO_2-1:1, mpg-C_3N_4/TiO_2-1:2, mpg-C_3N_4/TiO_2-2:3, mpg-C_3N_4/TiO_2-1:4 and TiO_2

References

[1] Wang, W.; Yang, D.; Yang, W.; Sun, J.; Hou, H., *Journal of Materials Science: Materials in Electronics* **2016,** 28 (2), 1271-1280.

[2] Liang, J.; Wang, Y.-Z.; Wang, C.-C.; Lu, S.-Y., *J. Mater. Chem. A* **2016,** 4 (25), 9797-9806.

[3] Chen, D.; Wang, K.; Ren, T.; Ding, H.; Zhu, Y., *Dalton transactions* **2014,** 43 (34), 13105-14.

[4] Liu, Z.; Zheng, H.; Yang, H.; Hao, L.; Wen, L.; Xu, T.; Wu, S., *RSC Adv.* **2016,** 6 (59), 54215-54225.

[5] Cao, S.; Low, J.; Yu, J.; Jaroniec, M., *Advanced materials* **2015,** 27 (13), 2150-76.

This work was financially supported by the Fundamental Research Funds for the Central Universities (2232015D3-17), the National Natural Science Foundation of China (51303022) and Industry-University - Institute Project (Booster Plan) of Shanghai Municipal Education Commission (15cxy55).

Study on Structures and Properties of Ultra-Hot Drawing UHMWPE Fibers Fabricated via Dry Spinning Method

Xinwei Wang, Han Zheng, Yongfei Sun

State Key Laboratory of Polyolefins and Catalysis, Shanghai 200062,

Shanghai Research Institute of Chemical Industry, Shanghai 200062

E-mail: w_xv@qq.com

Ultra High Molecular Weight Polyethylene (UHMWPE) fiber is well known as one of high performance fibers together with carbon fiber and Kevlar. It shows high impact resistance, wear resistance, self-lubrication, non-toxic, alkali resistance, acid resistance, low temperature endurance and anti-bacteria, which lead varieties applications in many fields. Currently, two spinning routes are applied in UHMWPE fibers production. Decalin is used as solvent in dry route, while mineral oil is used in wet route. Comparing to wet route, dry route has advantages of short process, fast spinning, good product quality, directly recycled solvent and environment friendly.

In this study, high tensile strength UHMWPE fibers were prepared via advanced dry spinning method (Figure 1), which was designed independently by Shanghai Research Institute of Chemical Industry (SRICI). Raw material was UHMWPE resin with 6 million viscosity-average molecular weight (Mv). The mixture of resin and decalin was fed to a twin-screw extruder and the solution was extruded into a spinneret to form filaments. The filaments under the spinneret was swept by hot nitrogen to have majority of solvent evaporated. At the same time, a roller drawing these fibers pass a thermos-stated corridor, where filaments would gradually shaped and crystallized. Then pre-drawing process was applied to form the final as-spun fibers and the draw ration from spinneret to as-spun fibers was of about 10. Then drew the as-spun fibers and changes of morphology, mechanical properties, thermal properties and crystallization process of the fibers in multistage ultra-hot drawing process were studied by SEM, tensile tester, DSC, and WAXD. We found that within drawing process, the surface of fibers gradually changes from the horizontal texture of the folded structure to the axial orientation of the groove structure on fibers. Also, results shown that there was a maximum value of fracture force at draw ratio of 40 and excessive draw ratio would destroy crystal structures in fibers and led to decreasing of fracture force (F) value, while fiber strength increase gradually with the progress of drawing and then up to the maximum value 35.04 cN/dtex at the draw ratio of 75. A shoulder peak appeared in DSC curves of ultra-drawing samples and its peak value maintained at 154°C in different samples. In addition, three peaks can be observed in all WAXD pattern of drawing samples at the 2θ of 19.3°, 21.6° and 24.1° respectively. Mechanism of microstructural changes during hot drawing process was elucidated (Figure 2).

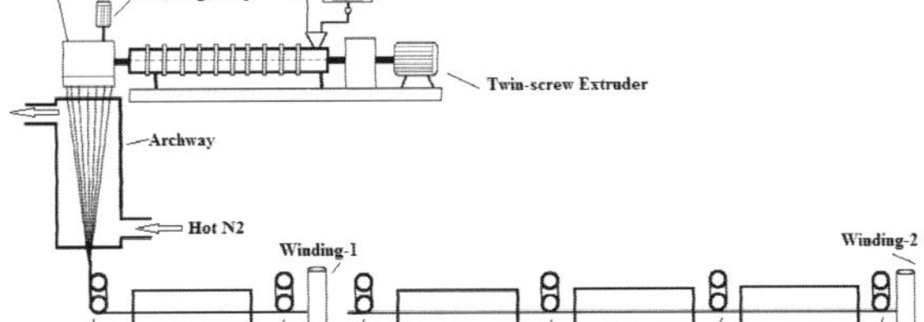

Figure 1. Dry spinning method of gel-spinning.

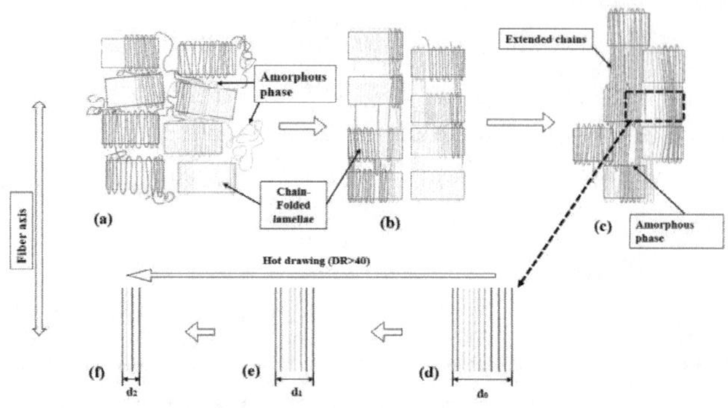

Figure 2. Model of the Fiber structure evolution in the super drawing process.

References

[1] Ismail Karacan, *J. APP. POLY. SCI*, **2006**, *101*, 1317-1333.

[2] Y. Jiang, J. Tang, Jens Tieger, etc., *Polymer*, **2009,** *16*, 4101-4111.

[3] H. Liu, F. Lv, J. Li, etc., *J. APP. POLY. SCI*, **2015**, 42823, 1-10.

[4] Sobieraj, M. C., Rimnac, C. M., *J. MECH. BEHAV. BIOMED*, **2009**, 2, 433-443.

[5] Uehara, H., Nakae, M., Kanamoto, T., etc., *Macromolecules*, **1999**, 32, 2761-2769.

[6] Xiao, M., Yu, J., Zhu, J., etc., *J. MATER. SCI.*, **2011**, 46, 5690-5697.

Air Drawing Effect of the Auxiliary Nozzle in the Melt Blowing Process

Lili Wu, **Jiabin Dong**, **Ting Chen**

College of Textile and Clothing Engineering, National Engineering Laboratory for Modern Silk, Soochow University, Suzhou, 215006, China

E-mail: tingchen@suda.edu.cn

Melt blowing is an important process for fabricating nonwoven fabrics of ultrafine fibers which can be used as fine filtering materials. Decrease of the fiber diameter will improve the filtration properties of the melt blown nonwoven fabrics. To produce much finer fibers is one of the development directions of the melt blowing technology. The polymer melt is drawn into ultrafine fibers by high velocity hot air stream. The air drawing force is the key force to stretch the polymer melt. Thus increasing the air drawing force is helpful to decrease the fiber diameter. The two streams of high velocity hot air flow out of the melt blowing dual slot nozzle and form an air flow field below the nozzle where the polymer melt is drawn. If the air flow field is changed properly, the air drawing can be increased and thus finer fibers can be obtained. In this research, an auxiliary nozzle with a converging-diverging cross section is installed below the melt blowing dual slot nozzle. Both the air velocity and air temperature along the spinnerete axis in the air flow field with the auxiliary nozzle are higher than those without the auxiliary nozzle, which is beneficial to the polymer drawing. Nonwoven fabrics are manufactured on the melt blowing nonwoven equipment with the auxiliary nozzle. Fiber diameters are measured using the image analysis method. A polymer drawing model is established to simulate the polymer drawing in the air flow field of the melt blowing process with the auxiliary nozzle. The predicted fiber diameters coincide with the measured fiber diameters. Fiber diameters with the auxiliary nozzle are also compared with those without the auxiliary nozzle. The results show that use of the auxiliary nozzle can produce finer fibers compared with those without the auxiliary nozzle. The results show great prospects for this research in the optimization design of melt blowing equipment.

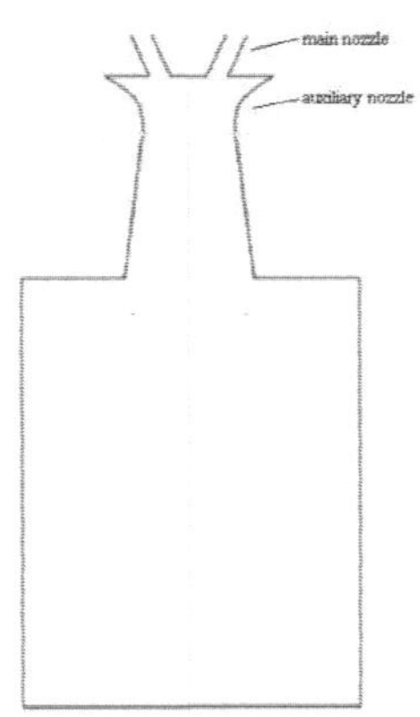

Figure 1. Auxiliary nozzle and air flow region.

References:

[1] Yuanling Cheng, Lili Wu, Ting Chen, *Heat Transfer Research*, **2012**, 44, 473-482.

[2] Ting Chen, Chong Zhang, Xia Chen, Liqing Li, *Journal of Applied Polymer Science*, **2009,** 111, 1775-1779

Design, Construction and Characteristics of Flexible Janus Nanofibers and Their Arrays

Endowed with Tuned Fluorescent Color-Electricity Bifunctionality

Xue Xi, Mingying Zhao, Xiangting Dong*, Qianli Ma, Dan Li, Wensheng Yu, Jinxian Wang, Guixia Liu

Key Laboratory of Applied Chemistry and Nanotechnology at Universities of Jilin Province, Changchun University of Science and Technology, Changchun, 130022, Jilin Province, China

E-mail: dongxiangting888@163.com

Since luminescent-electrical bifunctional nanomaterials have been applied in fluorescent display, electronic parts, biomedical fields, *etc*, many researchers have focused their attention on the preparations and properties of these novel nanomaterials in recent years. Recently, our research group has prepared a series of luminescent-electrical bifunctional nanomaterials *via* electrospinning technique [1, 2]. From the previous work of ours and other researchers, it has been proven that PANI will greatly decrease the luminescence of rare earth (RE) compounds if they are directly blended with RE compounds. Therefore, luminescent and electrical materials should be effectively isolated to avoid direct contact if the strong luminescence of the multifunctional nanofibers is achieved. Herein, we designed and fabricated luminescent-electrical bifunctional [Eu(BA)$_3$phen/Tb(BA)$_3$phen/PVP]// [PANI/PVP] Janus nanofibers *via* electrospinning technique using a homemade parallel spinneret. The flexible Janus nanofiber consists of two different sides assembled together in parallel, including an Eu(BA)$_3$phen/Tb(BA)$_3$phen/PVP nanofiber (luminescent side) and a PANI/PVP nanofiber (conductive side), and average diameter of each strand of the Janus nanofiber is *ca.* 450 nm. Janus nanofibers arrays were successfully constructed by using Janus nanofibers as construction units *via* electrospinning applying rotary drum as collector.

Compared with [Eu(BA)$_3$phen/Tb(BA)$_3$phen/PVP]//[PANI/PVP] Janus nanofibers non-array, Eu(BA)$_3$phen/Tb(BA)$_3$phen/PANI/PVP composite nanofibers array and composite nanofibers non-array, it is found that the Janus nanofibers array has much stronger fluorescent intensity than the contrast samples, and possesses tuned luminescent color by modulating added amounts of Eu(BA)$_3$phen and Tb(BA)$_3$phen. The conductivity along with the length direction of Janus nanofibers array reaches up to maximum six orders of magnitude higher than that along with perpendicular direction owing to its unique nanostructure, namely, the array has excellent electrically conductive anisotropy, and further, conductive anisotropy of the array can be tunable by changing added amounts of PANI. For the Janus nanofibers array, because the conductive sides of the Janus nanofibers are separated by the plenty of insulative-luminescent sides in perpendicular direction, which further blocks the movement of electrons. Therefore, a strong anisotropy in conduction is achieved in the Janus nanofibers array. It is very gratifying to see that the Janus nanofibers array simultaneously possess excellent color-tunable luminescence and electrically conductive anisotropy. The novel luminescent-electrical bifunctional Janus nanofibers arrays have potential applications owing to their excellent luminescent and electrical performance. More importantly, the design conception and construction technology are of universal significance to fabricate other bifunctional Janus nanofibers nanomaterials.

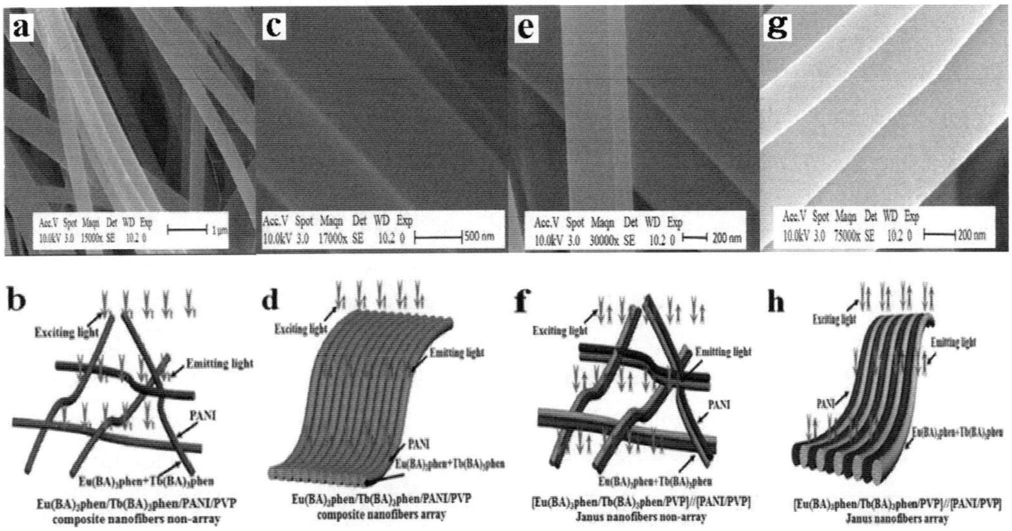

Figure 1. SEM image and schematic diagrams of composite nanofibers non-array (a, b), composite nanofibers array (c, d), Janus nanofibers non-array (e, f) and Janus nanofibers array (g, h)

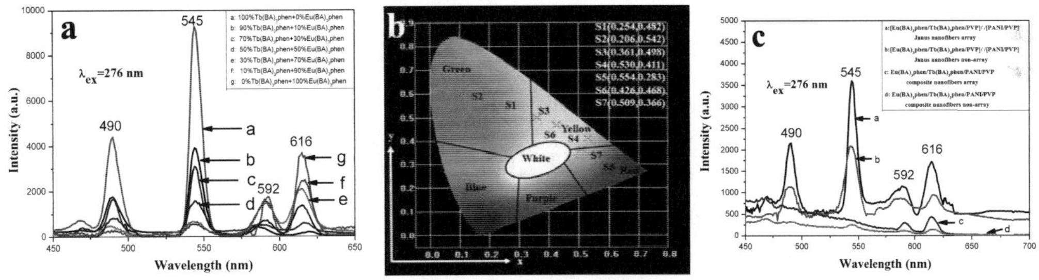

Figure 2. Emission spectra (a) and CIE chromaticity coordinates diagram (b) of Janus nanofibers array containing different mass ratio of $Eu(BA)_3phen$ to $Tb(BA)_3phen$, Emission spectra of Janus nanofibers array and contrast samples (c)

Table 1. Electrical conductivity values of Janus nanofibers array

Samples	Length direction ($S \cdot cm^{-1}$)	Perpendicular direction ($S \cdot cm^{-1}$)	Ratio of length and perpendicular directio
ANI/PVP: 15 %	1.65×10^{-3}	0.97×10^{-6}	1.70×10^{3}
ANI/PVP: 30 %	4.84×10^{-3}	1.67×10^{-7}	2.89×10^{4}
ANI/PVP: 50 %	6.28×10^{-2}	2.86×10^{-7}	2.19×10^{5}
ANI/PVP: 70 %	0.91	5.11×10^{-7}	1.78×10^{6}

References

[1] K. Lun, Q. L. Ma, X. T. Dong*, W. S. Yu, J. X. Wang and G. X. Liu, *Journal of Materials Science: Materials in Electronics*, **2014**, 25, 5395–5402.

[2] S. J. Sheng, Q. L. Ma, X. T. Dong*, N. Lv, J. X. Wang, W. S. Yu and G. X. Liu, *Luminescence,* **2015**, 30, 26-31.

Study on Preparation of Carbon Fiber Felt by Melt Melting

Wenchi Xue, Jun Song, Bowen Cheng, Lanyan Liu

Key Laboratory of Tianjin City Advanced Fiber and Energy Storage, Department of Materials Science and Engineering, Tianjin Polytechnic University, Tianjin 300387, China

E-mail: sjhb2000@163.com

Lignin is the highest natural aromatic polymers and as a by-product of the pulp industry is directly burned, which is a waste of resources. Because of its high carbon content, low price and other advantages, it can be used to prepare low-cost carbon fiber in order to improve the utilization of lignin. However, carbon fiber felt prepared by meltblowing are rarely reported. In this paper, the lignin raw materials were modified by phenolic as well as the insoluble and non-fused softwood lignin was transformed into fusible lignin. Then lignin-based fibers were prepared by meltblowing, and carbon fiber was obtained by curing, pre-oxidation and carbonization. The lignin before and after modification was characterized by FTIR, TG, Py-GC/MS and HAKEE rheometer. The optimum meltblowing temperature of modified lignin was between 100 °C and 160 °C, and the lignin-based carbon fiber felt with carbon content of 95.15% was prepared under this condition.

Keywords: Softwood lignin; Meltblown; Carbon fiber

Interfacial Complexation of Cellulose Nanocrystals with Synthetic and Bio-based Polyelectrolytes

Hajo Yagoub[a], Mahmoud H. M. A. Shibraen[a], Songmei Ma[a], Vincent Mukwaya[b], Shuguang Yang[a*], Jian Xu[a,c]

[a] *State Key Laboratory for Modification of Chemical Fibers and Polymer Materials, College of Material Science and Engineering, Donghua University, Shanghai 201620, China;*

[b] *State Key Laboratory of Textile Science & Technology, Ministry of Education, Donghua University, Shanghai, P. R. China201620;*

[c] *Laboratory of Polymer Physics and Chemistry, Institute of Chemistry, Chinese Academy of Sciences, Beijing 100190, China*

E-mail: shgyang@dhu.edu.cn

Composite thin films containing cellulose nanocrystal (CNC) and synthetic cationic polyelectrolyte poly allylamine hydrochloride (PAH) and poly(diallyldimethylammonium chloride) (PDDA), bio-based cationic polyelectrolyte cationic guar gum (CGG) and chitosan (CH) were fabricated using the layer-by-layer (LBL) deposition technique as shown in Figure 1. This study examines the effects of pH, ionic strength and temperature on the growth and properties of CNC films. As a result bio-based polyelectrolyte had the strongest dependence on the pH value, whereas, PAH exhibited the strongest dependence on the temperature at pH6, and CH was also found to be highly dependent on temperature on the addition of 0.2 M NaCl. The CNC, PAH, PDDA, CGG, and CH film growth in situ were monitored in real time by using the quartz crystal microbalance (QCM). The water uptake behavior, anti-fogging and anti-frosting properties were also investigated.

Keywords: Bio-based polyelectrolyte, Layer-by-layer deposition, Cationic guar gum, Chitosan.

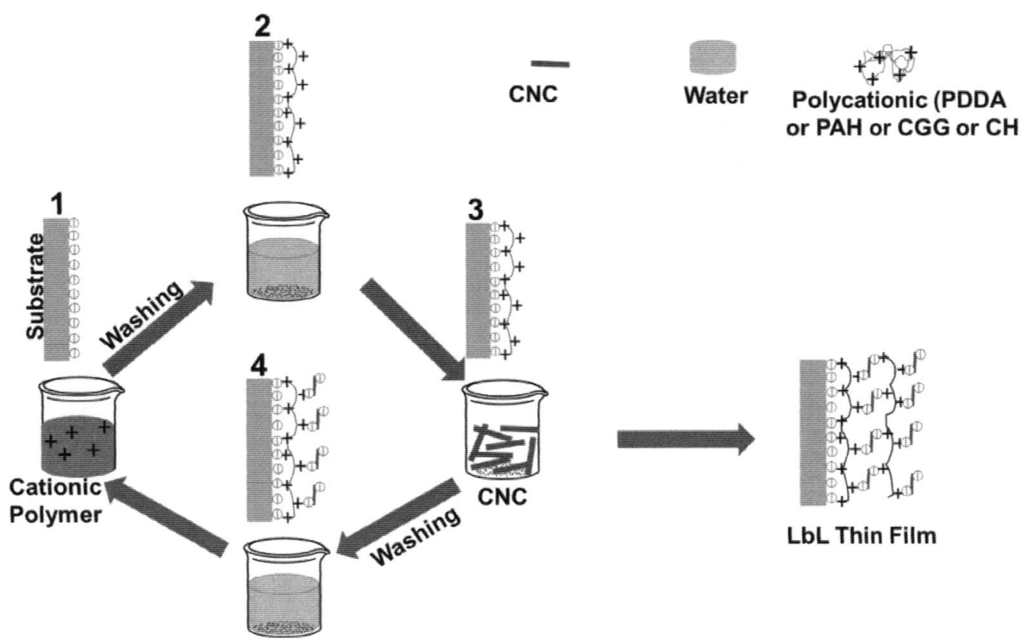

Figure 1. The layer-by-layer (LBL) deposition technique CNC with cationic polyelectrolyte.

One-dimensional $Mg_xTi_yO_{x+2y}$ Nanostructures: General Synthesis and Enhanced Photocatalytic Performance

Guorui Yang [a, b], **Ling Wang**[a], **Shengjie Peng**[b], **Wei Yan** [a] *, **Seeram Ramakrishna** [b] *

[a] *Department of Environmental Science & Engineering, State Key Laboratory of Multiphase Flow in Power Engineering, Xi'an Jiaotong University, Xi'an 710049, China;*

[b] *Department of Mechanical Engineering, National University of Singapore, 117574, Singapore*

E-mail: serram@nus.edu.sg; yanwei@xjtu.edu.cn

The photocatalytic H_2 evolution from water splitting is a promising process for generating renewable energy. In regard of the fascinating properties of the one-dimensional (1D) nanostructure, when used as photocatalyst, we report the synthesis of a series of $Mg_xTi_yO_{x+2y}$ (including Mg_2TiO_4, $MgTiO_3$, $MgTi_2O_5$) 1D nanofibers by the electrospinning technique and the investigations of their photocatalytic performance. Scanning electron microscopy (SEM), transmission electron microscopy (TEM), X-ray diffraction (XRD) and X-ray Photoelectron Spectroscopy (XPS) corporately confirmed the successful fabrication of pure $Mg_xTi_yO_{x+2y}$ nanofibers. By contrast, the $Mg_xTi_yO_{x+2y}$ nanoparticles obtained via sol-gel method still contains impurities because of the inhomogeneous crystallization. The $Mg_xTi_yO_{x+2y}$ nanofibers exhibited enhanced efficiency and stability in photocatalytic H_2 generation under ultraviolet light, compared with the $Mg_xTi_yO_{x+2y}$ nanoparticles and P25. The photoelectrochemical measurements further revealed that $Mg_xTi_yO_{x+2y}$ nanofibers facilitated the transport and separation of the photoinduced charge carriers, mainly resulting from their special 1D structure, unique mesh morphology, large specific surface area and pure phase. Herein, we take $MgTi_2O_5$ as a delegate to elucidate this topic. As shown in Fig. 1a, the electrospun $MgTi_2O_5$ demonstrates a typical nanofibrous morphology. Fig1 b obviously indicates that the electrospun $MgTi_2O_5$ was composed pure phase, while impurities existed in the sample obtained via sol-gel reaction. The photocatalytic investigation suggests the higher photocatalytic performance of the nanofibers, which was also confirmed by the EIS measurements.

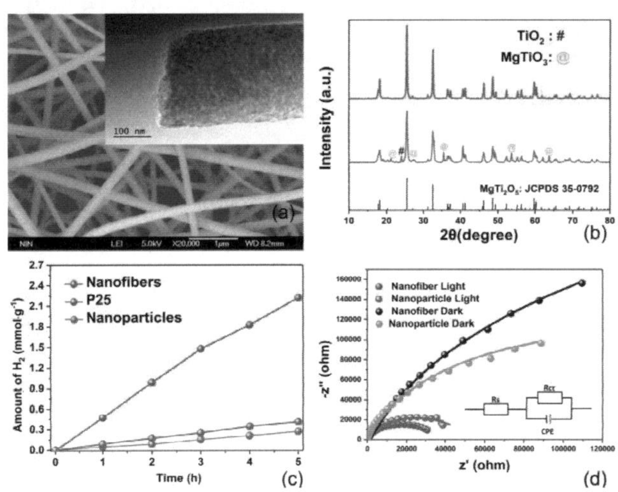

Figure 1. (a) SEM image of $MgTi_2O_5$ nanofibers, insert is the TEM image; (b) XRD patterns of $MgTi_2O_5$ nanofibers and nanoparticles; (c) H2 evolution efficiency; (d) EIS measurements

Stab -Resistant Performance of Fabrics with Hard Particles

Wanqiu Yang[1, 2], Xiao-Yan Liu[1,2]
[1]*Key Laboratory of Textile Science & Technology, Ministry of Education, Donghua University, Shanghai, 201620, China;*
[2]*College of Textiles, Donghua University, North Renmin Road, Shanghai, 201620, China*
E-mail: xiaoyanliu@dhu.edu.cn

In recent years, individual protection have attracted many attention in personal safety area, especially the stab resistant clothing. Stab-resistant fabric mainly include woven fabric, knitted fabric, and nonwoven fabric. Usually these fabrics were consisted of high performance fibers, such as ultra high molecular weight polyethylene fiber(UHMW-PE) , aramid fiber, and so on. There were many ways to prevent puncturing. In ancient days, people used metal sheets as an anti stab layer, it can play the protection role but they were so heavy. With high performance fiber development, some new technology such as the ceramic, metal powders were coated on the high performance fabric. They can make stab resistant fabric lighter and flexible. In this paper, the woven fabric of UHMW-PE fiber was taken as the sample, and B_4C hard particles were coated on them. Here, B_4C different particle sizes were compared on the fabric. The best particle size was elected by comparing stab-resistant performance of coated fabric. Then, different number of coating layers was compared on the fabric with the best size particle. The best number of coating layers was found. The coating liquid at different temperature was used, and the influence of temperature on stab resistance was discussed. Different ratio of SiC powder and binder was compared, and stab resistant performance of coated fabric under the same coating thickness was discussed. Finally, a best experimental scheme was found. Lightweight and soft anti stab clothing can be manufactured with this experimental scheme. This stab resistant clothing was suitable for people who need to wear individual protective armor for a long time, policemen, security men, and so on. The stab resistant fabric can also be applied to many other areas where stab resistant were needed.

Keywords: stab resistant; fabrics; hard particles; coating

Polypyrrole Assisted Cu NPs Deposition on Papers for Energy Storage

Yang Yang, Quanbo Huang, Xiaohui Wang*

State Key Laboratory of Pulp & Paper Engineering, School of Light Industry and Engineering, South China University of Technology, Guangzhou, 510640, P. R. China.

E-mail: fewangxh@scut.edu.cn

High-performance supercapacitors (SCs) are promising energy storage devices to meet the pressing demand for future wearable applications. Here we report a porous, flexible and conductive paper with polypyrrole and copper nanoparticles by using in-situ polymerization and electroless deposition. The conductive polypyrrole was used as interfacial layer, which not only assisted the electroless deposition of copper nanoparticles on the paper, but also enhanced the adhesion strength between metal nanoparticles and paper substrates. The obtained polypyrrole coated papers were high conductivity with a low resistance of 10 $\Omega\ sq^{-1}$ with polymerization time for 60 min and showed enhanced thermal and mechanical stabilities. It was found that the conductivity was increased when the polypyrrole paper was coated with copper NPs, and the resistance reduced by 0.2 $\Omega\ sq^{-1}$ by electroless deposition for 10 min. Notably, the electrochemical performance of the Cu coated papers was improved remarkably, and they reached an area capacitance up to 257 mF cm^{-2} at an area current density of 1mA cm^{-2}. Furthermore, this method could be easily scaled up to large-scale fabrication of conductive paper and opened up new opportunities for flexible energy storage.

Keywords: polypyrrole, copper nanoparticles, polymerization, electroless deposition, supercapacitors

FeCo Nanoparticles-embedded Carbon Nanofibers as Robust Peroxidase Mimics for Sensitive Colorimetric Detection of L-cysteine

Zezhou Yang, **Ce Wang, Xiaofeng Lu***

Alan G. MacDiarmid Institute, College of Chemistry, Jilin University, Changchun, 130012, China

E-mail: xflu@jlu.edu.cn

Artificial enzymes, which served as alternative of natural enzymes, have aroused increasingly interests in recent years.[1,2] Various materials have gradually come into our sights and presented their potentials in diverse enzyme-like catalytic fields. In this work, a facile strategy was established which combines electrospinning technique and calcination treatment to decorated FeCo nanoparticles in carbon nanofibers. We prepared a series of FeCo-CNFs nanocomposites under different calcination temperature, these nanofibers can still maintain uniform fibrous morphology with the increment of temperature. Additional, it can be observed that the sizes of FeCo nanoparticles are augment with the increasing of calcination temperature. XRD data show that the crystallinity of the FeCo nanoparticles are promote under the higher temperature. Moreover, we investigated the catalytic properties of these obtained samples as peroxidase mimics, the optimum activity of FeCo-CNFs under 550 °C can be observed. The superior catalytic properties of the as-prepared FeCo-CNFs over individual FeCo nanoparticles as well as CNFs suggest the synergistic effect between the two components. Furthermore, based on the prominent catalytic performance of FeCo-CNFs nanocomposites, we developed a colorimetric assay to sensitive detect H_2O_2 and L-cysteine. The low detection limits and high selectivity can be realized. In summary, this work provides promising potentials of FeCo-CNFs nanocomposites in sensors, biomedicines and environmental monitoring.[3]

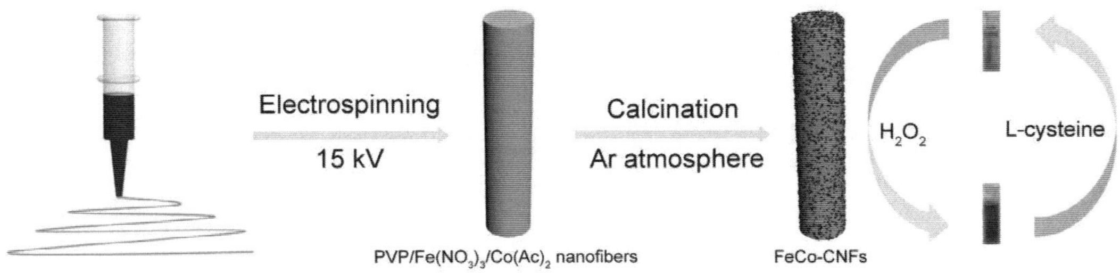

Figure 1. FeCo nanoparticles embedded in carbon nanofibers as efficient peroxidase-like mimics for the sensitive detection of L-cysteine have been developed.

References

[1] Y. H. Lin, J. S. Ren, X. G. Qu, *Acc. Chem. Res.*, **2014**, 47, 1097-1105.

[2] H. Wei, E. K. Wang, *Chem. Soc. Rev.*, **2013**, 42, 6060-6093.

[3] Zezhou Yang, Yun Zhu, G. D. Nie, M. X. Li, C. Wang, X. F. Lu, *Dalton Trans.*, **2017**, 46, 8942-8949.

Establishment and Application of a Test Method for the Fibrillation of Regenerated Cellulose Fibers

Wei Yuan, Tiehan Wang, Yumei Zhang *, Huaping Wang
State Key Lab. for Modification of Chemical Fibers and Polymer Materials,
College of Materials Science and Engineering, Donghua University, Shanghai, 201620
E-mail: ywqdu@126.com

In the test methods for the fibrillation of Lyocell fiber, the equipment of ultrasonic oscillations method is simple and the condition is easy to control. Based on the ultrasonic oscillations method, through the study of factors affecting the production of fibrillation, we obtain stable conditions with repeatability. That is, when ultrasonic power is 500W, ultrasonic time is 30min, the probe height is 10mm and the fiber sample amount is 1mg, we can get a clear picture of the fibrillation effect, and the degree of fibrillation is easy to distinguish, it can be used to calculate the fibrillation index; The regenerated cellulose fibers from different sources were treated with this condition, and we can distinguish the conventional Lyocell fiber, antifibrillation Lyocell fiber, viscose fiber and Model fiber by the fibrillation degree. It can be illustrated that the optimized ultrasonic oscillation condition can be used as the basis for the establishment of the test standard of Lyocell fiber fibrillation.

Key words: Lyocell; fibrillation; ultrasonic treatment; Fibrillation index

One-Pot Synthesis and Characterization of Graphene with Grafted Phosphorus-Nitrogen Flame Retardant

Fanxin Zeng, Zongyi Qin*, Yao Xiao

State Key Laboratory for Modification of Chemical Fibers and Polymer Materials, College of Materials Science and Engineering, Donghua University, Shanghai 201620, China.

E-mail: zengfx@mail.dhu.edu.cn; *phqin@dhu.edu.cn

ABSTRACT: Functionalized reduced graphene oxide (FRGO) wrapped with a P-N flame retardant was successfully synthesized by a simple one-pot method and well characterized. The chemical structure of FRGO was confirmed by using FT-IR and the thermal properties was characterized by TGA.

INTRODUCTION

Polymers containing the phosphoramide structure are families of P–N compounds, and have been used as flame retardant materials due to their good thermal stability, low flammability, high Tgs, and more importantly inherent synergistic effect between phosphorus and nitrogen[1-3]. Moreover, it was reported that phosphoramide structure could be more effective than phosphate structure in promoting the char formation and reducing the pyrolysis gases[4].

In this work, Functionalized reduced graphene oxide (FRGO) wrapped with a nitrogen-phosphorus flame retardant was successfully synthesized by a simple one-pot method.

RESULTS AND DISCUSSION

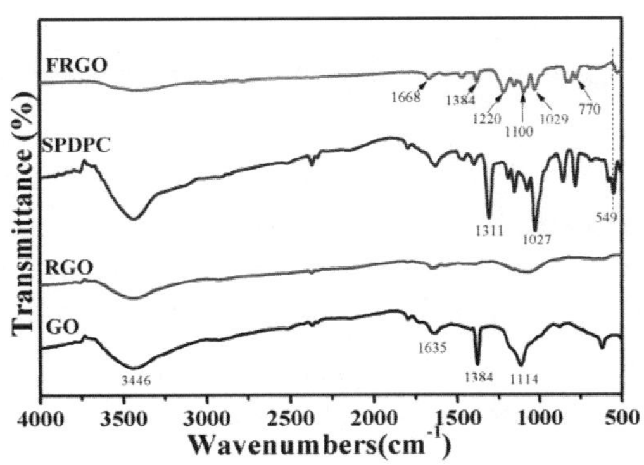

Figure 1. FTIR spectra of GO, RGO, SPDPC and FRGO.

Figure 1 shows the FTIR spectra of graphene and its derivatives. In the FTIR spectrum of GO, the typical absorption peaks of oxygen-containing groups at 3446, 1384 cm^{-1}(-OH), 1635 cm^{-1}(C=O), 1114 cm^{-1}(C-O) are observed[5]. And there is a clearly decreased intensity of these peaks in the spectrum of RGO. In the FTIR spectra of SPDPC, the characteristic bands at 1311cm^{-1}(P=O), 1027cm^{-1}(P-O-C), 549cm^{-1}(P-Cl) are found[6]. The characteristic bands of FRGO are similar to SPDPC, and there are new absorption peaks at 1100, 770cm^{-1}(P-N-C), furthermore, the peak at 549cm^{-1}(P-Cl) disappears for FRGO, indicating that RGO has been successfully functionalized by the P–N flame retardant.

The thermal stability and thermal degradation behaviours of GO, SPDPC and FRGO were investigated by TGA under nitrogen. Fig. 2 shows the TGA curves of these thermosets. At the temperature range of 150-250 °C. GO exhibits an obviously mass loss stage which is attributed to the decomposition of the unstable oxygen-containing groups. And there are two-step degradation process of SPDPC in the regions of 200-300 °C and 450-550 °C. What's more, GO exhibits a higher char residue (68.86%) than SPDPC (36.35%) at 700 °C. In comparison, the char residue at 700 °C for FRGO is 64.52, and FRGO shows a sharp degradation behaviour at the temperature range of 400-500 °C, corresponding to the degradation of the flame retardant. These results demonstrate that the FRGO has a good thermal stability.

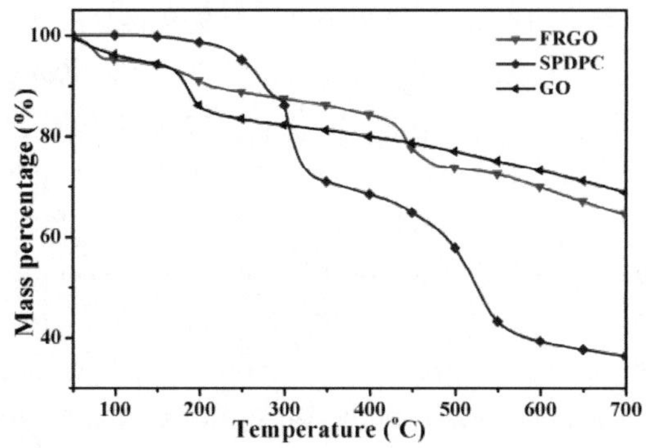

Figure 2. TGA curves of GO, SPDPC and FRGO under nitrogen

CONCLUSION

In this study, FRGO was successfully synthesized from SPDPC, diamino diphenylmethane and GO via a simple one-pot method. The chemical structure of FRGO was confirmed using FT-IR and the thermal properties was characterized by TGA. The results from TGA demonstrate that the FRGO with good thermal stability and the char yield at 700 °C for FRGO is 64.52%.

ACKNOWLEDGEMENT

This work was financially supported by the Fundamental Research Funds for the Central Universities (CUSF-DH-D-2017040).

References

[1] Veen I, Boer J. *Chemosphere*, **2012**, *88*, 1119-1153.

[2] Tai Q, Hu Y, Yuen R, et al. *Journal of Materials Chemistry*, **2011**, *21*, 6621-6627.

[3] Ma H, Wang J, Fang Z. *Polymer degradation and stability*, **2012**, *97*, 1596-1605.

[4] Gaan S, Rupper P, Salimova V, et al. *Polymer Degradation and Stability*, **2009**, *94*, 1125-1134.

[5] Tran D, Kabiri S, Losic D. *Carbon*, **2014**, *76*, 193-202.

[6] Wang X, Xing W, Feng X, et al. *Polymer Chemistry*, **2014**, *5*, 1145-1154.

Flexible Asymmetric Fiberlike Supercapacitors based on Ternary $NiCo_2Se_4$ Nanosheet and $NiCo_2O_4$@PPy Electrodes

Daohong Zhang*, **Yun Ma, Qiufan Wang, Yunlong Wu**

Key Laboratory of Catalysis and Materials Science of the State Ethnic Affairs Commission & Ministry of Education, Hubei Province, South-Central University for Nationalities, Wuhan, Hubei Province, 430074, China

E-mail: zhangdh27@163.com

Flexible fiberlike supercapacitors needed for wearable electronic devices are attracting more and more attention in recent years. Here, we reported a high performance flexible asymmetric all-solid-state fiberlike supercapacitor (ASC) with a $NiCo_2Se_4$ positive electrode and a $NiCo_2O_4$@PPy negative electrode (coated with KOH PVA). The fabrication procedure of the ASC is demonstrated in Figure 1(a). The fiber-shaped ASC can reached a high volumetric capacitance of 14.2 F/cm^3 (111.5 mF/cm) at a current density of 63.7 mA/cm^3 owing to their nanosheet and nanosphere structures confirmed by scanning electron microscopy (SEM) and transmission electron microscopy (TEM). The higher electrochemical performance of $NiCo_2Se_4$ was shown in our work, which is attributed to the superiority of its interconnected nanostructure compared with $NiCo_2O_4$. The voltage window of the fiberlike ACS device was expanded from 0-1.0 V to 0-1.7 V and the energy density was increased to 5.18 mWh/cm^3 at a power density of 51.6 W/cm^3. The fiber-shaped ASC device exhibited an excellent long-term cycling stability (94% retention of the initial capacitance after 5000 cycles at 0.6 mA). Furthermore, the fiber-shaped ASC also shows superior flexibility under different bending conditions and bending test indicating the outstanding fatigue durability of our ASC device. Figure 1(b) shows Galvanostatic charge/discharge (GCD) curves for the ASCs connected in series and parallel. The operating potential was increased to 3.4 V when two ASCs was connected in series and the total capacitance of two ASCs connected in parallel was twice than a single. Inset in (i) shows the light-emitting diode (LED) can be powered by connecting three such ASCs in series. This work provides a promising method for fabricating high electrochemical performance flexible supercapacitors from easily available materials for wearable and portable electronic devices.

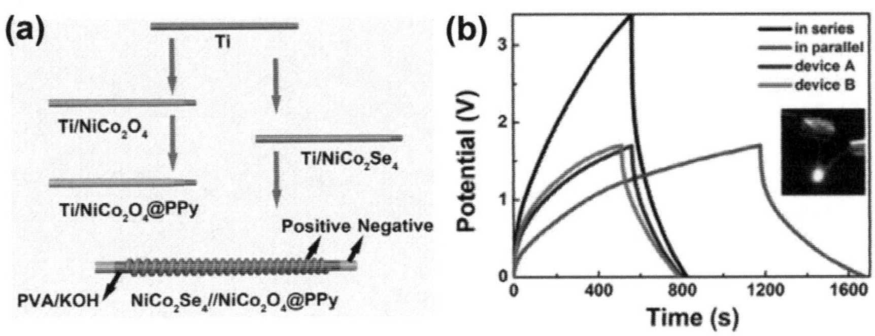

Figure 1. (a) Fabrication of fiber-shaped ASC based on $NiCo_2Se_4$ nanosheets//NiCo2O4@PPy. (b) GCD curves for the ASC connected in series and parallel, the picture inserted in (b) shows three ASCs connected in series powering a LED.

Corresponding author: Prof. Dr Daohong Zhang, zhangdh27@163.com.

Graphene-oxide Modified Sodium-alginate / Krill-protein Composite Fiber

Rui Zhang, Jing Guo, Jing Wu, Sen Zhang

Department of high polymer, College of Material Science and Engineering,
Dalian Polytechnic University, Dalian, 201620, China

E-mail: 28994703@qq.com

The graphene-oxide (GO), sodium-alginate (SA), krill-protein (AKP) were blended to get a functional fiber, And the GO/SA/AKP composite fibers were obtained by the conventional wet-spinning used 5% calcium-chloride as coagulation solution. The intermolecular interaction of GO/SA/AKP(SA/AKP as shown in Figure 1-(a), GO/SA/AKP as shown in Figure 1-(b))composites was characterized by Fourier Transform infrared spectroscopy (FT-IR). The morphology, crystallinity, thermal stability and mechanical properties of GO/SA/AKP composite fibers were investigated by scanning electron microscope(SEM), X-ray diffraction(XRD),Thermogravimetric Analysis(TG) and Differential Scanning Calorimeter (DSC). The results show that the addition of GO promotes the increase of intermolecular hydrogen bonding and the decrease of intramolecular hydrogen bonding in the composites, Which lower the crystallinity of the composites; simultaneously, the thermal stability and fracture strength of the composite increased. When GO added, the number of grooves raised and the width of the fiber reduced.

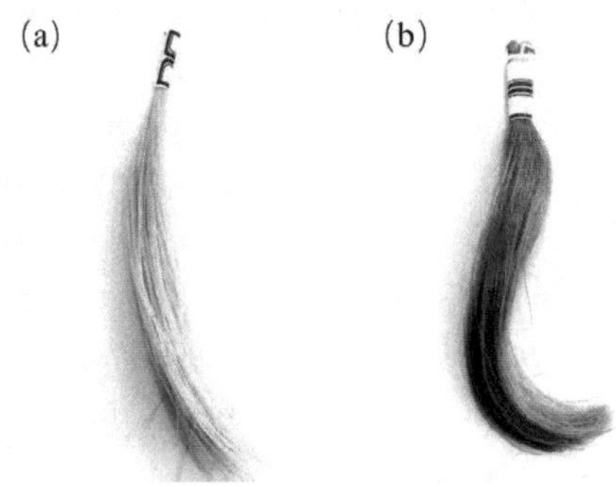

Figure 1. SA/AKP and GO/SA/AKP composite fibers.

Fabrication of Core-spun Yarn Containing a Metal Wire and Study of its Structure and Properties for Intelligent Wearable Device

Yan Zhang, Zeguang Pei, Ge Chen

College of Mechanical Engineering, Donghua University, Shanghai 201620, *China*

E-mail: 2150656@mail.dhu.edu.cn

Conductive traces work for connecting the electronic devices on the smart apparels. The core-spun yarn which consists of metal wire as core surrounded by short-staple fibers is an excellent candidate material for the conductive trace. In order to improve the characteristics of metal wire-based conductive traces on fabrics, a modified vortex spinning method is proposed for fabricating conductive traces-oriented core-spun yarns with an ultra-fine metal wire of 50 μm in diameter as the core. The structure and electro-mechanical properties of the core-spun yarns are analyzed. The results show that the proposed vortex spinning apparatus and method are capable of continuously fabricating core-spun yarns containing the ultra-fine metal wire at a speed 250 m/min. The fabricated vortex core-spun yarn containing a metal wire is composed of the metal wire, the untwisted parallel fibers, and the wrapper fiber from the interior to the exterior of the yarn. The positions of the metal wire in the cross-section of the core-spun yarn are mainly center~$^{1}/_{3}$R type and $^{1}/_{3}$R~$^{2}/_{3}$R type. The breaking strength of the vortex core-spun yarn containing a metal wire is increased by 100%～200% compared to the metal wire. Wrapping characteristics and amount of the short-staple fibers have insignificant effects on the resistance variation - elongation relations of the metal wire core in core-spun yarn, indicating that the vortex core-spun yarn containing a metal wire has similar electrical properties to the bare metal wire.

Keywords: smart textiles; conductive yarn; core-spun yarn; metal wire; vortex spinning

Construction and 3D Printing of Fiber Reinforced Biomimetic Hybird Materials for Dental Restoration

Menglu Zhao, Huili Shao, Xuechao Hu, Yaopeng Zhang*

State Key Laboratory for Modification of Chemical Fibers and Polymer Materials,

College of Materials Science and Engineering, Donghua University, Shanghai, 201620, China

E-mail: zyp@dhu.edu.cn

Composite resin materials for dental restoration have been widely used due to the excellent aesthetic performance, simple and quick clinical operation, good binding ability with tooth, and low security risks. The performance of the composite materials can be improved by many parameters, such as organic monomer selection, inorganic filler composition optimization, regulation of dimension and function, the compatibility of the organic/inorganic components, and the optimization of the compound performance. As a new type of processing molding, 3D printing has been widely used in biomedical field due to the features of precise formation of complex structure, personalized and mass production. Dental model can also be printed effectively and ultra-accurately. 3D printing is therefore promising to construct composite resin for dental restoration and personalize dental repair. In this study, Bis-GMA derivatives and cholic acid derivatives are mixed with inorganic nanoparticles to configure the crown material, which are 3D printed for personalized tooth repair.

The process of 3D printed tooth is shown in the figure. The human teeth were scanned by the micro CT and the data was reconstructed to obtain its three-dimensional model. Then it was imported to the 3D printer (Envision TEC bioplotter) to print. Printed materials limit the development of 3D printing technology. An organic and inorganic mixed slurry filled with different proportions of micro-nano silica powders were prepared as print ink. They are all shear thinning fluids from the rheological performance test which are easier to extrusion. Therefore, they can be adapted to print. The slurries were printed into cylinders, lumps, strips, and the true proportions of reduced teeth for further testing. By changing the print speed, extrusion pressure, different sizes of print needles, UV curing condition and print spacing, the stable internal structure and high mechanical strength of the dental repaired materials were obtained. The ratio of micron silica, nanometer silica and organic monomer at 55:15:30 wt% were better to shape as a suitable printed materials. The microstructure and the mechanical properties of the printed stents were observed respectively by SEM and instron machine, and the biocompatibility of the printed samples were characterized.

The materials can be used successfully for printing. We will choose different print materials to achieve personalized repaired teeth according to the different parts of the tooth in the further research.

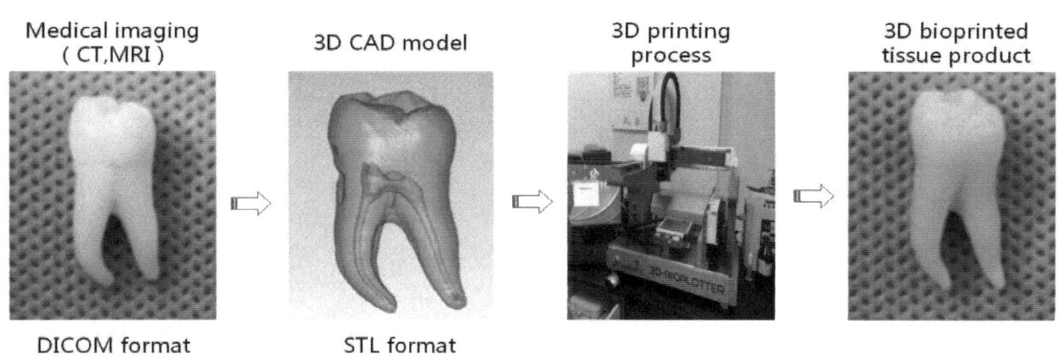

Figure 1. The process of 3D printed teeth

Tunicate Cellulose Nanofiber-Starch-Lignin Composite Film: Strong, 100% Bio-based and Completely Green Composite Film Best Suitable for Coating and Packaging of Food and Medicine

Yadong Zhao and **Jiebing Li**

Department of Fibre and Polymer Technology, School of Chemical Science and Engineering, Royal Institute of Technology, KTH, Teknikringen 56-58, 10044 Stockholm, Sweden

E-mail: yadong@kth.se

Facing problems from declining petroleum-related resources and ever-increasing environmental pollutions, there is a global development trend to replace synthetic polymers/plastics with recyclable and biodegradable materials. Cellulose and its derived cellulose nanofibers (CNF) are promising choices as a substitute that offers similar or better properties and also opens up possibilities to introduce novel functional properties.[1-3] To date, however, no large scale or practical breakthrough has been reported mostly due to shortcomings from obviously weaker mechanical strengths of the produced cellulose (mostly woody cellulose) polymer/plastic than the synthetic counterparts as well as from judgments in terms of being waterborne processing technique and 100% free from synthetic additives or components.

In this study, we have developed the composite film concept by using tunicate CNF as the basic matrix to overcome all the shortcomings primarily thanks to the excellent tunicate CNF microfibrillar network structures. The composite film includes two other components being also 100% natural polymers, namely starch (commercial starch) and lignin (industrial softwood kraft lignin), which reinforces the CNF network and supplies further properties respectively to the final composite films.

Comprehensive characterizations have revealed that the tunicate CNF-starch-lignin composite films produced are structurally starch-cemented CNF networks with well-distributed lignin as nanoparticles both inside and at the surface. Strikingly, after optimizations of the processing operation and composition formulation, the mechanical strengths of the composite films have reached the levels stronger than most of the synthetic polymers/plastics in terms of tensile strength and Young's modulus values (**Figure 1**). In addition, the starch reinforcement renders the composite films with optical transparency and oxygen barrier properties and the lignin modification supplies hydrophobicity or water repellent and UV screening properties to the composite films. Moreover, the preparation procedure is completely based on waterborne processing technique and could thus be considered 100% green. These films are promising green alternatives to synthetic plastics especially for food and medical packaging applications.

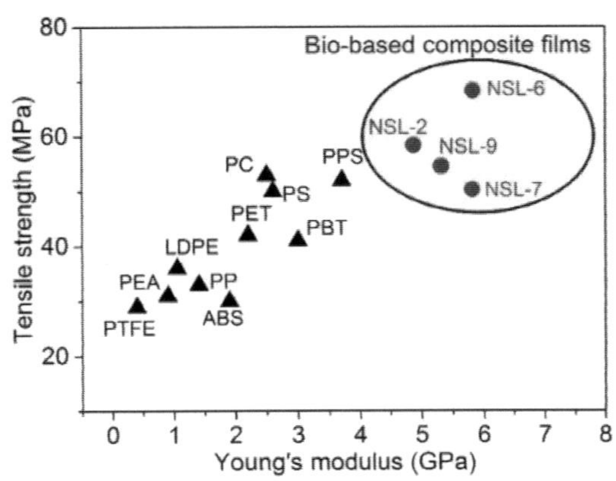

Figure 1. Mechanical properties of tunicate CNF-starch-lignin (NSL) composite films in comparison to common synthetic plastics (NSL-2, 6, 7 and 9 being with different formulations)

References

[1] Yadong Zhao, Carl Moser, Mikael E. Lindström, Gunnar Henriksson, and Jiebing Li, *ACS Applied Materials & Interfaces*, **2017**, 9, 13508-13519.

[2] Yadong Zhao, Yujia Zhang, Mikael E. Lindström, and Jiebing Li, *Carbohydrate Polymers*, **2015,** 117, 286-296.

[3] Yadong Zhao and Jiebing Li, *Cellulose*, **2014**, 21, 3427-3441.

Oil-water Interface Self Assembly of Bowl-like Magnesium Nickel Oxide Monolayer Nanofilm for Enhanced Tetrahydrofuran Sensing

Yan Zhao, Yunpeng Huang, Huaming Li

Institute for Energy Research, Jiangsu University, Zhenjiang 212013, China.

E-mail: yanzhao@ujs.edu.cn

High quality film-based nickel oxide and magnesium nickel oxide (10%, 20%, and 30% Mg loading) with bowl-like structure have been prepared via a facile oil-water interface self-assembly strategy and annealing process. It was found that the morphologies of the as-obtained products can be tuned by changing the concentrations of the precursors. The pristine NiO nanofilm presents a closely-packed unites and the incorporation of Mg still keep the bowl-like structure. The NiO based sensor shows efficient response to ethanol, methanol, acetone, and benzene. Besides, the sensor based on pristine NiO shows relatively low response at the operating temperature in the range from 200°C to 350°C, with the maximum response of 3.58 at 325°C. Compared with the pristine NiO, the presence of Mg can largely enhance the responses to these gases. Moreover, the 30% Mg loading of NiO exhibits an excellent response to tetrahydrofuran gas and EG gas at 325°C, which is extraordinary superor than other chemical gases. The 30% Mg-loaded NiO based gas sensor displayed a response value of 0.5 to 50 ppm of tetrahydrofuran at 325°C with a maximum response of 15.

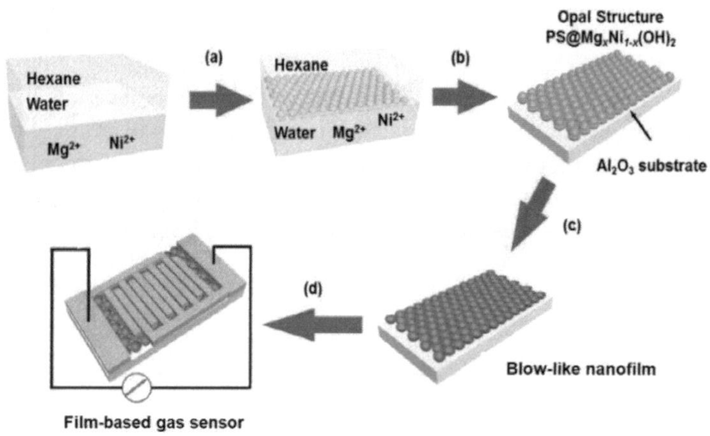

References

[1] L.F. Hu, M. Chen, X.S. Fang, L.M. Wu, *Chem. Soc. Rev.*, **2012**, 41, 1350–1362

[2] N.D. Hoa, S.A. El-Safty, *Chem. Eur. J.* **2011**, 17, 12896–12901

Preparation of Waterborne Epoxy Resin Emulsion by Middle-Temperature Radical Graft Polymerization

Wankai Zheng, **Kemei Pei***

Department of Chemistry, Zhejiang Sci-Tech University, Hangzhou 310018, China

E-mail: peikemei@zstu.edu.cn

The epoxy resin E-20 was modified by middle-temperature radical graft polymerization. In the preparation experiments, benzoyl peroxide(BPO) is used as an initiator and methacrylic acid(MAA), styrene(St) and butyl acrylate(BA) were used as comonomers. The modified epoxy resin can be solved in water well for acrylate monomers with the strong hydrophilic group –COOH were introduced into the polymer chains. Meanwhile, ketone and hydrazine cross-linking reaction was applied in this work for it can increase the dry rate significantly. The performance of the modified waterborne expoxy resin has been investigated in detail, and the resin shows very good surface, adhesion, impact resistance, water-resistance characters. This work supplies an excellent synthesis method for a waterborne resin emulsion that can be used as a green coating material in many cases.

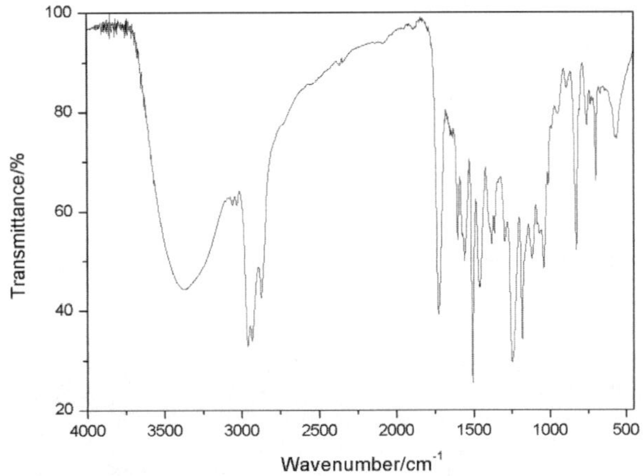

Figure 1. FTIR spectra of grafting product.

References

[1] L Li, PH Wang, CH Liu, et al. *Paint & Coatings Industry*, **2011,** 41(2), 17-20.

[2] XL Lv, *Guangxi University*, **2008.**

[3] AM Deng, Q Dang, R Mu. *Chemistry and Adhesion*, **2013,** 35(1):22-2.

Preparation and Characterization of Temperature Stimulus-Responsive Hollow Nanofiber

Xie Zheng, Yifan Zhou, Jianfeng Zhou, Liusheng Zha*
State Key Laboratory for Modification of Chemical Fiber and Polymer Materials& College of Materials Science and Engineering, Analysis and Measurement Center, Donghua University, Shanghai 201620, P. R. China
E-mail: lszha@dhu.edu.cn

Temperature stimulus-responsive nanofiber, a kind of smart nanofiber that can respond to environmental temperature change, has been paid considerable attention due to their potential applications in sensor, actuator, drug delivery system, separation and purification [1]. If the smart nanofiber has hollow structure, its application range will be broadened. In our work, poly(N-isopropylacrylamide-co-N-hydroxymethylacrylamide-co-stearyl acrylate) (PNNS) was firstly synthesized via solution free radical copolymerization of N-isopropylacrylamide, N-hydroxymethylacrylamide and stearyl acrylate, and its chemical structure and composition were analyzed by nuclear magnetic resonance spectrometer. Its lowest critical solution temperature determined by ultraviolet visible spectrometer is ca. 36℃. The nanofibers with core-shell structure were prepared by coaxial electrospinning process using the ethanol and water solution of PNNS as shell spinning solution and mineral oil as core spinning solution. After heat treatment and extraction of mineral oil with toluene, the hollow nanofibrous membrane with good stability in aqueous medium was produced. The hollow structure of its constituent nanofibers was confirmed by scanning electron microscope (Figure 1). As the aqueous medium temperature was increased from 20 °C to 40 °C, the area contraction ratio of the hollow nanofibrous membrane was up to 50%.

(a) (b)

Figure 1. SEM images of the obtained hollow nanofibrous membrane (a) and the cross sections of its constituent hollow nanofibers (b)

References

[1] Huang C, Soenen S J, Rejman J, et al. *Chemical Society Reviews,* **2011**, *40,*2417-2434.

This work was supported by the national natural science foundation of China (51373030).

Numerical Simulation of Melt Flow Behavior of ABS/CNT Composite in FDM Process

Jin Zhu, **Biao Wang***

State Key Laboratory for Modification of Chemical Fiber and Polymer Materials, College of Material Science and Engineering, Donghua University, Shanghai, 201620, China

E-mail: wbiao2000@dhu.edu.cn

Fused deposition modelling (FDM) is one of additive manufacturing techniques in which a thermoplastic filament is fed through a heated barrel to achieve molten state and then extruded layer by layer to create three dimensional parts. The present work is focused on analyzing the melt flow behavior of multi-walled carbon nanotube (MWCNT)/acrylonitrile butadiene styrene (ABS) composite in the extrusion liquefier of FDM process. Experiments were carried out to characterize the physical properties like thermal conductivity, specific heat, viscosity and shear thinning property of MWCNT/ABS composite. These values were used for simulating in the extrusion liquefier and the melt flow behavior was studied using accurate channel geometry by varying filament velocity at the entry. The results showed that the temperature gradient, pressure drop and velocities of the MWCNT/ABS composite melt depended on the filament velocity. These simulation results provide promising information in optimizing the FDM parameters for better part quality with such composite and can also be used to predict the melt flow behavior of other ABS based composites.

Keywords: Fused deposition modelling (FDM); multi-walled carbon nanotube (MWCNT); acrylonitrile butadiene styrene (ABS); melt flow behavior

Isolation and Application in Polyurethane Foam of Lignin from Enzymatic Hydrolysis Residues

Jun Xu[1,2], Shiyun Zhu[1], Peng Liu[1], Jun Li[1], Kefu Chen[1*]

[1]State Key Laboratory of Pulp and Paper Engineering, South China University of Technology, Guangzhou 510640, China;

[2]Key Laboratory of Pulp and Paper Science & Technology of Ministry of Education of China, Qilu University of Technology, Jinan 250353, China.

E-mail: ppkfchen@scut.edu.cn

Enzymatic hydrolysis residues (EHR) from corn stalks are by-products in the process of producing cellulosic ethanol, which mainly consists of lignin and few cellulose, hemicelluloses and ashes. Lignin can replace part polyether polyol to produce polyurethane foam materials due to existence of large amounts of phenolic hydroxyl in the structure. In this paper, lignin was isolated as raw material from EHR to prepare polyurethane forms, which could not only improve additional economic benefits of cellulosic ethanol industry but also protect atmospheric environment to some extent. Single factor experiment and response surface method (RSM) of four factors three levels were applied to determine the optimal isolation conditions such as NaOH concentration, liquid-solid ratio, reaction temperature and reaction time. Basic structure and functional groups of isolated lignin were analyzed by FT-IR and ^{13}C-NMR. Then, polyurethane foams were synthesized by lignin, MDI, polyether polyol and catalysts with different addition of isolated lignin. Surface morphology, water adsorption rate, thermogravimetry (TG), FT-IR and mechanics performance of polyurethane foam were investigated. Results showed that the optimal conditions for NaOH concentration was 39.90 g/L, and liquid-solid ratio ratio of 38.75:1 ml/g, reaction temperature of 61.33 ℃, reaction time of 2 h. Simultaneously, it was found that the interaction effect of NaOH concentration and liquid-solid ratio, NaOH concentration and reaction temperature, liquid-solid ratio and reaction temperature had significant influence on the isolation rate of lignin. FT-IR analysis exhibited that there were many phenol hydroxyl in isolated lignin, and the proportion of guaiacyl was obviously increased. ^{13}C-NMR analysis showed that guaiacyl was the primary basic structural unit in isolated lignin, but p-hydroxylphengl and syringyl were in the minority, indicating lignin isolated from HER was GHS type. Results showed that isolated lignin can take the place of polyether polyol in some degree to manufacture polyurethane foam and its mechanics performance improved. Results of TG and water adsorption test indicated that the heat resisting property of lignin-polyurethane foam was better than pure polyurethane foam, and water adsorption rate of lignin-polyurethane foam exhibited an increasing trend with increasing the amount of lignin. Thus, this research could provide a potential and promising method for complete utilization of lignin from HER in consideration of their abundant supply.

Keywords: enzymatic hydrolysis residues, lignin isolation, response surface method, polyurethane foam

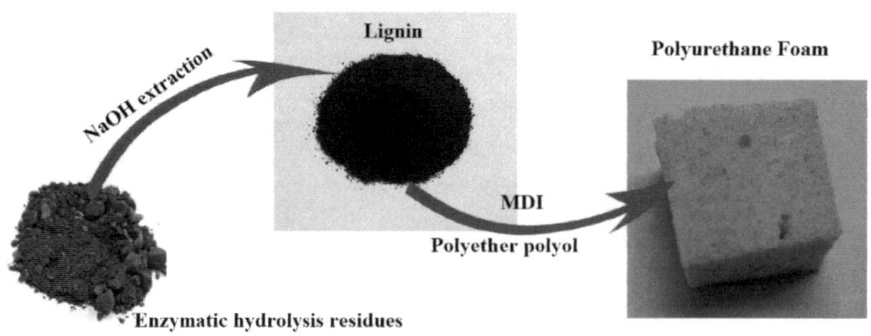

Figure 1. Diagram of research

References

[1] He K, Chen K K, Guo M. *Journal of Zhejiang A& F University*, **2012**, 29, 203-209.

[2] Liu Q X, Li J H, Tian J L, et,al. *China Plastics*, **2002,** 16, 37-39

[3] Tana T, Zhang Z Y, Moghaddam L, et al. *ACS Sustainable Chemistry & Engineering,* **2016**, 4, 5483-5494

[4] Lu Q F, Huang Z K, Liu B, et al. *Bioresource Technology*, **2011**, 104, 111-118

[5] Toledano A, Serrano L, Garcia A, et al. *Chemical Engineering Journal*, **2010**, 157, 93-99

[6] Hao Y M, Chen M, Hu Z B. *Journal of Hazardous Materials*, **2010**, 184, 392-399

[7] Ho Y S. *Water Research*, **2003**, 37, 2323-2330.

Preparation and Characterization of Electrospun Fiber Mats from Regenerated *Antheraea pernyi* Silk Fibroin Aqueous Solutions

Shengzhi Zou[a], Xinru Wang[a], Wenqing Feng[a], Yaopeng Zhang*[a], Jiaming Zhang*[b], Huili Shao[a], Xuechao Hu[a]

[a] *State Key Laboratory for Modification of Chemical Fibers and Polymer Materials, College of Material Science and Engineering, Donghua University, Shanghai 201620, China;*

*E-mail: zyp@dhu.edu.cn.

[b] *Department of Geriatric Cardiology, Huashan Hospital, Fudan University, Shanghai 200040,China.*

* E-mail: zhangjm73@126.com.

Antheraea pernyi is China's unique wild silkworm and in *Antheraea pernyi* silk fibroin (ASF), there are arginine-glycine-asparagine tripeptides with good cell-adhesion ability[1,2]. However, till now, the regenerated ASF fiber mat was generally electrospun from its toxic organic solution[3,4]. The research on electrospinning of regenerated ASF from its nontoxic aqueous solution has been little reported. Therefore, in this work, the regenerated ASF aqueous solutions with high ASF concentration were prepared firstly by dissolving water-soluble regenerated ASF lyophilized porous material into deionized water. Then, the regenerated ASF fiber mats were prepared by electrospinning from these solutions and post-treated by different methods such as being soaked in 90% (v/v) ethanol-water solution and heat-moisture treatment at 65°C and 90% relative humidity. The viscosities of regenerated ASF aqueous solutions with different concentrations were measured by Haake rheometer. The micro-morphologies and molecular configurations of electrospun regenerated ASF fiber mats were characterized by scanning electron microscopy (SEM), Fourier transform infrared (FT-IR) and wide-angle X-ray diffraction (WAXD). Furthermore, Schwann cells were cultured on the electrospun regenerated ASF fiber mats. The cell proliferation ability of the mat was measured by MTT, and the cell growth morphology on the mat was observed by SEM. The results showed that with the increase of regenerated ASF concentration in the spinning dope, the viscosity of the dope and the average diameter of electrospun ASF fiber increased. The regenerated ASF fiber mat electrospun from the spinning dope with a concentration of 38 wt% had the largest average micro-pore size. The as-spun regenerated ASF fibers presented mainly α-helix/random coil configuration and the concentration of spinning dope had little effect on configuration. However, the post-treatment could make the molecular configuration of ASF fiber mats change significantly. Being soaked in 90% (v/v) ethanol-water solution only for 30 min could result in secondary structure of ASF with the dominant β-sheet configuration. However, even if being heat-moisture treated for 96 h, the main configuration of ASF in the electrospun fiber was still α-helix. The above phenomenon indicated that post-treatment by 90% (v/v) ethanol-water solution was more suitable to prepare insoluble electrospun regenerated ASF fiber mats for cell culturing. The results of cell culturing further showed that the cytocompatibility of ASF fiber mat was better than that of *Bombyx mori* silk fibroin fiber mat and the ethanol treated regenerated ASF fiber mat electrospun from a 38 wt% ASF spinning dope had relatively higher cell proliferation ability.

References

[1] Hwang J, Lee J, Goo T, et al. *BIOTECHNOL LETT*, **2001**, 23, 1321-1326.

[2] Sezutsu H, Yukuhiro K. *J MOL EVOL*, **2000**, 51, 329-338.

[3] Wang J, Sun B, Bhutto M A, et al. *FRONT MATER SCI*, **2017**, 11, 22-32.

[4] Li X, Zhang Q, Ye D, et al. *POLYM ENG SCI*, **2017**, 57, 206-213.

PLGA Fiber Leads into the "Big Fiber Era"

Hong Song *

Shanghai Library (Shanghai Institute of Science and Technology Information), Shanghai 200031, China

*E-mail: hsong@libnet.sh.cn

Poly-lactic-glycolic-acid copolymers (PLGA) fiber has generated great interest because of their excellent degradation performance, similar properties of thermoplastic, soluble, low-temperature melting, good processing and excellent strength. The present article studies the innovation trend of PLGA fiber, and reflects the change of investment trend of patentee from three different technical development periods. Combined with the technical hot spots cloud, scientific research structure analysis, development pattern, government guidance and the active participation of enterprises, the article indicates that the cross-disciplinary integration of new areas are the technical trends of PLGA fiber. It also shows the strength and gap between each country through patent activity by source country. Analyzing major players by activity, the paper points out the leading institutions in the field. Based on promoting the development of new pharmaceutical preparations and tissue engineering disciplines construction, the industrialization of PLGA fiber is likely to form a new high-paying technology industry. All the information above on PLGA fiber can provide cognitive assistance to professionals. On the other hand, PLGA fiber and industry have multi-domain integration characteristics towards the intelligent, super and green directions. It is the representatives of large fiber era.

Keywords: PLGA fiber; innovation trend; hot technology cloud map; scientific research structure analysis

Introduction

Poly-lactic-glycolic-acid copolymers (referred to as PLGA) have more excellent degradation, slow releasing and processing performances than PLA. PLGA molecular contains a large number of water-unstable ester bond, can be hydrolyzed into a single lactic acid and a glycolic acid[1]. The degradation process has nothing to do with the role of enzymes in the body, having no species difference. Meanwhile, the lactic acid and the glycolic acid are the inherent material of human being that can participate in the body's metabolic cycle, and ultimately be degraded into water and carbon dioxide[2]. Compared to polylactic acid, PLGA has some of the more hydrophilic characters[3], so degradations occurs relatively faster. As a synthetic biodegradable polymer, PLGA can be designed to provide a wider range of properties, having more advantages than natural biodegradable polymers[1]. Therefore, the biodegradable PLGA macromolecule fiber can be directly used in medical practice and can be directly in contact with the living body, such as the implants, the drug delivery systems, organ replacement, and human tissue systems.

PLGA has similar properties of thermoplastic, soluble, low-temperature melting, and good processing properties than PLA[4]. It can not only form films, be drawn to wires, weaven to web to make any geometric shape, but also can be a composite product with other materials together. The key is to use modern technology to achieve its mechanical properties. So far, scientists have been improving PLGA fiber preparation technology[5]. PLGA can be successfully post-processed (stretching, knitting, or weaving,) into desirable textile structure. Various studies have investigated the preparation and applications of PLGA fiber but few studies that investigated the patents of PLGA fiber are reported. The present article reveals the evolution of PLGA fiber technologies in the emerging, growing, mature and outbreak phases, helping to judge the new research and development trends, forecast the next development path of the technologies, and inspire technological innovation ideas. The application and market demand of PLGA fiber are also discussed.

Overall innovation trend within PLGA fiber

In recent years, the international number of patents in the field of PLGA fiber increased rapidly year by year (Figure 1, data from 101 countries and organizations based on the Orbit World Patent Database for data sources), indicating that the field of research and development is very active. As the patent from the application to the public usually takes some time, even up to 18 months, and if the applicant through the PCT procedures for international applications, then it enters the national phase, the cycle can be extended to 30 months. Therefore, some of the patent applications in 2014 and later are not included.

From the patent time series, PLGA fiber began to appear since the 1970s. Because of its excellent performance, PLGA quickly attract the researchers' attention. In the 1970s, PLGA fibers were mainly used for medical sutures. After entering the 21st century, with the integration of biological activity and degradation, PLGA fiber began to apply to tissue regeneration and tissue engineering projects. Some PLGA fibers, which reduce the cost and improve the processing performance, have been developed, and the demand for PLGA fiber has been gradually increased. There are three different periods that reflect the change in the investment trend of the patentee in this technology.

Industrial germination period: from the 1970s to 1998, the global number of patent applications is small;

Industry growth period: the number of patent applications in 1999 began to show a steady growth until 2006;

Industrial outbreak: since 2007, the speed of publication for the patent has been significantly accelerated and the number has increased significantly.

Hot research on PLGA fiber

Figure 2 shows researching hotspots for PLGA fiber .This cloud map is derived from the "Technology Radar" put out by Orbit. We evaluated the current suitability of each technology for people, and the level of "crowd readiness" to respond to work in a given technology area. The PLGA fibers or yarns or textiles can be integrated with biodegradable polymers, collagen, growth factors, hydrogels, implants, microcapsules, stents, therapeutic drugs, and fertility sectors. This is attributed to the growing usage of PLGA textiles in medical applications, primarily due to the textiles' excellent biocompatibility, biodegradability and mechanical strength. Helping users stay protected.

Cross-disciplinary integration of new areas are the technical trends

The Orbit patent map shown in Figure 3 highlights major growth areas of 7099 patent families in the field of PLGA fiber technology. Each technical field is distinguished by the Orbit technology concept. Each point in the figure 5 represents a patent family, being classified. The largest red cluster in the figure 5 is nerve cell treatment. From the 12 o'clock clockwise, the clusters are followed by expandable support, end effector, polymerizable liquid, pulmonary drug delivery, ophthalmic drug, implant, composite scaffold, implanted pharmacies, fiber (nanofibers / microfibers), active blood vessels, cardiac myocardium and so on. Each cluster reflecting the industry's most concentrated areas of patent and technology investment is a hot research front and a hot market spot around the world today. But industry participants may also face great risk of intellectual property in the field. The map aggregates key risk and opportunity factors from patents across technology segments/categories.

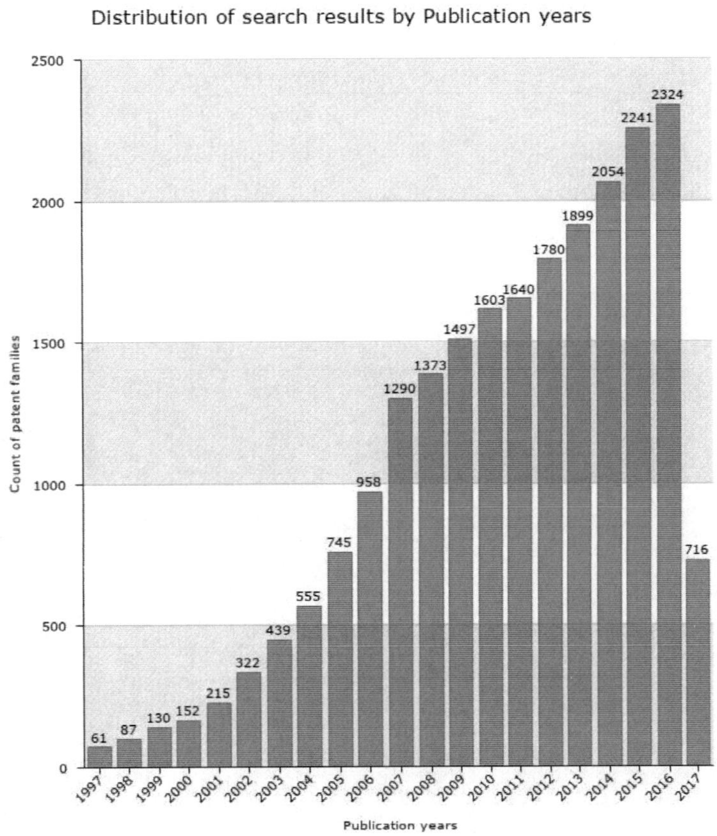

Figure 1 The changing number of PLGA fiber patent applications in the past 20 years

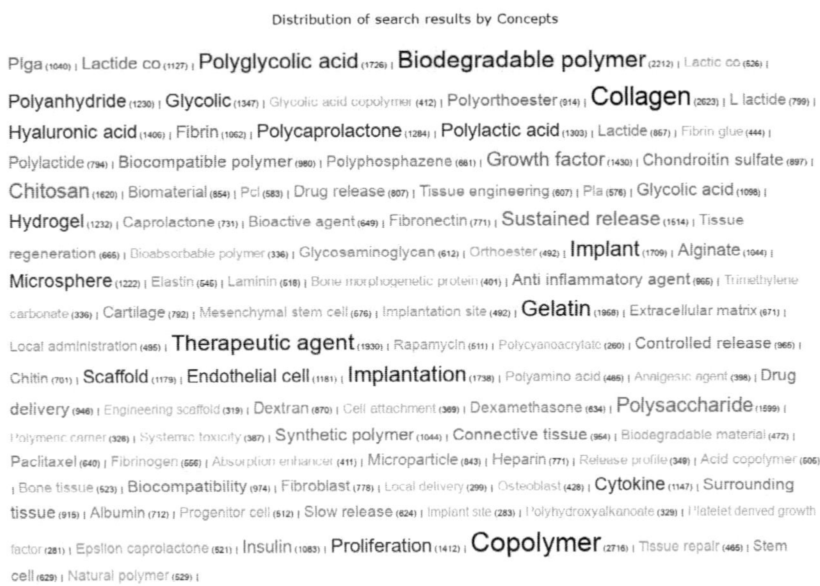

Figure 2 Hot technology cloud map of PLGA fiber

Figure 3 also reflects the basic stability of the scientific research structure of PLGA fiber. Although the scale of PLGA research (cluster) will continue to grow, the layout structure would remain stable outline on the whole. However, it should be noted that the cross-disciplinary integration of new areas becomes more and more widely, especially being prominent around the life of biological science cross-integration. According to the evolutionary trajectory analysis in Figure 5, the cross of chemistry, nanotechnology and life sciences will gradually expand. More researches have been increasingly applied to clinical medicine.

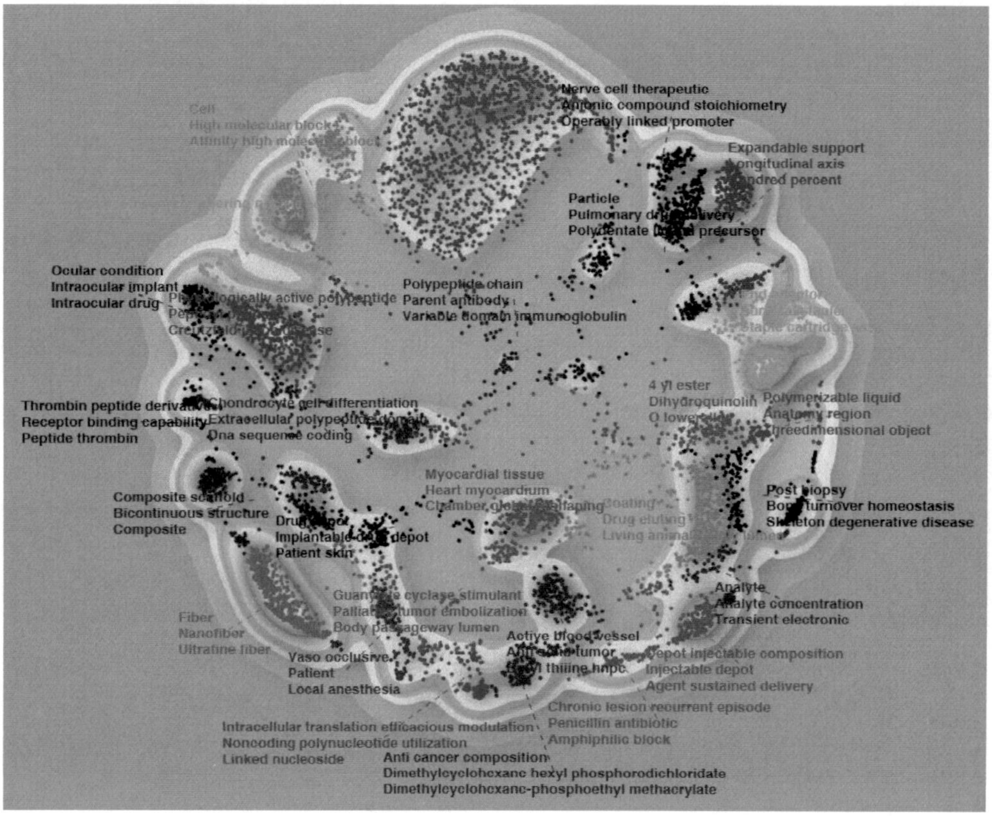

Figure 3 Orbit patent map of PLGA fiber

Patent activity by source Country

Many countries are involved in PLGA fiber research and development (Figure 4).The United States dominated the industry. The number of patent applications in the United States is the largest, reaching 4644; while the integration number of which in China （ranked second） and South Korea（ranked third） only accounts for about 30% of the United States . The gaps between the first level three countries are obvious, showing the difference of technological forces among United States, China and South Korea in the PLGA fiber research. From the intercontinental region, Asia takes the leading role in the study of PLGA fiber. China, South Korea, Japan and China Taiwan have seen considerable growth in the last few years. Europe, including Germany, France and the United Kingdom also has a certain group advantage.

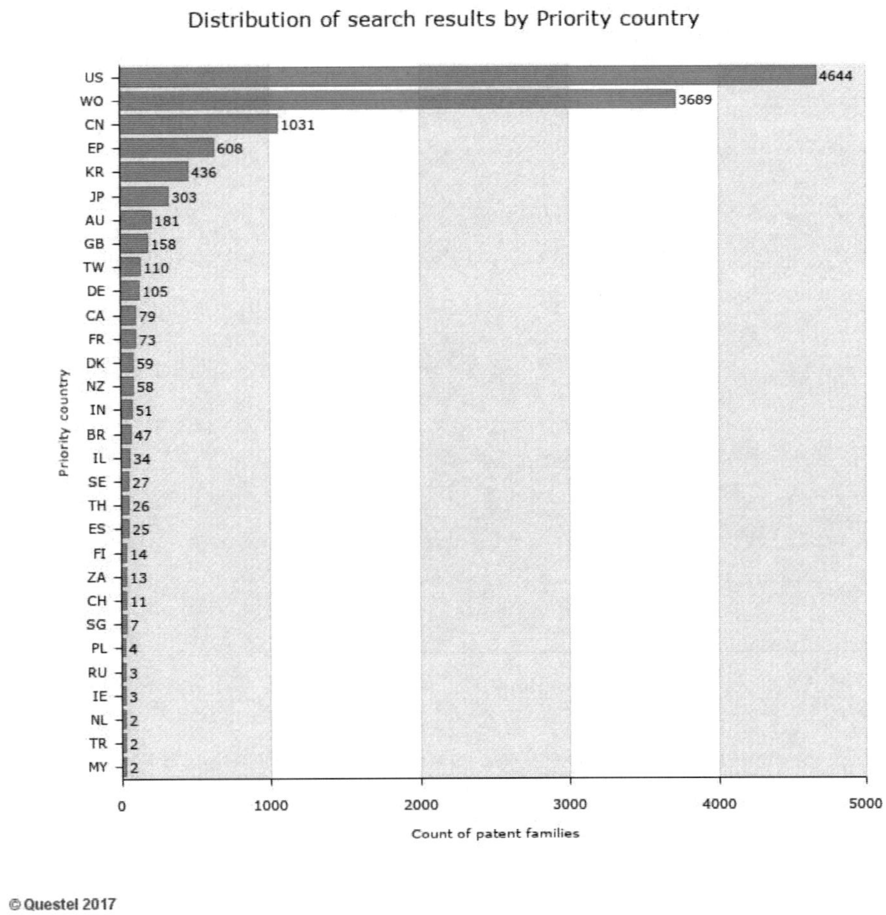

Figure 4 National and regional distributions of PLGA fiber patents in 20 years

Major players by activity

As shown in Figure 5, PLGA fiber research and development are mainly concentrated in the famous companies, colleges and research institutes, where are also the industry's leading investments in intellectual property rights. Among them, the US companies and research institutes further reflect the leading position of the U.S. in the PLGA fiber research in turn. The top 50 R & D institutions, among which the United States accounts for 41, reaches 82%. WARSAW ORTHOPEDIC， INC.,Abbott Cardiovascular Systems Inc., Boston Scientific scimed, AbbVie (leader on medical equipment and pharmaceutical research and development), Allergan(a multidisciplinary health care company), and Cordis(a vascular medical device manufacturing company) are among the top ten companies. In addition, Japan's Shinshu University, China Jinan Shuaihua Pharmaceutical Technology Co., Ltd., China Jinan Kangquan Pharmaceutical Technology Co., Ltd.,and South Korea's electrostatic fiber nanofiber manufacturer Finetex EnE also rank among the top 20.

It is noteworthy that the number of patent applications between the top two corporations, WARSAW ORTHOPEDIC INC. and Abbott Cardiovascular Systems Inc. is very close, but the two companies have different R & D priorities.

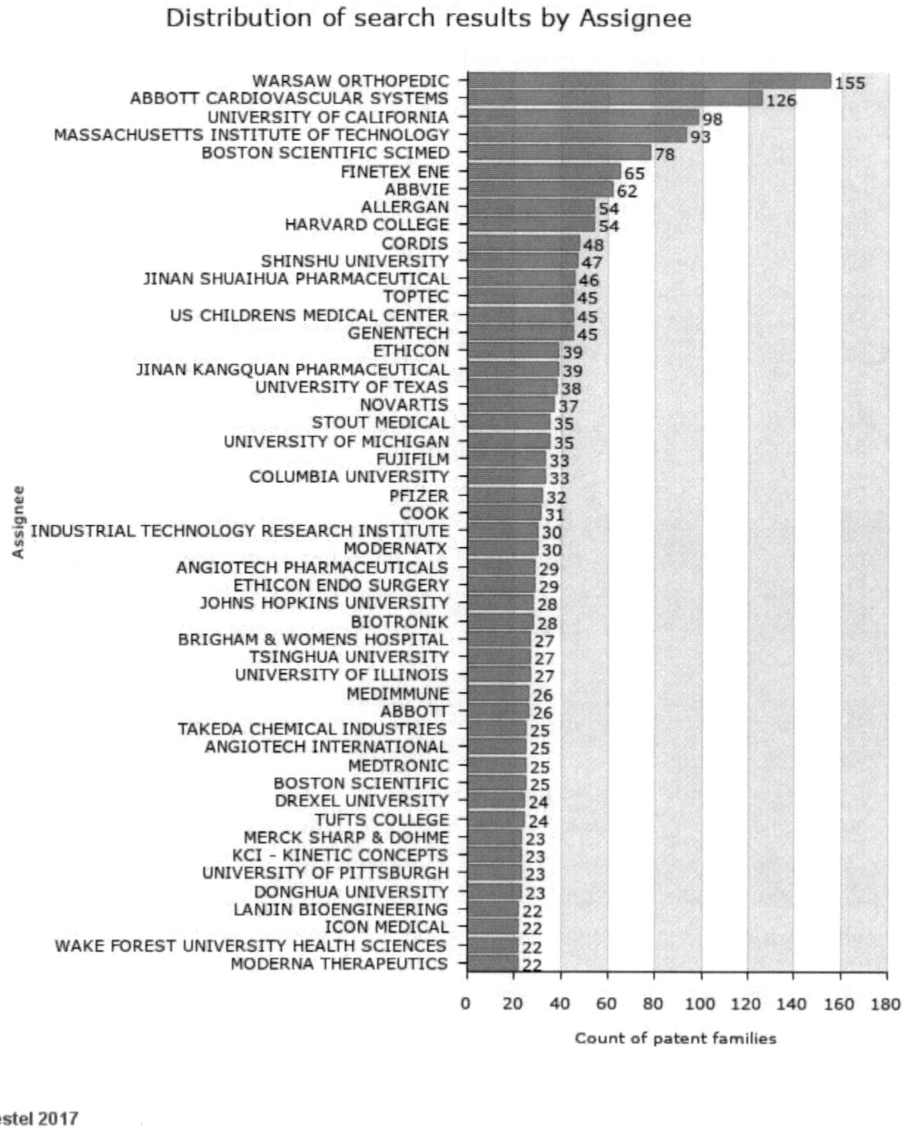

Figure 5 The first 50 patent assignees of PLGA fiber

Development pattern of PLGA Fiber

PLGA Fiber is successful in terms of technology maturity and success. Since the 1970s, PLGA has been used for surgical suture and body implantation, such as artificial joints, guard, bolts and so on. After the 1980s, PLGA is widely used in the field of biomedicine, such as drug control release system, bone fixation, tissue repairing and cell culture matrix, medical suture and anti-adhesion film. At present, PLGA fibers are widely used in pharmaceutical, medical engineering and modern chemical industry[6]. FDA has approved PLGA as a medical surgical suture and injection microcapsules, microspheres, implants and other preparations of the carrier. PLGA was adopted as a medicinal material by the US Pharmacopoeia[6-8]. Chinese Pharmacopoeia 2000 has listed lactide - glycolide copolymer (including PLGA) as the preferred biodegradable synthetic medicinal accessories.

PLGA fiber has some formal products commercialized, some are in the clinical trial stage, but a large number of which are still in the laboratory research stage. In 1975[9], ETHICON company developed PLGA (LA / GA: 90/10) surgical suture (trade name Vicryl) for the first time on the market, that was favored by doctors. At present, its upgraded version of VicrylRapid ® has also been listed[10]. The upgraded version being radiated degrades faster than before. PANACRYL® is another PLGA suture product commercially available[10]. Shanghai Tianqing Biomaterials Co., Ltd. has used a technology investigated by Donghua University for PLGA surgical sutures that have been sold nationwide for many years[11]. The Vicryl PLGA Guided Tissue Regeneration Mask (referred to as periodontal regenerative tablets) produced by Johnson & Johnson, USA, has good biocompatibility, no any discomfort, no local inflammation, and good bioabsorbability for patients[12]. Besides, such firms as U.S. Surgical, Instrument Makar, and Surgical Dynamics provide PLGA surgical sutures, named as Polysorb, Biologically Quiet Biosphere, Biologically Quiet Mini-screw, and SD sorbs 2mm and 3mm respectively. Some PLGA orthopedic fixtures are developed and used. Instrument Makar provides Biologically Quiet Interference Screw for interface screw. Biomet provides Lactosorb Screw and Plates for craniomaxillofacial fixation. Instrument Makar provides Biologically Quiet Staple for anterior cruciate ligament reconstruction. Surgical Dynamics provides SD sorb Meniscal Staple for meniscus repair. Bionx provides BiosorbPDX Screws and Tacks for craniomaxillofacial fixation. Bionx also provides SmartPinPDX for fracture fixation[6].

Donghua University has studied a new type of ureteral stent tube, cooperated with the hospital[13] . The membrane and fiber bicomponent structure stents were prepared by weaving and post-treatment process and using polyglycolic acid (PGA) or PLGA as the raw materials. The wall thickness was 1/3 ~ 2/9 of the traditional non - degradable stent. At the same time, the proportion and distribution of bicomponent in the stent tube were optimized, and the mechanical properties were even better than that of the conventional ureteral stent. With PLGA as raw material, Donghua University makes fibers, followed by a special core weaving, plus core and reinforced weaving process and chitin, low temperature plasma protein coating process. Nerve regeneration catheter thus can be made, and nerve growth factor coated to the catheter to provide nerve growth of nutrients is used to help nerve regeneration of the full biological activity, which can promote the peripheral nerve along the catheter and the catheter wall regeneration recovery. This artificial nerve catheter product has entered clinical trials.

Kim K et al. used the electrospun PLGA nanostructures to successfully blend and release the hydrophilic antibiotic drug Mefoxin[14]. In a culture dish filled with Staphylococcus aureus, the drug released by the drug-loaded fiber causes more than 90% of the strains to disappear and the bactericidal effect is very pronounced. This kind of fiber combines the mechanical properties of nonwoven nanofibers with their efficacy in the release of antibiotics and has a good application prospect in biomedicine. Casper CL acquied PLGA fiber contained liver phospholipids by spinning. The release of heparin in the membrane can be maintained for 14 days. Stents made of such fibers can release growth factors in long-term stable release of tissue healing and repairing until tissue and organs are fully recovered. PLA / PLGA (30/70) nanofibers containing cisplatin have a drug entrapment rate of 90% and show sustained release for up to 75 days with no significant burst release.

In nearly 10 years, reports on PLGA fibers as microsphere or microcapsule skeletons, capsule drug deliveries and drug release systems are more and more. Scientists have made great achievements on PLGA fiber applications. In the future, PLGA fiber can be used in various biomedical domains.

Government guidance and the active participation of enterprises

The United States "revolutionary fibers and textiles institute for manufacturing innovation ", referred to as RFT-IMI, supports the unique fibers and fabric innovation. These innovative fibers and textiles are used in a wide range and can be easily removed when an injured soldier needs to use an antibacterial tape. Companies participating in the project contain many industries, including Chip makers Intel and ADI, textile and apparel producers Milliken, Nike and VF Corp, materials producers Corning and DuPont, as well as Medtronic, a health and medical device maker[15]. PLGA fibers as a renewable and green biological fibers, meet the requirement of such revolutionary fibers. In the Western Europe and China, the government and many manufacturers pay high attention to them.

For example, Evonik, the creative industrial group from Germany, is one of the world leaders in specialty chemicals. On November 24, 2016, Evonik announced that it was expanding its production facilities in Birmingham (Alabama, USA) and Darmstadt (Germany). This would create additional capacity for the production of biodegradable polymers marketed globally under the brand names RESOMER® and RESOMER® SELECT. These poly-lactic-glycolic-acid (PLGA) copolymers are primarily used to manufacture bioresorbable medical devices and controlled-release formulations for parenteral drug delivery.

In 2016, the State Food and Drug Administration issued a "Notice on the Relevant Matters Concerning the Examination and Approval of Drug-related Materials and Drugs for Pharmaceutical Packaging Materials No. 134 of 2016" and strengthened the supervision of pharmaceutical excipients. In 2015 China's pharmaceutical excipients industry market size of 50 billion yuan, is expected to increase to 100 billion yuan in 2020, a huge space for growth. Hunan Erkang Pharmaceutical, as a leading domestic enterprise, has been the layout of high-end research and development of pharmaceutical excipients, such as PLGA and other high-end varieties have been associated with the review. Erkang will seize the high-end accessories market high ground in the future.

Conclusion

We have shown means of intelligence analysis of key issues in PLGA fiber development. By introducing the technology roadmap of PLGA fiber, we successfully acquired the information on PLGA fiber which provide cognitive assistance to professionals. The method must also be developed in hand with patent professionals considering their practices and expectations. Our study confirms the PLGA fiber can be applied to drug controlled releasing carriers, tissue engineering scaffolds, artificial organs, surgical procedures, and to increase the variety of pharmaceutical excipients, to promote the development of new pharmaceutical preparations and tissue engineering disciplines construction. In the United States, organizational engineering has now formed a $ 6 billion industry and is growing at an annual rate of 25 percent. The industrialization of PLGA fiber is likely to form a new high-paying technology industry. On the other hand, PLGA fiber and industry have multi-domain integration characteristics towards the intelligent, super and green directions. It is the representative of large fiber era.

References

1. Hirenkumar K. Makadia and Steven J. Siegel. Poly Lactic-co-Glycolic Acid (PLGA) as Biodegradable Controlled Drug Delivery Carrier. Polymers (Basel). 2011 Sep 1; 3(3): 1377–1397.

2. Lakshmi S. Nair, Cato T. Laurencin .Biodegradable polymers as biomaterials. Prog. Polym. Sci. 32 (2007): 762–798.

3. Vijay Kumar Thakur, Manju Kumari Thakur , John Wiley & Sons. Handbook of Polymers for Pharmaceutical Technologies, Bioactive and Compatible Synthetic/Hybrid Polymers.2015.10.20

4. Christopher J. Bettinger , Zhenan Bao. Organic Thin Film Transistors Fabricated on Resorbable Biomaterial Substrates. Adv Mater. 2010 Feb 2; 22(5): 651–655.

5. Piergiorgio Gentile , Valeria Chiono, Irene Carmagnola , Paul V. Hatton . An Overview of Poly(lactic-

co-glycolic) Acid (PLGA)-Based Biomaterials for Bone Tissue Engineering. Int. J. Mol. Sci. 2014, 15, 3640-3659.
6. Bahareh Azimi, Parviz Nourpanah, Mohammad Rabiee, Shahram Arbab. Poly (lactide -co- glycolide) Fiber: An Overview. Journal of Engineered Fibers and Fabrics .Volume 9, Issue 1 – 2014:47-66.
7. Li Junnan. Research Status and Development of Frontier Field of Biomedical Materials. http://www.sohu.com/a/123645351_465915.
8. PLGA. https://baike.baidu.com/item/PLGA.
9. CHENNAKKATTU KRISHNA SADASIVAN PILLAI , CHANDRA P. SHARMA. Review Paper: Absorbable Polymeric Surgical Sutures: Chemistry, Production, Properties, Biodegradability, and Performance. JOURNAL OF BIOMATERIALS APPLICATIONS Vol. 25 — November 2010:291-366.
10. Biodegradable Polymer Materials in Medical Devices. http://y.tt/39375000070en.html.
11. Lu Changchun. The High - end of Medical Textiles: Game in Technology and Market. http://qk.laicar.com/Home/Content/311498.
12. De Sanctis M, Zucchelli G. Guided tissue regeneration with a resorbable barrier membrane (Vicryl) for the management of buccal recession: a case report. Int J Periodontics Restorative Dent. 1996 Oct;16(5):435-41.
13. Wang Lu, Guan Guoping, Wang Fujun, Lin Jing, Gao Jing, Hu Jiyong. Research progress on biomedical textile materials and devices. Journal of Textile Research. 2016,37(2):133-140.
14. Kim K, Luu YK, Chang C, Fang D, Hsiao BS, Chu B, Hadjiargyrou M. Incorporation and controlled release of a hydrophilic antibiotic using poly(lactide-co-glycolide)-based electrospun nanofibrous scaffolds. J Control Release. 2004 Jul 23;98(1):47-56.
15. David L. Chandler. New institute will accelerate innovations in fibers and fabrics. April 1, 2016. http://news.mit.edu/2016/national-public-private-institute-innovations-fibers-fabrics-0401.

Donghua University – College of Materials Science & Engineering
2999 North Renmin Road
Songjiang, Shanghai
China, 201600

ISBN 978-1-7138-0623-3
9 781713 806233

CURRAN ASSOCIATES INC.
proceedings.com